Janine Cossy (Ed.)

Organozinc Derivatives and Transition Metal Catalysts

Also of Interest

Organocatalysis.
Stereoselective Reactions and Applications in Organic Synthesis
Maurizio Benaglia (Ed.), 2021
ISBN 978-3-11-058803-3, e-ISBN 978-3-11-059005-0

Engineering Catalysis
Dmitry Yu. Murzin, 2020
ISBN 978-3-11-061442-8, e-ISBN 978-3-11-061443-5

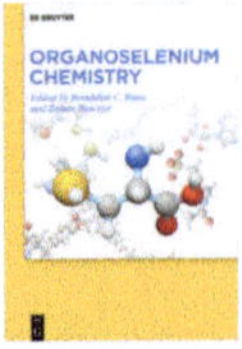

Organoselenium Chemistry
Brindaban C. Ranu and Bubun Banerjee (Eds.), 2020
ISBN 978-3-11-062224-9, e-ISBN 978-3-11-062511-0

Chemistry of the Non-Metals.
Syntheses – Structures – Bonding – Applications
Ralf Steudel, 2020
In cooperation with: David Scheschkewitz
ISBN 978-3-11-057805-8, e-ISBN 978-3-11-057806-5

Organozinc Derivatives and Transition Metal Catalysts

Formation of C-C Bonds by Cross-coupling

Edited by
Janine Cossy

DE GRUYTER

Editor
Prof. Janine Cossy
ESPCI Paris, PSL University
Molecular, Macromolecular chemistry
and Materials, C3M
10 Rue Vauquelin
75005 Paris
France
Janine.Cossy@espci.fr

ISBN 978-3-11-072877-4
e-ISBN (PDF) 978-3-11-072885-9
e-ISBN (EPUB) 978-3-11-072894-1

Library of Congress Control Number: 2022951402

Bibliographic information published by the Deutsche Nationalbibliothek
The Deutsche Nationalbibliothek lists this publication in the Deutsche Nationalbibliografie; detailed bibliographic data are available on the internet at http://dnb.dnb.de.

Cover image: Antoine2K/iStock/Getty Images Plus
Typesetting: Integra Software Services Pvt. Ltd.
Printing and binding: CPI books GmbH, Leck

www.degruyter.com

This book is dedicated to the Memory of Prof. Ei-ichi Negishi

Contents

Preface —— IX

Mae Féo and Guillaume Lefèvre
1 Organozinc reagents and palladium —— 1

Olivier Piva
2 Organozinc reagents and nickel —— 63

Julien Legros and Bruno Figadère
3 Organozinc reagents and iron —— 117

Pauline Schiltz, Mengyu Gao and Corinne Gosmini
4 Organozinc reagents and cobalt —— 141

Armelle Ouali and Marc Taillefer
5 Organozinc reagents and copper —— 179

List of contributors —— 207

Index —— 209

Preface

Organozinc reagents proved to be of great interest in synthesis as they are much more chemoselective than the corresponding lithium and Grignard reagents. Their involvement in cross-coupling reactions as nucleophiles, catalyzed by transition metals, has been revealed to be very useful as these organometallic reagents can be utilized in the late-stage functionalization of complex molecules. *Organozinc Derivatives and Transition Metal Catalysts: Formation of C-C Bonds by Cross-coupling* provides an overview of the progress, trends and accomplishments in the cross-coupling reactions using organozinc reagents.

This book is for university teachers and researchers, researchers in industry and advanced students. Instead of covering the entire subject in full detail, it gave a flavor of what can be done and obtained from these cross-coupling reactions using organozinc reagents. A description of the reactions and mechanistic rationalization of the obtained results are reported. This book is divided into five chapters, each of which corresponds to the cross-coupling reactions catalyzed by a transition metal: **palladium** (Chapter 1), **nickel** (Chapter 2), **iron** (Chapter 3), **cobalt** (Chapter 4) and **copper** (Chapter 5). A parallel can be done between this book and the book *Grignard Reagents and Transition Metal Catalysts: Formation of C-C Bonds by Cross-coupling*, published by De Gruyter in 2016, and readers can see the similarities and differences between the use of organozinc reagents and Grignard reagents. One can notice that in this book, contrary to the *Grignard Reagents and Transition Metal Catalysts: Formation of C-C Bonds by Cross-coupling*, there are no reported cross-coupling reactions from organozinc compounds catalyzed by manganese. This could be explained by the fact that transmetallation between organozinc reagents and manganese salts doesn't seem to proceed efficiently. Attempts to extend manganese-catalyzed cross-coupling reactions from Grignard reagents to the corresponding organozinc compounds have been failed. For example, in 2015, concerning the possibility of the cross-coupling reaction of vinylbromide with organozinc reagents in the presence of manganese chloride, Madsen wrote, "β-bromostyrene was selected because it had served as a successful substrate for coupling with aryl Grignard reagents. However, unexpectedly, the coupling did not occur with organozinc reagents" (see: A. Solvhoj, A. Ahburg, R. Madsen, Chem. Eur. J., 2015, 21, 16272–16279). Since then, no real cross-coupling reactions have been reported between organozinc reagents and electrophiles catalyzed by manganese salts.

Acknowledgments: I would like to warmly thank all the authors for their participation in the adventure of this book and for their contribution. I would like also to thank Dr. Alban Moyeux and Dr. Olivier Gager (Université Sorbonne Paris Nord) for all their efforts to find publications related to the cross-coupling reactions catalyzed by manganese. My thanks to all the team at De Gruyter, especially Stella Müller, for her helpful assistance during the preparation of this book.

https://doi.org/10.1515/9783110728859-203

We hope that this book will be useful in my respects: for preparing lectures and seminars, in finding the right conditions to solve synthetic problems to form C-C bonds in a very chemo-, regio-, diastereo- and enantioselective way, or to be a good source of inspiration to develop new reactions, or just have fun with cross-coupling reactions catalyzed by transition metals.

Janine Cossy
Paris, 6 December 2022

Mae Féo and Guillaume Lefèvre

1 Organozinc reagents and palladium

1.1 Introduction

The choice of main-group nucleophiles as cross-coupling partners in transition-metal-catalyzed cross-coupling reactions is a central question in synthetic chemistry. In particular, the use of a cross-coupling reaction as a key step in a total synthesis will strongly depend on the reactivity and functional tolerance displayed by the nucleophiles. Soon, the organozinc reagents proved to be of high interest in this synthetic context, since they usually are more chemoselective than their lithium or magnesium analogues.

Owing to the high importance of the use of organozinc derivatives as nucleophilic partners in transition-metal-catalyzed cross-coupling reactions, several reviews have been devoted to this topic in the recent past, with a particular focus on Ni- and Pd-mediated transformations [1, 2]. Herein, we will exclusively focus on the development of cross-coupling methods involving an organozinc reagent as the nucleophilic partner and an organic halide or pseudo-halide as the electrophile, in the presence of a palladium catalyst – e.g. on the palladium-mediated Negishi cross-coupling. The Negishi cross-coupling led to a plethora of synthetic developments during the last 40 years, and was awarded in 2010 with the Nobel Prize to Ei-ichi Negishi, along with Richard Heck and Akira Suzuki for the breakthrough brought by palladium catalysis in synthetic organic chemistry [3].

In this chapter, we will briefly discuss the seminal work which paved the way for the use of organozinc derivatives in palladium-catalyzed cross-coupling reactions (Section 1.2), as well as several classic and more recent mechanistic insights (Section 1.3). Several families of important ligands, displaying a particularly efficient reactivity in Pd-catalyzed Negishi cross-coupling reactions, will be introduced (Section 1.4). An overview of the possible cross-coupling patterns accessible, depending on the hybridization of the coupling partners, will be given (Section 1.5). Finally, several recent Negishi cross-couplings in modern or high-scale processes will be discussed (Section 1.6) as well as some classical total syntheses featuring such cross-coupling reaction as a key step (Section 1.7).

1.2 Discovery, early times and examples

In a seminal work, Negishi and Baba reported that the use of palladium catalysts could efficiently achieve the stereoselective synthesis of conjugated dienes using organoaluminium nucleophiles [4]. The use of palladium as an alternative to nickel

Mae Féo, Guillaume Lefèvre, CNRS, Chimie ParisTech, UMR8060 i-CleHS, 11 rue Pierre et Marie Curie, 75005 Paris, France

https://doi.org/10.1515/9783110728859-001

catalysts suppressed the loss of stereoselectivity observed with palladium in conjugated diene synthesis, which opened the door to the exploration of the Pd-based cross-coupling reactions. Following this result, the quest for soft nucleophilic coupling partners which allow a good functional group tolerance soon led to the use of organozinc reagents. The early examples of cross-coupling reactions involving organozincs as nucleophiles, and palladium catalysts, were reported in 1977 by Negishi et al. [5, 6] and Jutand and Fauvarque [7], respectively, in the aryl-aryl and in the alkyl-aryl cross-coupling reactions (Fig. 1.1).

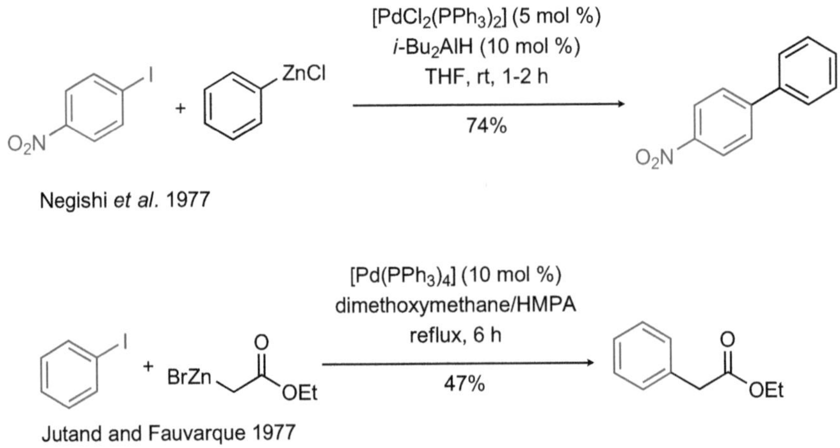

Fig. 1.1: Early reports by Negishi and Jutand and Fauvarque [5–7].

The Negishi group was particularly involved in the development of new cross-coupling reactions involving the main group organometallics as nucleophilic partners in the late 1970s. Between 1976 and 1978, numerous seminal publications related to Pd- or Ni-catalyzed cross-coupling reactions were reported, using a variety of the main-group organometallic nucleophiles. Among this distribution of organometallic reagents, organozinc species proved to be more efficient and selective than the other organometallic partners. An early example of a selective alkynyl-alkenyl cross-coupling reaction was thus reported using an *in situ* generated alkynylzinc nucleophile, to produce the conjugated enynes with a stereoselectivity higher than 97%. This transformation is efficiently mediated by the $Pd(PPh_3)_4$ catalyst, whereas the use of the nickel analogue $Ni(PPh_3)_4$ only affords 50% of the expected product (Fig. 1.2) [6].

Following this strategy, the Negishi group reported several procedures illustrating the success of the association of organozinc reagents with Pd-based catalysts. In the case of the alkynyl-aryl cross-coupling reactions, generation of a catalytically active palladium(0) species, by reduction of $PdCl_2(PPh_3)_2$ with i-Bu_2AlH, was more efficient than the direct use of the classical tetrakis complex $Pd(PPh_3)_4$ [8]. Similarly, the use of similar conditions to produce bisaryls, by an aryl-aryl cross-coupling reaction using arylzinc derivatives and aryl halides, was reported by the same group [5].

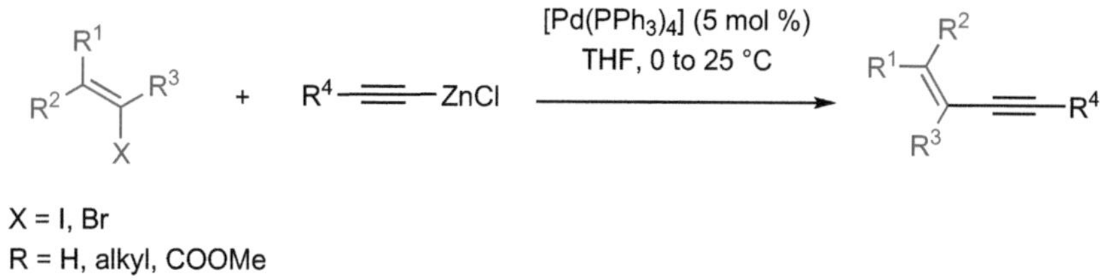

Fig. 1.2: Formation of conjugated enynes by Negishi et al. [6].

Due to the success and the incredible breakthrough brought by transition-metal-catalyzed cross-coupling reactions in organic chemistry, the following decades witnessed an impressive variety of improved methods, making accessible new coupling patterns and opening the way to the incorporation of Pd-mediated cross-coupling reactions between organozinc and electrophiles in total syntheses of complex targets.

Concomitantly, the progresses made in comprehensive mechanistic features of transition-metal-catalyzed transformations contributed to the improvement of the cross-coupling methods. In particular, this field of research strongly benefited from the mechanistic findings related to palladium redox chemistry. The next section of this chapter (Section 1.3) provides some details about the important milestones met in mechanistic understanding of Negishi cross-coupling reactions mediated by palladium catalysts.

1.3 Mechanistic insights

In first approximation, the Negishi cross coupling reaction can be described with a conventional catalytic cycle involving the classical oxidative addition/transmetallation/reductive elimination sequence (Fig. 1.3). However, this simplified pathway fails to reflect the intrinsic complexity of this reaction. Indeed, it is worth mentioning that the reactivity of a transition-metal-based catalyst is governed by the stereoelectronic properties of the coordination sphere, which are shaped by the nature of the ligands. Pd-Complexes are not an exception, so, recently, efforts have been made to use the knowledge provided by the coordination chemistry in the design of new efficient organometallic catalysts. The underlying philosophy is that accurate and deep mechanistic studies of a given catalytic transformation can help to identify the key energy-demanding elementary steps, thus leading to a useful bill of specifications for more efficient catalyst design. In addition to the ligand influences, which will be extensively discussed in the following sections, by-products or additives such as metal salts can affect the catalytic activity [9], either hampering or improving the cross-coupling process. In order to fully understand the Negishi cross-coupling mechanism, a close attention is thus given to the

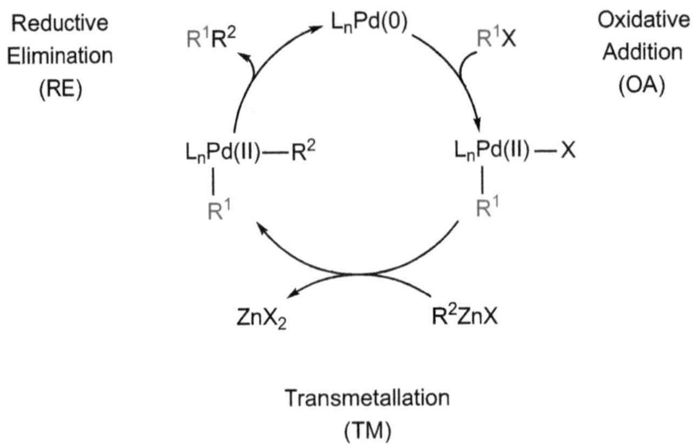

Fig. 1.3: Conventional catalytic cycle for a Negishi cross-coupling.

cross-coupling process by realizing mechanistic studies, both from experimental reactions and from computational studies.

1.3.1 Rate determining step

On one hand, the Negishi cross-coupling is an extremely versatile method, regarding the nature of the electrophiles, the numerous catalysts or the reaction conditions that can be encompassed. On the other hand, such an impressive diversity makes the identification of the Rate Determining Step (RDS) extremely difficult. Examples are reported in Tab. 1.1.

Tab. 1.1: Proposed RDS for several cross-coupling systems for SPhos structure, see Fig. 1.5; OA = oxidative addition, TM = transmetallation, NHC = N-heterocyclic carbene).

Electrophile	Nucleophile	Catalyst	RDS	Ref.
ArX	AlkylZnX	Pd-SPhos	OA	[10]
VinylX	AlkylZnX	$Pd(PMe_3)_2$	OA	[11]
AlkylX	AlkylZnX	Pd-NHC	TM	[12]

1.3.2 Role of additives

Focusing on the transmetallation step with RZnX reagents, formal identification of the catalytically active nucleophile is challenging. Indeed, the versatility of the method does not help to propose a unique active species, especially because of the reactivity variations induced by all the possible hybridizations of the substrates. In addition, the organozinc species are involved in the Schlenk equilibrium (Fig. 1.4), which questions the true form of the nucleophile in solution. It is also important to note that zinc-based compounds easily evolve toward tetracoordinated species. Given this specificity, combined with the strong affinity of zinc with halide ions, which are by-products formed stoichiometrically in the reaction, the conventionally proposed RZnX nucleophile is possibly not always the true transmetalating agent [13].

$$R_2Zn + ZnX_2 \rightleftharpoons 2\ RZnX$$

Fig. 1.4: Schlenk equilibrium.

For all these reasons, the use of halide salts as additives has been intensively investigated. The idea behind that is that halide ions can be associated with RZnX to form $RZnX_2^-$ complexes. In this new species, the metal-carbon bond is supposed to be weaker than in the initial organometallic reagent. Thus, transmetallation of the R group on the catalyst would be easier. This hypothesis has been supported by numerous experimental observations for several cross-coupling reactions, for example with ArX and AlkylZnX (Tab. 1.2) [10]. In a very polar solvent, like a THF/NMP mixture, this hypothesis is, however, overly simplistic, at least for alkyl-alkyl Negishi cross-coupling. Indeed, on the basis of their previous work on BuZnBr [14], Organ et al. observed that when 2 equivalents of halide salts are added, EtZnBr is converted into $EtZnBr_2^-$ and further to $EtZnBr_3^{2-}$, which was found to be the active transmetalating agent in the alkyl-alkyl cross-coupling reaction (Tab. 1.2). Formation of the later double-charged species requires very polar solvent in order to stabilize the zincate by complexation and a close attention should also be given to the positive counter ion. The same group also highlighted the very different effect of halide salts on the aryl-aryl Negishi cross-coupling, since the additive does not participate in the high-order zincate formation but rather impacts the solvent properties (Tab. 1.2) [15].

Tab. 1.2: Proposed role of additives for several cross-coupling systems.

Electrophile	Nucleophile	Additive	Solvent	Active species	Additive effect	Ref.
ArX	ArZnX	$MgCl_2$ (1 equiv)	THF	ArZnX	Breaking aggregates	[15]
ArX	AlkylZnX	LiCl (1 equiv)	THF	$(AlkylZnX_2)^-$	Zincate formation	[10]
AlkylX	AlkylZnX	Bu_4NBr (2 equiv)	THF/DMI	$(AlkylZnX_3)^{2-}$	High-order zincate formation	[13]

Given the affinity of the zinc(II) oxidation state for halides, formation of $ZnX_3^-M^+$ or $ZnX_4^{2-}M^{2+}$ salts in the presence of MX additives has also been investigated since ZnX_2 is a formal stoichiometric by-product of the reaction [13]. In addition, because ZnX_2 is much more electrophilic that organozinc reagents, it is more likely able to capture a halide anion than the organozinc. This is consistent with the inhibiting effect of ZnX_2 when the latter is used as an additive in the alkyl-alkyl or aryl-alkyl Negishi cross-coupling reaction [10, 14] and matches the stability of $ZnX_4^{2-}M^{2+}$ edifice highlighted by calculations [16]. Consequently, the halide ion delivered by the MX additive should first react with ZnX_2 instead of participating in the formation of the charged organozincates previously mentioned. However, since halide ions are also formal stoichiometric by-products of the reaction, it is reasonable to think that ZnX_2 is, thanks to either the additive MX or the X^- by-product, kept saturated with anions so there is no accumulation of ZnX_2 during the process whereas, at the same time, there is still enough halide ions to participate in the formation of the active nucleophile [13].

1.3.3 Bimetallic interactions

In considering the Negishi cross-coupling mechanism, it is usually accepted to neglect the interaction between the two metallic centers, Pd and Zn, except during the transmetallation step. However, several groups have suggested that the metal centers interact during all the catalytic cycle, involving either the nucleophile RZnX (or R_2Zn) or ZnX_2. These hypotheses are mostly supported by computational studies.

In particular, the full mechanistic investigation of alkyl-vinyl coupling using $Pd(PMe_3)_2$ and MeZnX by Aurrecoechea et al. showed that the formation of the bimetallic species $[(PMe_3)_2Pd\text{-}ZnMeX(solvent)]$ prior to the oxidative addition disfavors this step, whereas $[(PMe_3)_2Me(vinyl)Pd\text{-}ZnMeX(solvent)]$ favors the reductive elimination process [11]. It is worth mentioning that the implication of organozinc reagent in

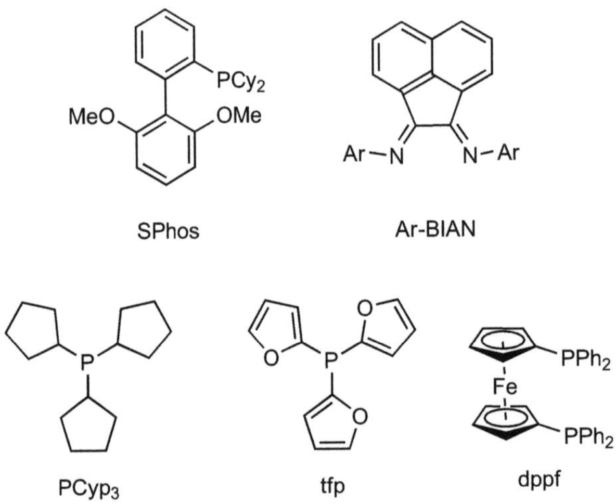

Fig. 1.5: A few usual ligands for Negishi cross-coupling.

the oxidative addition step of alkyl-aryl cross-coupling reaction using ArX, AlkylZnX and the SPhos ligand (Fig. 1.5) has been highlighted by kinetic measurement, thus supporting the idea of bi-metallic species formation early in the process [10].

Regarding the transmetallation step, several reports highlighted the possible existence of Pd-Zn bond interactions which seems to significantly lower the transmetallation energetic barriers, thus explaining, somehow, why the transmetallation step is often so easy in the Negishi cross-coupling reaction [11, 17, 18]. Simplified conclusions are summarized in Fig. 1.6.

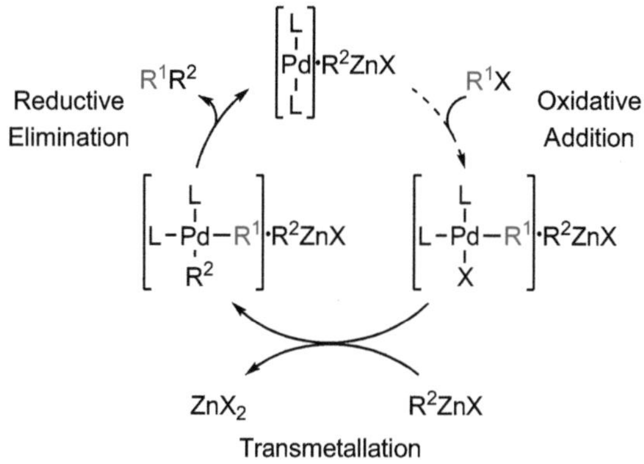

Fig. 1.6: Proposed bimetallic interactions between the Pd-catalyst and R_2ZnX (dashed arrows represent disfavored pathway compared to the corresponding monometalic pathway) [10, 11].

Apart from the organozinc reagent, formation of formal $ZnCl_2$ through the catalytic cycle also leads to possible Pd-Zn interactions. As a representative example, the theoretical role of ZnX_2 was depicted by Pidko and Polynski for the cross-coupling reaction between ArX and AlkylZnX using phosphines or NHC ligands [16]. Calculations revealed that prior to the oxidative addition, interactions between the Pd catalyst and ZnX_2 are very unlikely to happen regarding the energetic profile of the system. On the contrary, after the oxidative addition, interactions between the two metallic centers are favored and lead to off-cycle resting states, thus resulting in the cross-coupling inhibition. Consequently, participation of ZnX_2 at this stage of the process should be avoided, for instance by trapping ZnX_2 with MX additives as mentioned above. Simplified conclusions are summarized in Fig. 1.7. On the other hand, calculations on the alkyl-alkyl cross-coupling reaction using NHC ligands showed the assistance of ZnX_2 in the reductive step through Pd-Zn interactions [12].

1.3.4 Homocoupling undesirable pathways

Although the Negishi conditions provide a selective access to the cross-coupling product, homocoupling products of the nucleophile RZnX (or R_2Zn) or the electrophile R'X are frequently observed. Aiming to hamper the homocouplings, substantial efforts have

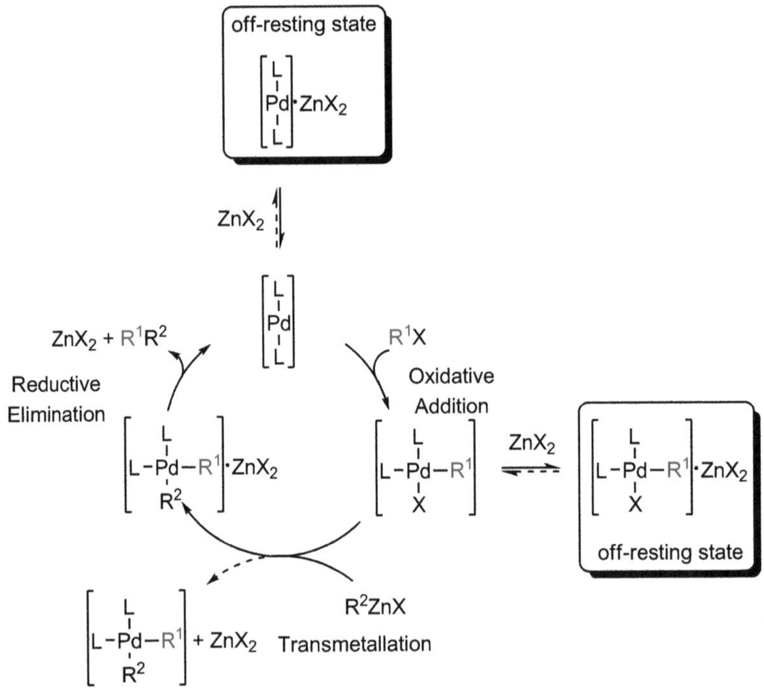

Fig. 1.7: Proposed bimetallic interactions between the Pd-catalyst and ZnX_2 (dashed arrows correspond to disfavored pathway) [16].

been made to identify the species involved in the formation of the homocoupling products. As a starting point, it is reasonable to think that they are produced by competitive transmetallations during the process through the simplified intermediates [PdR_2] or [PdR'_2] respectively. In order to explain their existence, several main scenarios are conceivable such as: (1) after the usual oxidative addition, unwanted organic group exchange between the catalyst and the nucleophile, followed by the usual transmetallation and reductive elimination steps (Fig. 1.8, eq. 1); (2) starting from the usual [PdRR'] intermediate, an unwanted second transmetallation with the nucleophile, followed by a reductive elimination (Fig. 1.8, eq. 2); (3) after the unwanted transmetallation between RZnX and whatever [PdR'] intermediate, transmetallation between the usual [PdRR'] intermediate and the new species R'ZnX, followed by a reductive elimination (Fig. 1.8, eq. 3); (4) starting from the usual [PdRR'] intermediate, retro-transmetallation giving species which can further evolve toward [PdR_2] or [PdR'_2] (Fig. 1.8, eq. 4). Neither pathway excludes the other, and the dominant mechanism seems to be extremely ligand and solvent dependent and furthermore, to differ with the nature of the organozinc reagent.

Among the systems investigated, it is worth mentioning that the early proposed pathway (Fig. 1.8, eq. 1), suggested for the coupling between AlkylX and ArZnX using Ar-BIAN ligand (Fig. 1.5) [19], was corroborated by kinetic experiments and computational

$$[\text{X-Pd-R'}] + \text{R-Zn-X} \rightleftharpoons [\text{X-Pd-R}] + \text{R'-Zn-X} \quad \text{(eq. 1)}$$

$$[\text{R-Pd-R'}] + \text{R-Zn-X} \rightleftharpoons [\text{Pd-R}_2] + \text{R'-Zn-X} \quad \text{(eq. 2)}$$

$$[\text{R-Pd-R'}] + \text{R'-Zn-X} \rightleftharpoons [\text{Pd-R'}_2] + \text{R-Zn-X} \quad \text{(eq. 3)}$$

$$[\text{R-Pd-R'}] + \text{ZnX}_2 \rightleftharpoons [\text{X-Pd-R}] + \text{R'-Zn-X} \quad \text{(eq. 4)}$$

Fig. 1.8: Possible homocoupling pathways.

studies on alkyl-aryl coupling using $ZnMe_2$ and PPh_3 ligands, when the concentration of the intermediate *trans*-$[PdArX(PPh_3)_2]$ is high [20, 21]. On the contrary, experimental evidence was provided by Lei et al. to support the pathway given in Fig. 1.8 (eq. 2) in the case of aryl-aryl cross-coupling with ArZnX, using dppf ligand (1,1′-Ferrocenediyl-bis(diphenylphosphine, Fig. 1.5) [22]. Formation of the electrophilic homocoupling product through pathway given in Fig. 1.8 (eq. 3) was also found to be competitive when it comes to ArX and Et_2Zn cross-coupling, using electron-withdrawing phosphine-based ligands [21]. Finally, the importance of retro-transmetallation pathways (Fig. 1.8, eq. 4) has been highlighted when reacting *trans*-$[PdArX(PMePh_2)_2]$ with MeZnX [17].

Since all the mentioned undesirable pathways involve R'ZnX or R'ZnR species, hampering their formation is crucial. One strategy is the use of RZnX reagents instead of R_2Zn nucleophiles. This idea was found to work on the *trans*-$[PdArX(PPh_3)_2]$ system, using MeZnX instead of Me_2Zn [20]. Focusing on pathway given in Fig. 1.8 (eq. 2), the cross-coupling final elimination should be fast enough so the presence of the intermediate is elusive, thus preventing undesirable transmetallations. Making this elimination step faster is a challenge in the ligand design which will be discussed in the next section (Section 1.4). Another way to suppress undesirable transmetallations is the use of an excess of ligand, when it is easily available. This hypothesis is supported by calculations on the *trans*-$[PdArX(PPh_3)_2]$ and Me_2Zn system, for which the transmetallation was found to proceed mainly through ligand-dependent associative substitution pathways, in consistence with the observed experimental influence of an excess of PPh_3 in the homocoupling inhibition [18].

1.3.5 β-hydride undesirable pathways

A classic drawback in Pd-catalyzed cross-coupling reactions, involving nucleophiles or electrophiles bearing β-hydrogens, is the production of alkenes by β-hydride elimination (Fig. 1.9). With these substrates, more tricky is the isomerization of the corresponding alkyl-Pd(II) chains, the alkyl part arising from either the nucleophilic or the electrophilic partner, respectively, after the transmetallation or the oxidative addition steps. The isomerization is induced by successive sequences of β-hydride elimination and 1,2-insertion, as represented in Fig. 1.10 for the alkyl-aryl cross-coupling, which leads to complex mixtures of regioisomers, featuring both linear and branched species.

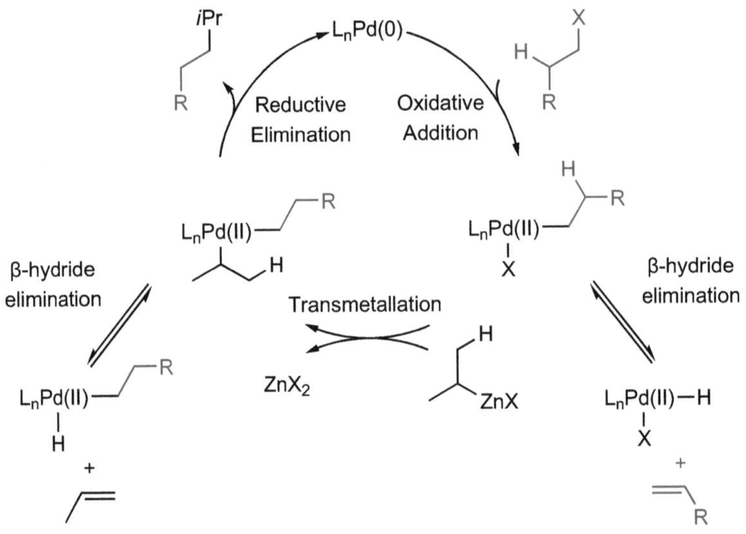

Fig. 1.9: Alkene formation through β-H elimination for the alkyl-*i*Pr cross-coupling reaction.

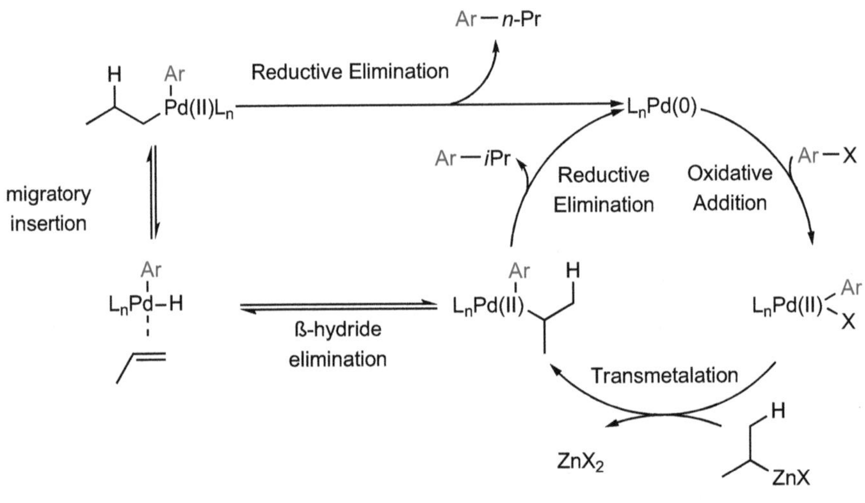

Fig. 1.10: Product isomerization through β-H elimination for the aryl-*i*Pr cross-coupling reaction.

To circumvent this β-hydride elimination, the rate of the cross-coupling reductive elimination should be significantly higher than that of the β-hydride elimination, thus hampering not only alkenes and RH formation, but also the isomerization pathway, thus ensuring a satisfying selectivity and regioselectivity of the cross-coupling process. Examples of tailor-made ligands designed to do so are discussed in the next section (Section 1.4).

1.4 Remarkable ligands

The first examples of Pd-mediated Negishi cross-couplings involved the association of palladium(0) precursors with PPh_3 as a stabilizing ligand (Section 1.2). Accessing a broad substrate scope, for either the nucleophilic or the electrophilic partner, requires the development of palladium catalysts able to activate a variety of C-X bonds (X = halide or pseudo-halide), and able to lead to intermediate organopalladium(II) species which can efficiently follow the catalytic cycles discussed in Section 1.3.

Among the different challenges required to achieve efficient cross-coupling transformations, two major classic limitations must be faced: (1) some electrophiles with strong C-X bonds (typically organic chlorides) may be hard to be activated and usually require more reactive catalysts than their bromides or iodides analogues and (2) the use of aliphatic partners (either the nucleophile or the electrophile) with a hydrogen atom in the β-position can easily lead to unstable alkyl-palladium(II) intermediates followed by decomposition pathways by β-hydride elimination (Figs. 1.9 and 1.10).

To circumvent those classical limitations, palladium organometallic chemistry keeps witnessing the development of more and more efficient ligands. Classical ligand families which were developed during the last 20 years have been successfully used in transition-metal catalysis. Important progresses were in particular made in the Negishi-type cross-coupling field thanks to those ligation platforms. An extensive review focused on tailor-made palladium-ligands and their applications in cross-coupling reactions has been published by Stradiotto and Lundgren [23].

1.4.1 Activation of strong C-X bonds: The quest for electron-rich ligands

To activate strong C-Cl bonds by a classic oxidative addition pathway, one must use an electron-rich palladium complex. Thus, stabilizing phosphines with strong electron-donating properties were associated with palladium(0) precursors to address this matter. Electron-rich trialkylphosphines, owing to the strong donating effects of their substituents, unlocked numerous activation pathways for electrophiles that are difficult to activate, e.g. with a strong C-X bond (X = halide, pseudo-halide).

Preliminary reports by Fu et al. demonstrated that bulky electron-rich $P(t\text{-Bu})_3$ associated with palladium(0) sources efficiently promoted cross-couplings involving aryl chlorides as electrophiles (Fig. 1.11 and see Section 1.5) [24, 25]. On the basis of the results reported by the Fu's group, more sophisticated trisubstituted monophosphines were developed, along with their well-defined Pd-ligated complexes. We can cite the example of di-adamantyl alkyl-phosphines developed by Beller et al. (used in Cata-CXium complexes), which are also particularly used in Suzuki-Miyaura cross-coupling systems [26], or of the Q-Phos ligand developed by Hartwig et al. (see Fig. 1.11) [27].

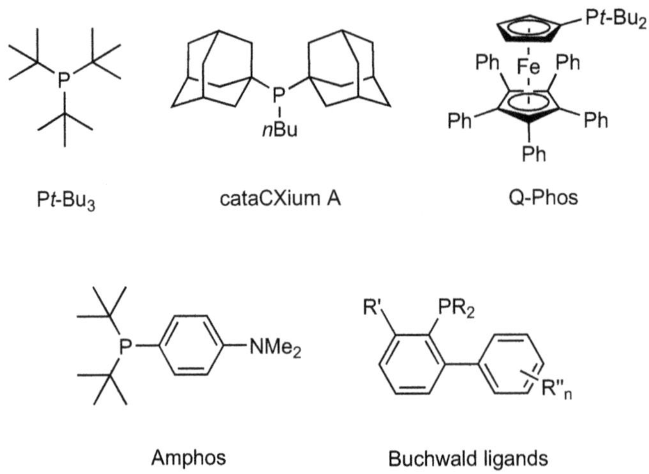

Fig. 1.11: Classical tailor-made ligands developed in palladium cross-coupling reactions.

However, the bulkiness of electron-rich alkyl substituents, such as *t*-Bu in P(*t*-Bu_3) or adamantyl groups in CataCXium ligands, can hamper the reactivity of the catalyst for steric reasons. To circumvent this matter, Guram et al. developed a new family of electron-rich monophosphines displaying a reduced bulkiness at the phosphorus atom, relying on the introduction of P-bound aryl substituents substituted with electron-donating groups in the *para* position. Bis-*tert*-butyl(4-dimethylaminophenyl) phosphine (denoted as Amphos) was thus designed for this purpose (Fig. 1.11). This ligand displays enhanced σ-donating properties, but brings a more modest steric pressure at the metal, the steric demand of the former being close to that of P(*t*-Bu)$_2$Ph. Association of the Amphos ligand to palladium soon proved to be particularly efficient in the promotion of Suzuki-Miyaura cross-coupling involving heteroaryl chlorides [28]. It was later demonstrated by Lipshutz et al. that Amphos-based palladium precatalysts such as $PdCl_2$(Amphos)$_2$ were excellent precursors to achieve Negishi cross-coupling reactions in water (micellar conditions) at room temperature under Barbier conditions (Fig. 1.12) [29]. Several applications of micellar Negishi cross-coupling reactions mediated by Amphos-Pd catalysis will be discussed in the next sections.

Dialkylbiaryl phosphines (Fig. 1.11) constitute another remarkable family of ligands allowing a fine control of the coordination sphere of the palladium(0) catalyst in cross-coupling reactions. Those ligands have been developed by Buchwald et al. in 1998 and have considerably been extended and used ever since [30]. A broad range of stereoelectronic effects can be modulated in this coordination platform, which makes these ligands applicable to a broad scope of transformations. For example, the substituents on the phosphorus atom can directly modulate the electronic properties at the metal center, whereas the bulkiness and the arene coordination provided by the lower aryl ring help to modulate the kinetic stability of the catalyst resting state.

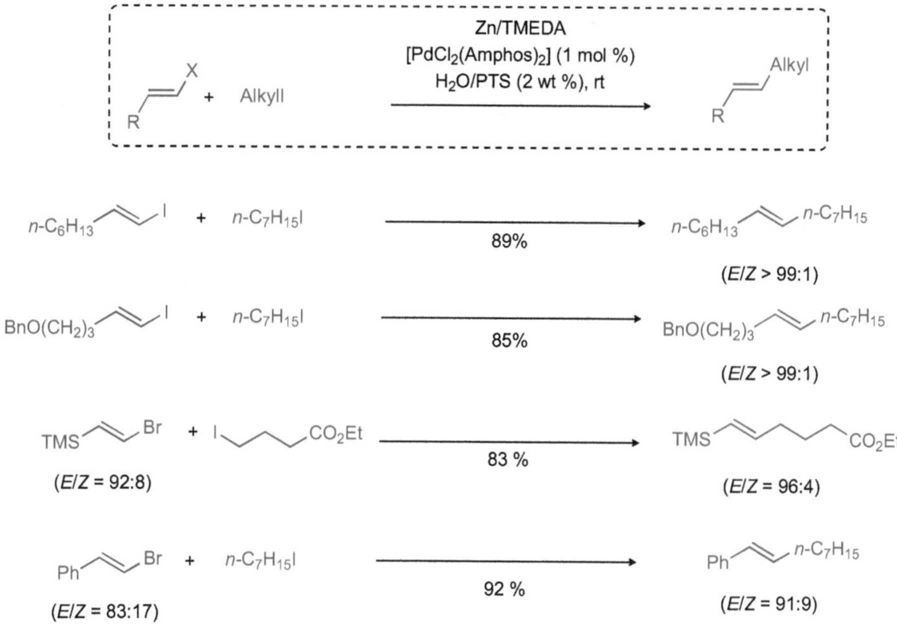

Fig. 1.12: Alkyl-vinyl coupling in Barbier conditions by Lipshutz et al. [29].

Elementary steps of the catalytic cycle (e.g. oxidative addition and reductive elimination) also proved to be possibly enhanced by the decoration of the arene rings.

1.4.2 Bypassing the hydride β-elimination

A second classic pitfall to be avoided during cross-coupling processes, involving aliphatic reagents, is the hydride β-elimination occurring on a lateral alkyl chain ligated to a palladium(II) center. Several groups addressed this problem using catalysts stabilized by bidentate phosphine ligands displaying a high bite angle, such as $PdCl_2$(dppf) (Fig. 1.5). The latter complex was used successfully by Hayashi et al. in selective Negishi-type cross-coupling reactions involving secondary aliphatic nucleophiles with benzene bromide [31].

Aiming at developing new coordination platforms which can be applied to a broader scope of substrates while decreasing the β-isomerization issues, the group of Organ deeply investigated the reactivity of Pd complexes ligated by *N*-Heterocyclic Carbenes (NHCs) displaying a strong bulkiness at the backbones [32]. The ligand design followed relies on two key points: (1) a high steric pressure at the metal will accelerate the reductive elimination compared to the β-hydride elimination process (*vide supra*), and

(2) according to DFT computation, increasing the bulkiness of the *N*-bound aromatic substituents will increase the positive charge on the palladium, thus making faster the reductive elimination process. Those general trends, induced by the stereoelectronic effects of various ligands onto the palladium(II)-to-palladium(0) reductive elimination rates, were moreover discussed by Hartwig [33]. With this in mind, the IPr ligand (Fig. 1.13) appeared as a good candidate for the tuning of the palladium stereoelectronic properties, and its use was successfully applied to highly regioselective alkyl-alkyl Negishi cross-couplings [32].

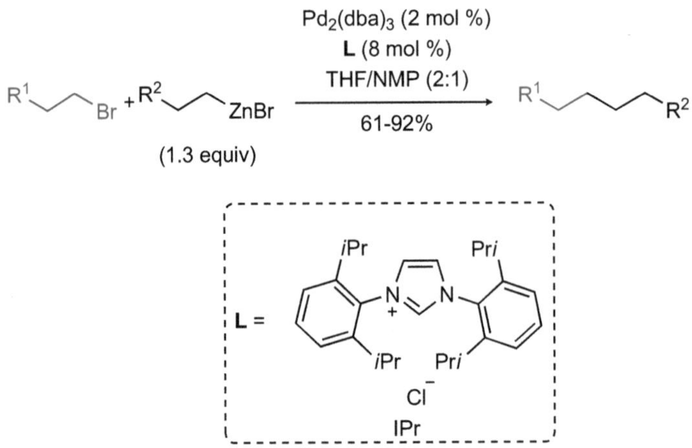

Fig. 1.13: Use of bulky NHCs in alkyl-alkyl coupling by Organ et al. [32].

Later on, the Organ group developed an elegant strategy to ensure the *in situ* formation of mono-coordinated NHC-palladium(0) species under the catalytic regime using easy to handle palladium(II) precatalysts stabilized by labile pyridine derivatives, namely the PEPPSI ligand family (Pyridine-Enhanced Precatalyst Preparation Stabilization and Initiation). As described in Fig. 1.14, the Pd-PEPPSI-NHC complexes can successfully be used as efficient pre-catalysts for Negishi cross-coupling reactions involving aliphatic nucleophiles, an excellent stereoselectivity being observed (see Section 1.5) [34].

The NHC-PEPPSI ligand family is thus considerably used in a variety of Pd-catalyzed Negishi cross-coupling reactions, the versatility of the structures accessible by NHC design allowing a fine-tuning of the stereoelectronic properties of the palladium. The principal performances of those systems were summarized in a review by Organ et al. [35].

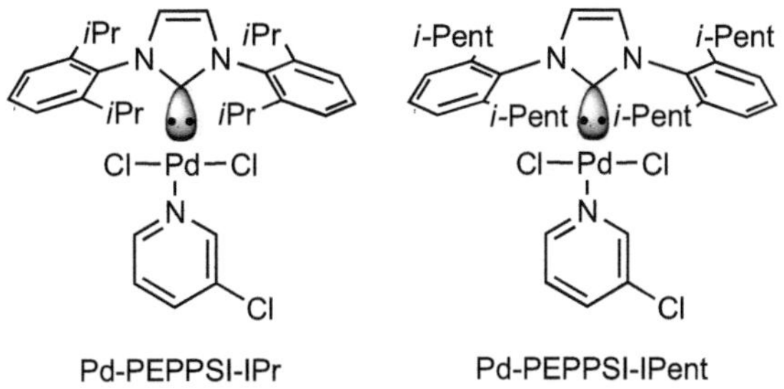

Fig. 1.14: Classic Pd-PEPPSI-NHC systems developed by Organ et al. [34, 35].

1.5 Applications

1.5.1 Arylzinc reagents

1.5.1.1 With aryl electrophiles

Early reports on Pd-mediated cross-coupling reactions by Negishi et al. described efficient aryl-aryl coupling procedures for activated aryl iodides, such as *p*-$NO_2C_6H_4I$ (Fig. 1.1). Later on, Knochel et al. also described selective conditions allowing the cross-coupling reaction of an arylzinc with a functionalized aryl iodide, this method tolerating a variety of functional groups (e.g. pivalates, esters or nitriles) [36]. The same group also reported efficient coupling methods of functionalized arylzincs and functionalized electrophiles displaying a high chemoselectivity (Fig. 1.15). Those results highlighted the crucial choice of the phosphine ligand associated to the $Pd(dba)_2$ precursor, tri-*o*-furylphosphine (tfp) or dppf (Fig. 1.5) showing promising activities [37].

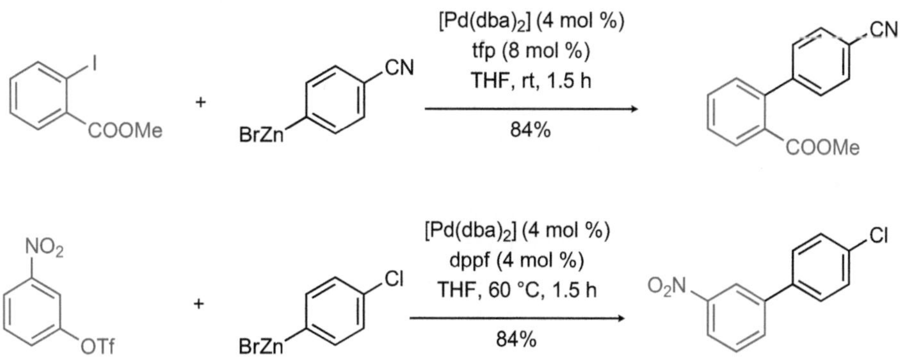

Fig. 1.15: Early examples of aryl-aryl Negishi cross-coupling by Knochel et al. [37].

Beyond the efficient procedures using aryl iodides or bromides, the use of chloroarenes in such cross-coupling reactions is of high industrial interest, since they are usually cheaper and more easily produced than their bromo and iodo analogues. However, non-activated chloroarenes are among the most difficult substrates to use in cross-coupling reactions, due to the strength of the Ar-Cl bond (BDE PhCl = 399.6 kJ/mol, BDE PhBr = 349.4 kJ/mol) [38]. This statement is particularly true when mild nucleophiles such as orgaonozinc are used as cross-coupling partners. By a fine-tuning of the substituents operated on monodentate phosphines, the group of Fu enabled the cross-coupling reaction of a variety of aryl and heteroaryl chlorides with RZnCl nucleophiles (R = aryl, alkyl) using a palladium(0) complex involving a sterically demanding and electron-rich trialkylphosphine, $Pd(P(t\text{-}Bu)_3)_2$ (Figs. 1.11 and 1.16) [24, 25].

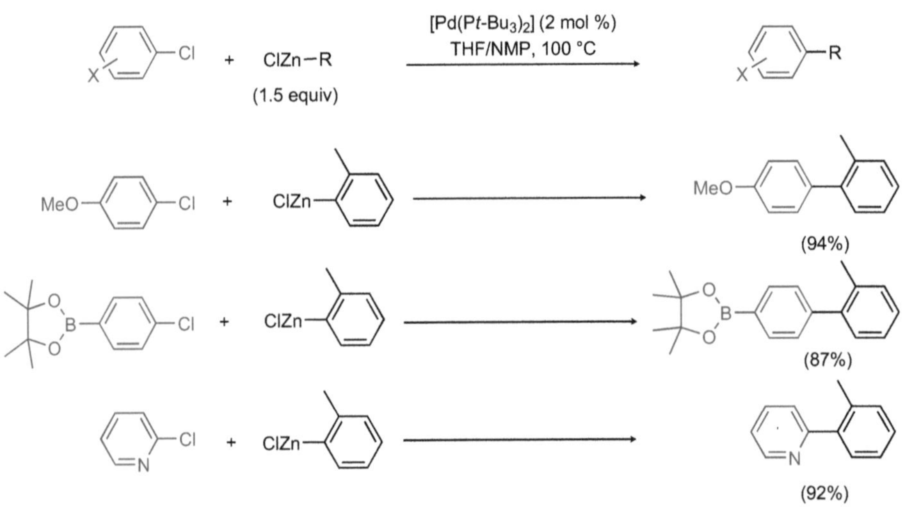

Fig. 1.16: Aryl-aryl cross-coupling reaction mediated by $Pd(P(t\text{-}Bu)_3)_2$ reported by Fu et al. [24, 25].

The high catalytic activity observed with $Pd(P(t\text{-}Bu)_3)_2$ catalyst in the arylation of non-activated electrophilic substrates soon witnessed various synthetic applications, both in natural product synthesis and in materials science. A representative example of the synthesis of hole-transporting materials relying on Negishi cross-coupling reactions has been reported by Tsuji et al. and Nakamura et al. (Fig. 1.17) [39].

As outlined in Section 1.4, PEPPSI ligand series developed by Organ et al. allows to hamper the β-elimination issues observed with aliphatic compounds. Another important milestone met with this family of catalysts deals with their remarkable performance in the promotion of aryl-aryl cross-coupling reactions between highly hindered substrates, allowing the synthesis of tetra-*ortho*-substituted bisaryls (Fig. 1.18) [40]. A high variety of substrates can be obtained with the Pd-PEPPSI-IPent catalyst (Fig. 1.14) in the aryl-aryl and aryl-heteroaryl series in particularly mild conditions, with temperatures ranking

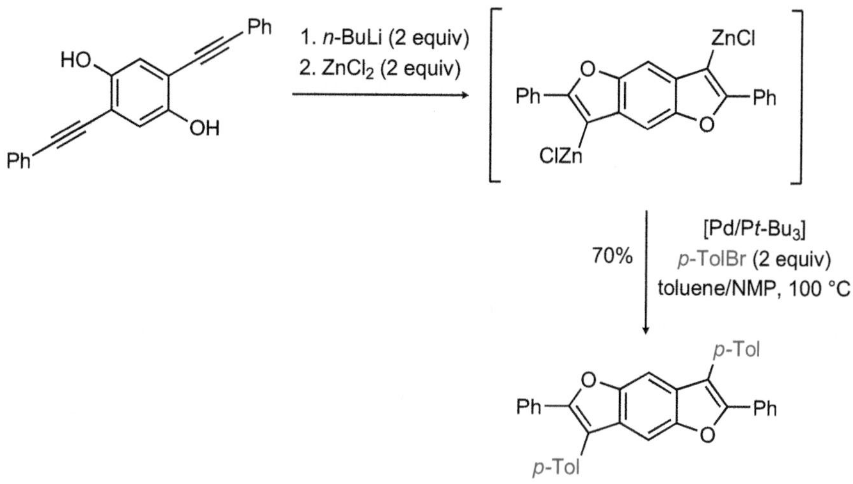

Fig. 1.17: Tandem 1,2-alkyne addition and aryl-aryl cross-coupling by Tsuji et al. and Nakamura et al. [39].

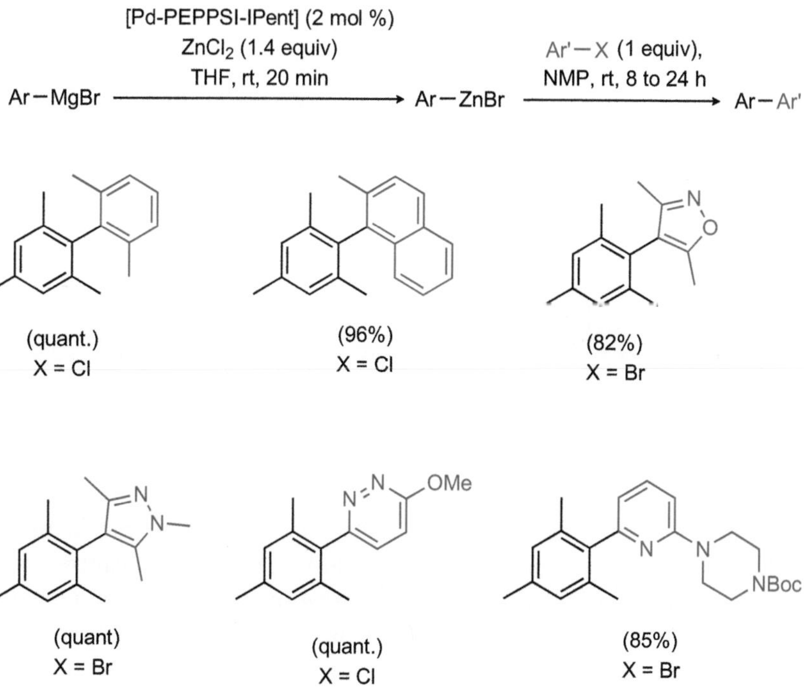

Fig. 1.18: Remarkable performances of Pd-PEPPSI-NHC systems in aryl-aryl coupling reactions involving bulky substrates [40].

between 0 °C and 20 °C. The key of the transformation is, again, an enhanced reductive elimination kinetics in the catalytic cycle allowed by those ligands.

Recently, Espinet and Ponce-de-León reported the challenging synthesis of fluorinated bisaryls by a Negishi aryl-aryl cross-coupling. This method involves a Pd-catalyst coordinated by the alkenylphosphine ligand PhPEWO-F (Fig. 1.19), and promotes the quantitative coupling of $Zn(C_6F_5)_2$ with aryl iodides. It is suggested that the active palladium species generated *in situ* is the mono-coordinated (PhPEWO-F)Pd complex [41]. The electron-withdrawing properties of the olefin moiety of PhPEWO-F ligand proved to facilitate the transmetallation of the nucleophile. The use of an electron-poor ligand to the palladium is thus of particular interest herein since Zn-to-Pd transmetallation of $C_6F_5^-$ requires a strong electrophilicity of the palladium(II) cation given the poor nucleophilicity of this electron-poor anion.

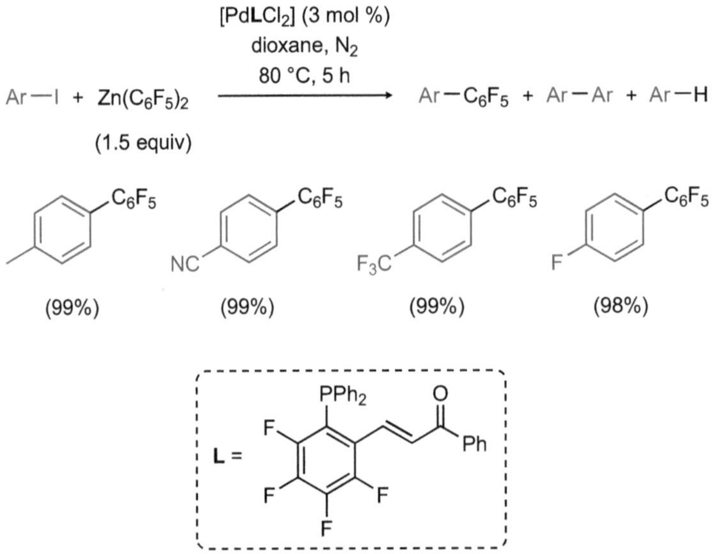

Fig. 1.19: Arylation of pentafluorophenyl nucleophiles by Espinet and Ponce-de-León [41].

Whereas the design of tailor-made ligands of palladium is crucial in order to unlock new reactivity patterns, it is important to remember that ligand-free methods are of particular interest since they are usually less expensive and generate lower quantities of chemical waste. In this regard, de Vries et al. have demonstrated that the use of ligand-free very small amounts of $Pd(OAc)_2$ (typically in the 0.01–0.1 mol % range) efficiently promoted several aryl-aryl Negishi cross-couplings. This remarkably low catalyst loading has been described as a “homeopathic” charge (Fig. 1.20) [42].

Formation of aryl-heteroaryl bond, involving Negishi cross-coupling reactions, is also broadly described. Given the intrinsic acidity of several heteroaryl-H bonds, higher than that of the arene analogues, several cross-coupling methods rely on an

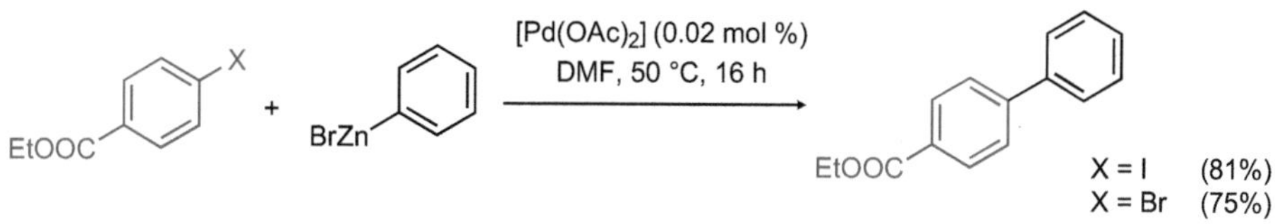

Fig. 1.20: Aryl-aryl Negishi cross-coupling using "homeopathic" palladium charge by de Vries et al. [42].

in situ preparation of the nucleophilic organozinc partner. A metallation of an acidic C-H bond followed by a transmetallation with a zinc salt is usually performed.

One-pot direct zincation of functionalized triazoles by the Knochel base, TMPZnCl•LiCl, followed by a subsequent Pd-mediated heteroaryl-aryl cross-coupling reaction has been reported by Zhang et al. (Fig. 1.21). In this case, a bulky electron-rich phosphine ($PPh(t\text{-}Bu)_2$) is used as an external ligand [43].

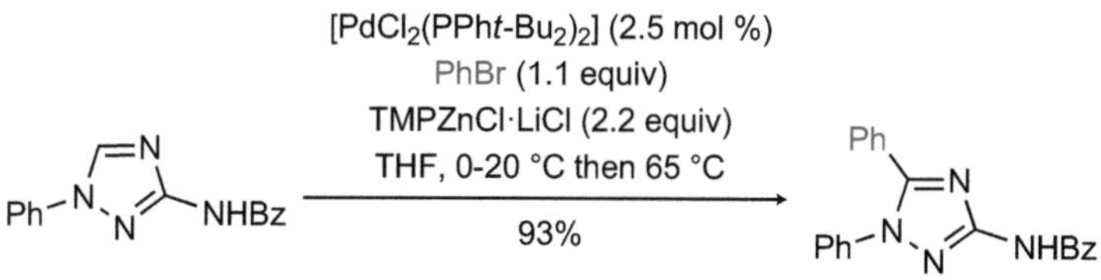

Fig. 1.21: Tandem metallation and heteroaryl-aryl cross-coupling of triazoles [43].

Following this strategy, a particularly elegant method for the α-arylation of functionalized pyrroles relying on a Negishi cross-coupling has been developed by Okano et al., and applied to the total synthesis of several lamellarins [44]. Metallation of α,β-dibromopyrroles by LDA followed by a selective halogen shift affords the more thermally stable β,β'-dibromopyrrole. A key electrophilic trap of the latter by $ZnCl_2$•TMEDA yields the expected α-metallated reagent, which can be involved in a one-pot subsequent Pd-catalyzed aryl-heteroaryl coupling with aryl iodides (Fig. 1.22).

A major drawback of the cross-coupling procedures involving the use of heteroarylzinc derivatives, especially in the case of pyridyl derivatives, is the intrinsic instability of this organometallic reagent, which is difficult to handle, precluding further easily applicable cross-coupling transformations. Buchwald et al. and Knochel et al. reported, in 2013, an original method allowing the preparation of solid-state pyridyl-based organozinc reagents. The key step is the obtention of a pivalate-coordinated zinc(II) center (Fig. 1.23). A variety of solid-state organometallic species could be prepared, which exhibited a moderate air-stability, making their further use particularly easy. Those reagents were used in subsequent cross-coupling procedures with various electrophiles, including activated aryl chlorides. Combination of $Pd(OAc)_2$ and X-Phos ligands (Fig. 1.24) was used as a catalytic system [45].

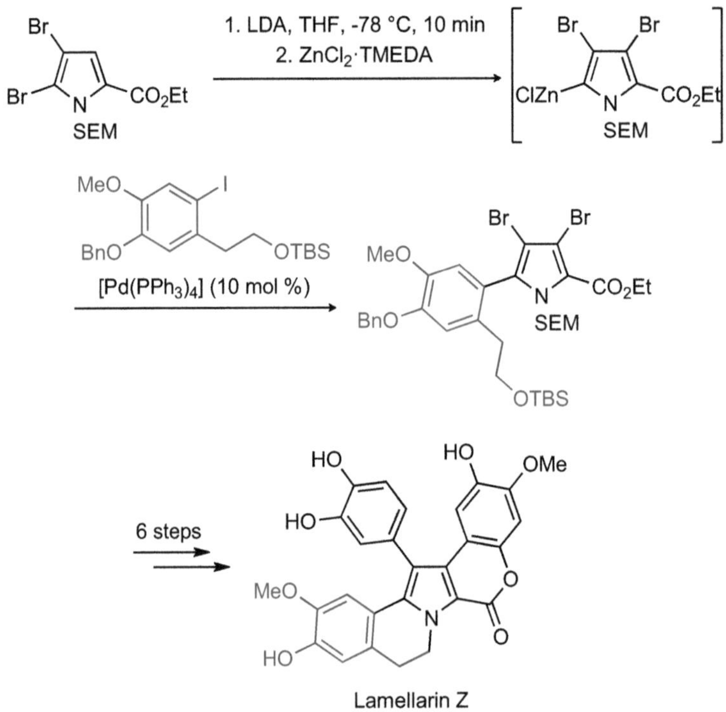

Fig. 1.22: Key heteroaryl-aryl Negishi cross-coupling in the synthesis of lamellarin Z [44].

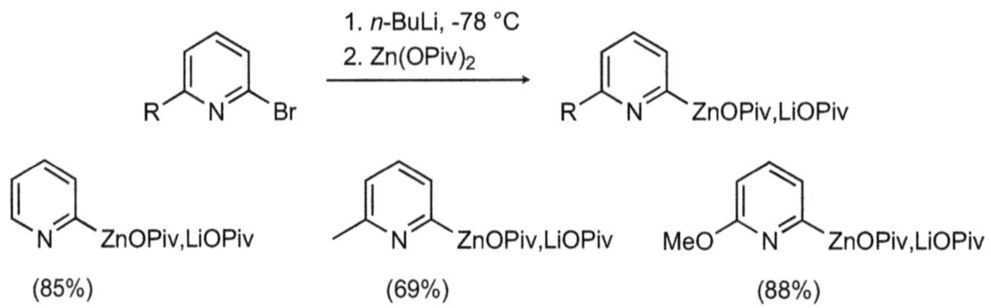

Fig. 1.23: Synthesis of solid-state pyridylzinc reagents stabilized by pivalate-ligated counter-cations by Knochel et al. [45].

Several one-pot procedures allowing the use of NHC-PEPPSI palladium catalysts in aryl-aryl cross-coupling reactions involving *in situ* generated organozinc complexes were also described. Those synthetic procedures are of practical interest since they do not require any manipulation of air-sensitive organozinc reagents [46].

In 2010, Luzung and Yin reported that heteroaryl-heteroaryl cross-coupling reactions between heterocyclic organozincs and heteroaryl chlorides could be mediated by association of palladium(0) precursors with the Buchwald-type ligand X-Phos [47].

In that case, the nucleophilic organozinc reagent is prepared in Knochel conditions, using fast halogen-metal exchange with *i*PrMgCl followed by *in situ* transmetallation with $ZnCl_2$ (Fig. 1.24) (for Knochel's metallation conditions using this strategy, see [48]).

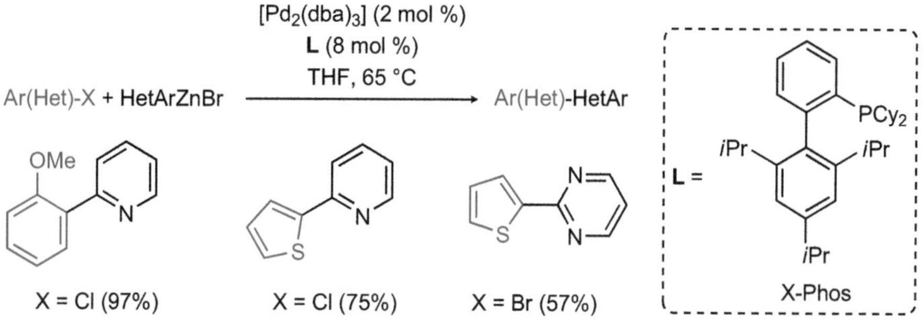

Fig. 1.24: Heteroaryl-heteroaryl coupling by Luzung and Yin [47].

1.5.1.2 With vinyl electrophiles

In line with the efficiency of the aryl-aryl cross-coupling discussed in Section 1.5.1.1, the cross-coupling reaction of arylzinc nucleophiles with electrophilic vinyl partners can easily be carried out using classic palladium sources. For example, highly stereoselective aryl-vinyl cross-coupling reactions involving the tetrakis $Pd(PPh_3)_4$ precursor have been reported, allowing the access to molecules of biological interest. A tamoxifen precursor was synthesized by a cross-coupling reaction between PhZnCl and a vinyl bromide (Fig. 1.25, eq. (1) [49]), and a UB-165 analogue could be synthesized by a cross-coupling reaction between a 3-pyridylzinc nucleophile and a vinyl triflate (Fig. 1.25, eq. (2) [50]).

Aryl-vinyl cross-coupling can also be performed with *in situ* generation of the organozinc partner when the starting pro-nucleophile displays acidic C-H bonds. For example, *ortho*-magnesiation of an aryl ester can be performed using $TMP_2Mg{\bullet}2LiCl$ base. Subsequent transmetallation with $ZnCl_2$ allows *in situ* generation of an arylzinc, which can be coupled with a vinyl iodide in the presence of $Pd(PPh_3)_4$. This sequence has been used to introduce an *ortho* chain in course of the synthesis of a natural product present in the essential oil of *Pelargonium sidoides DC*, a plant from South Africa (Fig. 1.26) [51].

The question of the stereochemical outcome of such cross-coupling reactions is central. Numerous systems proved to operate with a total retention of the stereochemistry with respect to the starting vinyl halide; however, Lipshutz et al. demonstrated that the stereoconvergence of the transformation could also be dramatically impacted by the nature of the combination of ligands and additives. Whereas cross-coupling reaction of $PhZnI{\bullet}LiCl$ with a stereochemically pure (*Z*)-iodoalkene in the presence of $PdCl_2(Amphos)_2$ leads to the formation of the *E* coupling product, addition of TMEDA allows a total retention of the configuration since the *Z* coupling isomer is obtained (Fig. 1.27) [52].

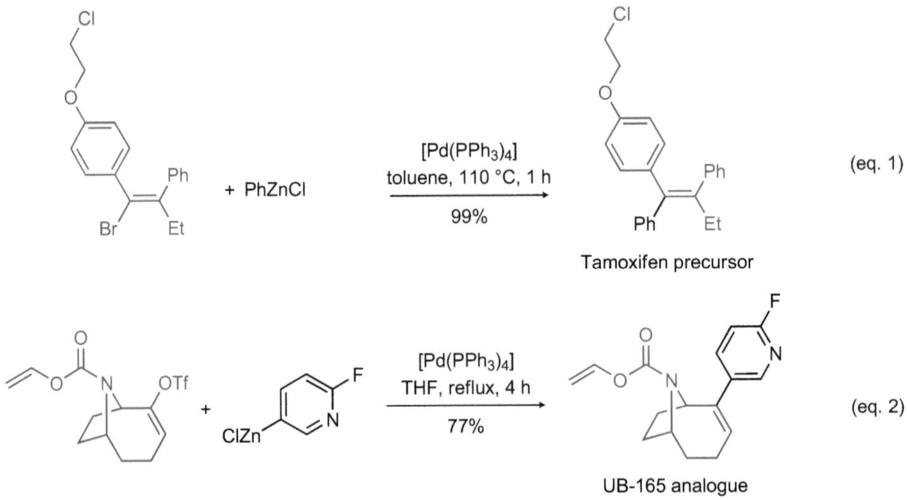

Fig. 1.25: Aryl-vinyl cross-coupling reactions applied to the synthesis of biologically active compounds such as tamoxifen [eq. (1)] and UB-165 analogue [eq. (2)] [49, 50].

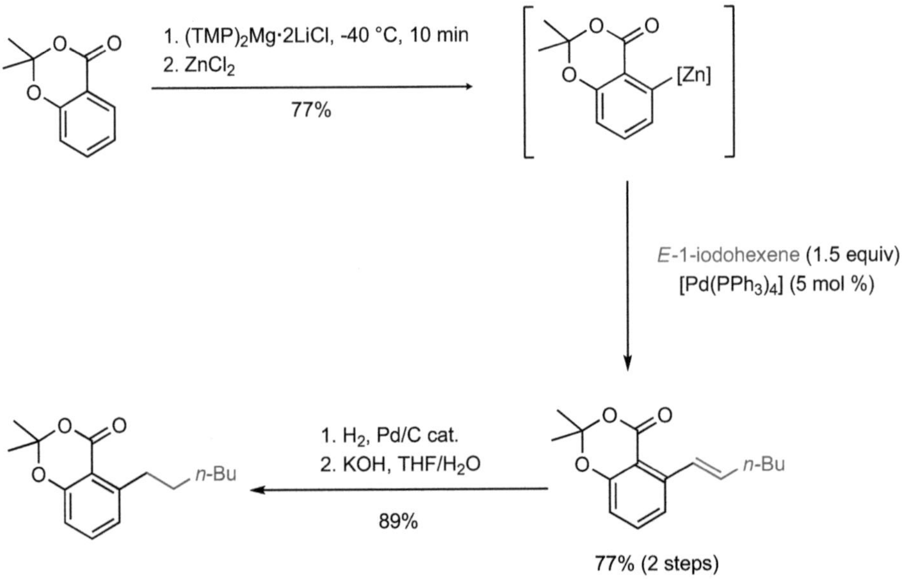

Fig. 1.26: Tandem zincation and aryl-vinyl cross-coupling sequence applied to the synthesis of a natural product [51].

An interesting example of regiodivergent cross-coupling process in the aryl-vinyl Negishi cross-coupling series was reported by Skrydstrup et al. in 2007 using vinyl phosphates as electrophiles. Depending on the nature of the phosphine-based stabilizing

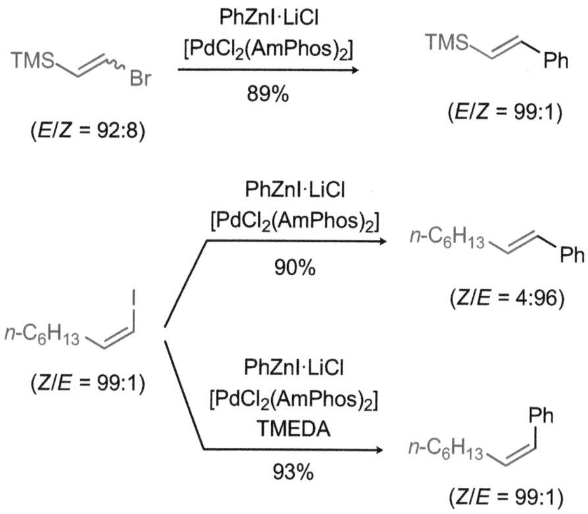

Fig. 1.27: Stereodivergence in aryl-vinyl cross-coupling reactions governed by the presence of TMEDA as additive by Lipshutz et al. [52].

ligand used in the cross-coupling system, two regioisomers can be formed with a high selectivity. When dppf ligand is used, the direct cross-coupling product is obtained, whereas the coupling product formed by a Heck-type 1,2-palladium migration is obtained when (*R*,*S*)-PPF-P*t*-Bu_2 is used (Fig. 1.28). In both cases, yields and selectivities are excellent [53].

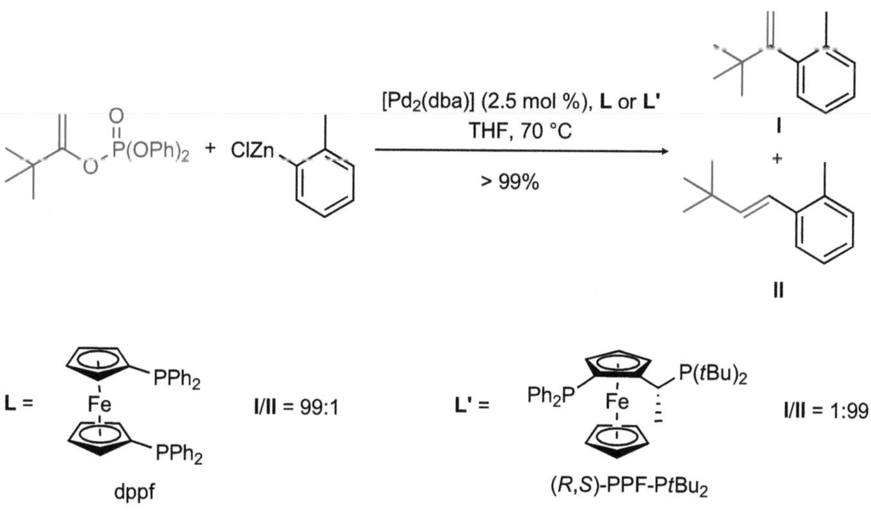

Fig. 1.28: Regioselectivity in aryl-vinyl cross-coupling reactions governed by the nature of the phosphine ligand using arylzincs and phosphate electrophiles by Skrydstrup et al. [53].

1.5.1.3 Alkyl electrophiles

The alkylation of aromatic rings is of high interest in organic synthesis and the alkyl substituted aryl compounds are thus appealing synthetic targets. Numerous methods involving a palladium-mediated Negishi cross-coupling allow the building of an aryl-alkyl bond; however, as it will be detailed in Section 1.5.3.1 of this chapter, the cross-coupling reaction between an alkylzinc nucleophile and an aryl electrophile is usually preferred. A classical reason invoked for explaining the reactivity of alkyl halides toward palladium, lower than that of their sp^2 or sp analogues, is that double or triple C-C bonds can enhance the oxidative addition rate by complexation with the palladium(0) center [54]. Therefore, few examples featuring a cross-coupling between an arylzinc and aliphatic electrophiles have been described. Fu et al. reported several examples of cross-coupling reactions between arylzinc reagents and alkyl electrophiles mediated by the association of Pd_2dba_3 with the electron-rich and sterically demanding phosphine $PCyp_3$ (Fig. 1.5, and 1.29) [55]. Moderate yields are obtained; the key of the efficiency of the cross-coupling reaction is that the strong donating properties of the ligand lead to a particularly nucleophilic palladium(0) catalyst able to activate the C-X bond.

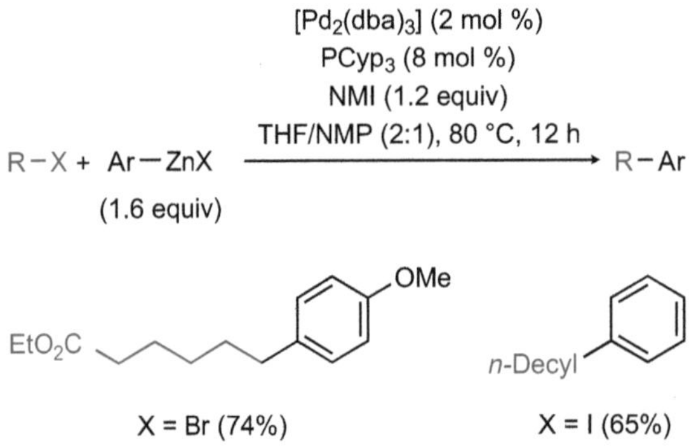

Fig. 1.29: Aryl-alkyl cross-coupling reaction reported by Fu et al., using palladium(0) and $PCyp_3$ system [55].

An interesting example of aryl-alkyl Negishi coupling featuring $[^{11}C]H_3I$ as an electrophile has been reported in 2012, affording ^{11}C-labeled products of interest in PET radiochemistry (see Section 1.7).

1.5.2 Vinylzinc reagents

1.5.2.1 With aryl electrophiles

The vinyl-aryl cross-coupling reaction in Negishi conditions usually affords satisfactory yields, but so far is scarcely explored in literature. The aryl-vinyl version is usually preferred (see Section 1.5.1.2), and a high variety of reactive vinyl halides can easily be prepared. However, several successful vinyl-aryl cross-coupling reactions were reported in the literature, but most of them involve Al- or Zr-based vinyl nucleophiles, these latter reagents being prepared *in situ* by carboalumination [56] or hydrozirconation of alkynes [57], and further subjected to a palladium-catalyzed cross-coupling.

1.5.2.2 With vinyl electrophiles

The stereocontrolled formation of conjugated dienes involving a vinyl-vinyl Negishi cross-coupling is a particularly reliable strategy, which can encompass an important functional group tolerance. This cross-coupling reaction has been used as a key step in several total syntheses of natural compounds (see Section 1.7). In order to illustrate the power and the mild synthetic conditions used to achieve the vinyl-vinyl Negishi cross-coupling, the total synthesis of the xerulin by Negishi et al. can be cited. Xerulin is a complex polyenynyl, inhibitor of cholesterol biosynthesis, which contains six double bonds and two triple bonds (Fig. 1.30). A key step of this synthesis involves a stereoselective vinyl-vinyl cross-coupling reaction between a vinylzinc, formed by transmetallation

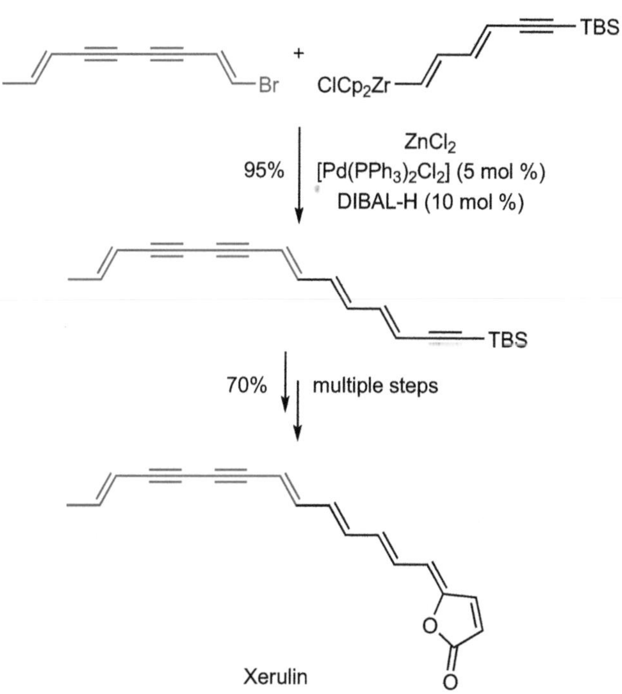

Fig. 1.30: Synthesis of xerulin with a key vinyl-vinyl cross-coupling reaction by Negishi et al. [58].

of an organozirconium with $ZnCl_2$, and a vinyl iodide. An excellent 95% yield was obtained for the cross-coupling step [58].

Illustrating again the high level of functional tolerance, which can be encompassed under palladium-mediated Negishi cross-couplings, a more recent total synthesis of xerulinic acid was performed using a Negishi cross-coupling as a key step featuring a difunctionalized conjugated dimetallotriene (Fig. 1.31). A highly stereo- and chemoselective cross-coupling reaction allowed to build a vinyl-vinyl linkage to access an all-(*E*)-tetraene key intermediate, the organotin moiety remaining unreacted under those conditions, and was further subjected to a Stille vinyl-alkynyl cross-coupling reaction [59].

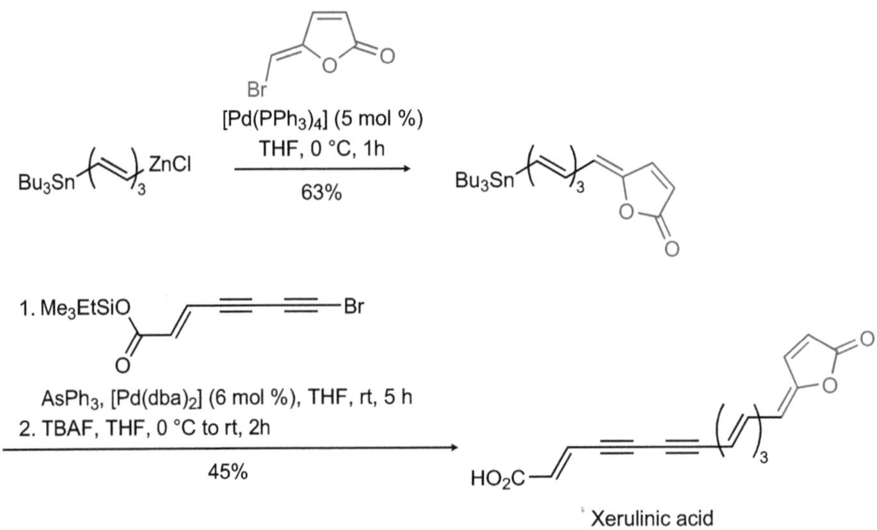

Fig. 1.31: Synthesis of xerulinic acid with a chemoselective vinyl-vinyl cross-coupling involving a dimetallo-Zn,Sn-triene as nucleophile by Brückner et al. [59].

Vinyl-vinyl cross-coupling reactions involving vinylzinc reagents usually proceed with a very good stereoselectivity. A representative example has been reported by Alexakis et al. in the synthesis of the four stereoisomers of methyl dimorphecolate, a natural metabolite of linoleic acid, using $Pd(PPh_3)_4$ as a catalyst [60]. The cross-coupling reaction of stereochemically pure vinylzincs and vinyl iodides affords the corresponding products with an excellent retention of stereochemistry. Two examples are reported in Fig. 1.32.

Similarly, a Negishi vinyl-vinyl cross-coupling was used as the key step in the construction of a diene moiety to produce the C16-C24 synthon of discodermolide by Schreiber et al. in 1996 (Fig. 1.33). In this case, the classic palladium tetrakis precursor $Pd(PPh_3)_4$ has also been used as catalyst [61].

An interesting double Pd-catalyzed Negishi cross-coupling on gem-dibromoalkenes has also been described by Negishi et al. A first cross-coupling reaction affords selectively the (*Z*)-bromoalkene, which then undergoes a second cross-coupling reaction with

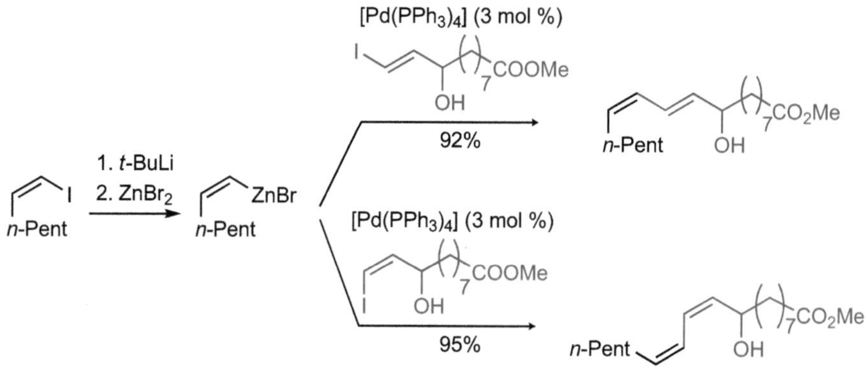

Fig. 1.32: Stereoselective syntheses of dimorphecolate isomers by Alexakis et al. [60].

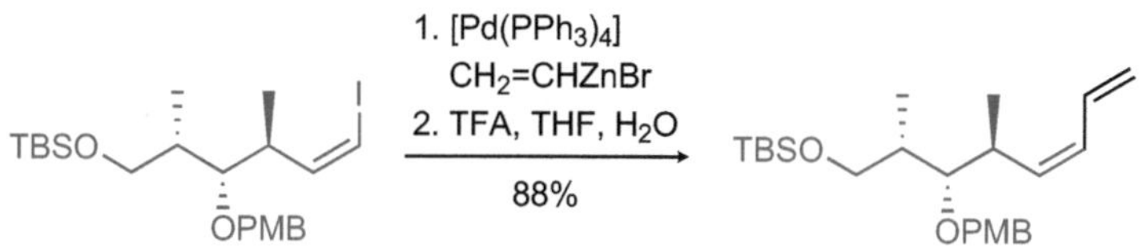

Fig. 1.33: Synthesis of the C16-C24 synthon of discodermolide involving a vinyl-vinyl Negishi cross-coupling by Schreiber et al. [61].

a total inversion of stereochemistry (Fig. 1.34). The full isomerization of the intermediate 2-pallado-1,3-diene was explained by the formation of a transient palladium carbene followed by rotation around the C(α)-C(β) bond. Generation of this carbene intermediate is precluded when double bonds or triple bonds are present in the γ position of the palladium. No isomerization was observed in that case [62]. The stereocontrolled generation of conjugated dienes can thus be easily accessible. This moiety is moreover a key fragment in numerous natural products. In this respect, examples of using the Negishi cross-coupling, leading to complex molecules of natural interest, will be discussed in Section 1.7.

1.5.2.3 With alkyl electrophiles

As discussed in Section 1.5.1.3 of this chapter, reactivity of alkyl halides toward palladium(0) catalysts is generally lower than that of their sp^2 or sp analogues. Therefore, strongly nucleophilic palladium(0) catalysts, stabilized by electron-rich alkyl phosphines, are usually required to activate such aliphatic substrates. Again, the procedure involving an alkylzinc nucleophile and a vinyl electrophile (see Section 1.5.3.2) is usually preferred for building alkyl-vinyl bonds by Negishi coupling procedure.

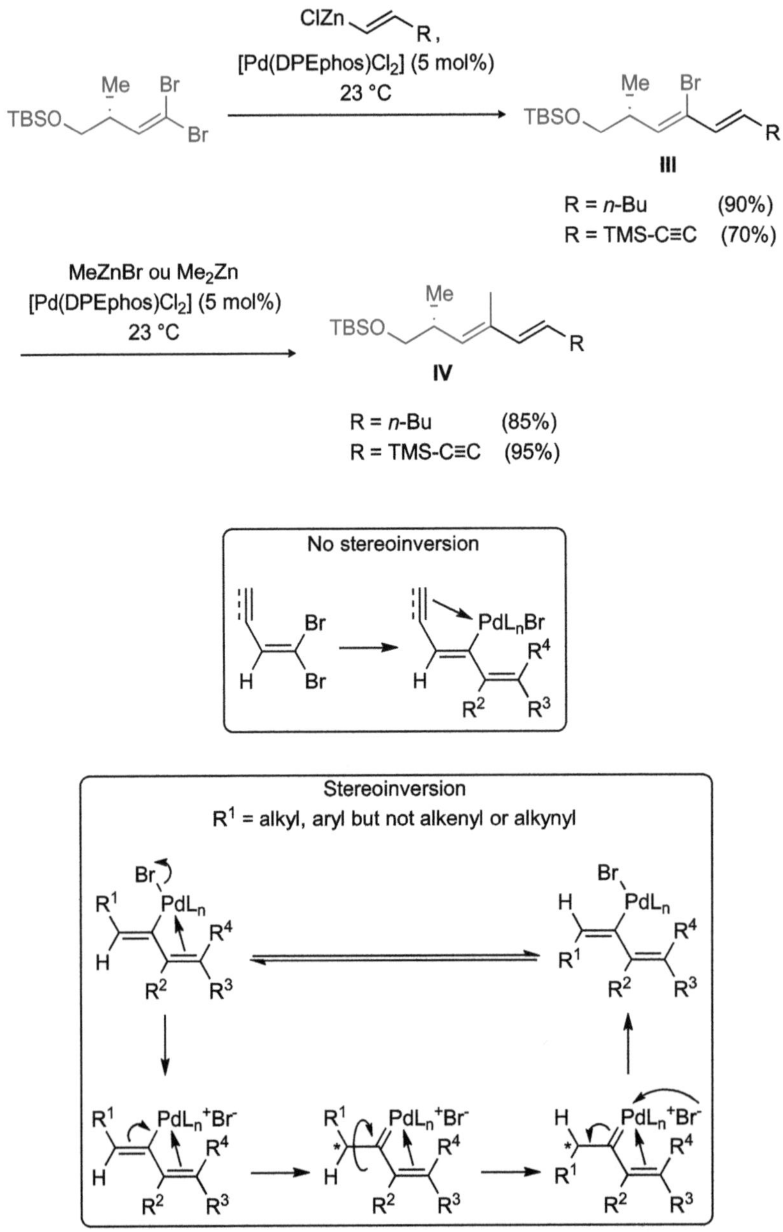

Fig. 1.34: Pd-catalyzed isomerization of 2-pallado-1,3-dienes in the course of vinyl-vinyl cross-coupling reactions [62].

Among the rare examples of vinyl-alkyl palladium-mediated Negishi couplings, Fu et al. reported that moderate to good yields could be achieved using combination of palladium(0) sources such as Pd_2dba_3 with the electron-rich bulky phosphine $PCyp_3$ (Fig. 1.5)

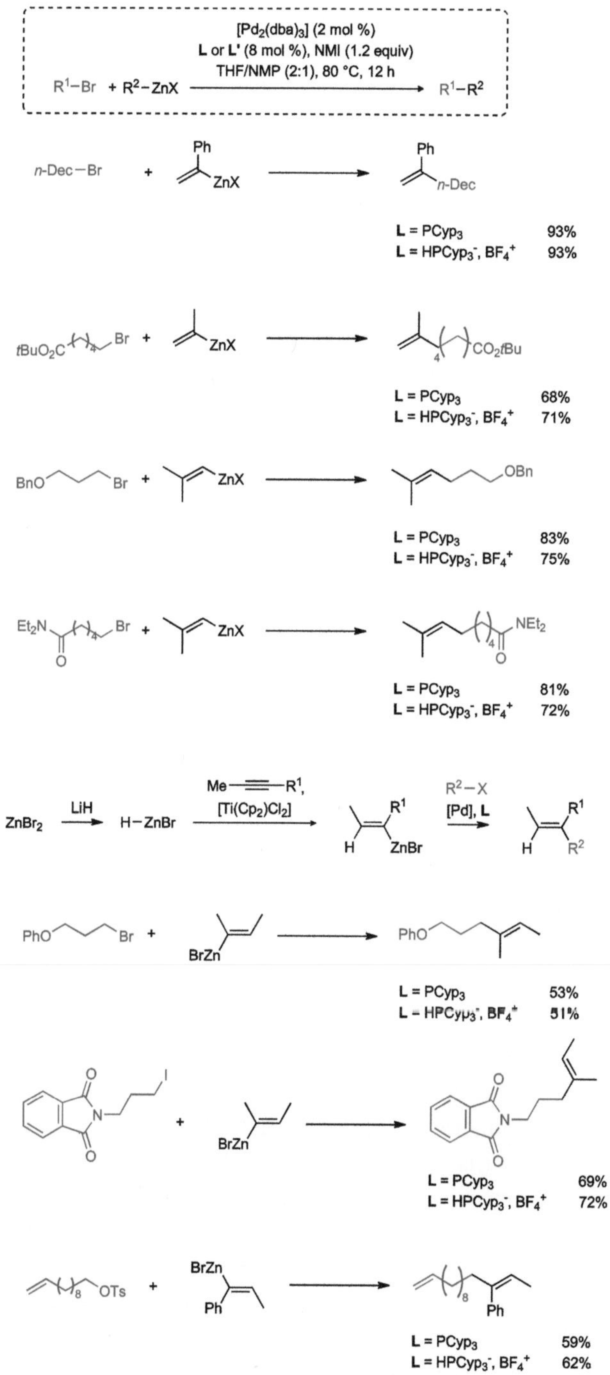

Fig. 1.35: Vinyl-alkyl cross-coupling reactions mediated by a palladium(0) source in the presence of $PCyp_3$ sources reported by Fu et al. [55].

[55]. This method has been applied to a broad range of alkyl electrophiles; vinylzinc reagents can either be used as a starting material, or be synthesized *in situ* by hydrozincation of acetylenic partners in the presence of a catalytic amount of Cp_2TiCl_2 (Fig. 1.35).

1.5.3 Alkylzinc reagents

1.5.3.1 With aryl electrophiles

In the quest for the development of palladium(0) catalysts ligated by two, or one monophosphine ligands, an inspiring preliminary result was reported by Lei et al. in 2007 [63]. In this work, the excellent catalytic performances of a palladium salt associated with a monophosphine bearing an electron-poor double bond is reported, applied to a difficult alkyl-aryl cross-coupling reaction of *ortho*-substituted aromatic substrates (Fig. 1.36). It is expected that the coordination of the palladium by the π-acceptor olefin ligand will lead to a less electron-rich metal, thus promoting the reductive elimination and transmetallation steps. This system was efficient enough for scalable conditions (up to 50 g-scale) and significantly low catalytic loading (0.005 mol %).

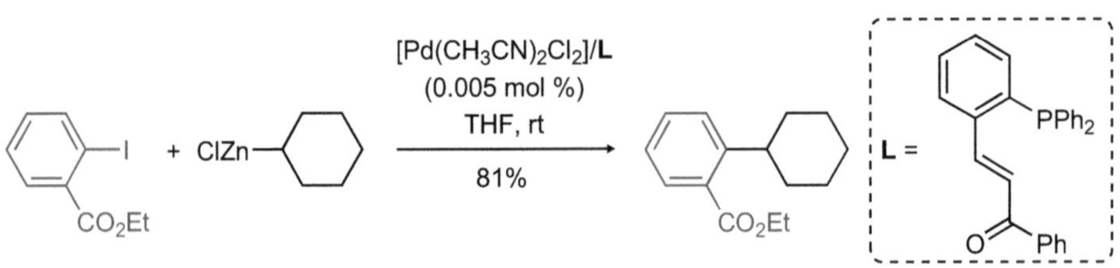

Fig. 1.36: Alkyl-aryl cross-coupling reaction of sterically hindered substrates involving electron-poor phosphines as Pd-ligands by Lei et al. [63].

The use of monophosphines substituted by electron-poor chelating groups is a particularly judicious choice for coupling systems involving energetically demanding reductive eliminations, such as *o*-arylester groups. Conversely, in such cases, electron-rich Fu-type phosphines such as $P(t\text{-Bu})_3$, tailored to enhance kinetically determining oxidative addition steps, do not ensure satisfying cross-coupling yields.

As outlined in Section 1.4 of this chapter, the PEPPSI ligand family displays a very good catalytic activity in cross-coupling reactions involving secondary alkylzincs as nucleophilic partners, hampering the β-elimination process of transient alkyl palladium(II) intermediates. Synthesis of functionalized aromatic cycles by alkyl-aryl cross-coupling reaction between either aryl chlorides or bromides with secondary alkylzinc partners using 2 mol % Pd-PEPPSI-IPent (Fig. 1.14) was reported by Organ et al. [34]. Again, the scope of this transformation shows a particularly broad functional group tolerance and can lead to very good yields with electron-poor aryl chlorides (up to 95%) (Fig. 1.37).

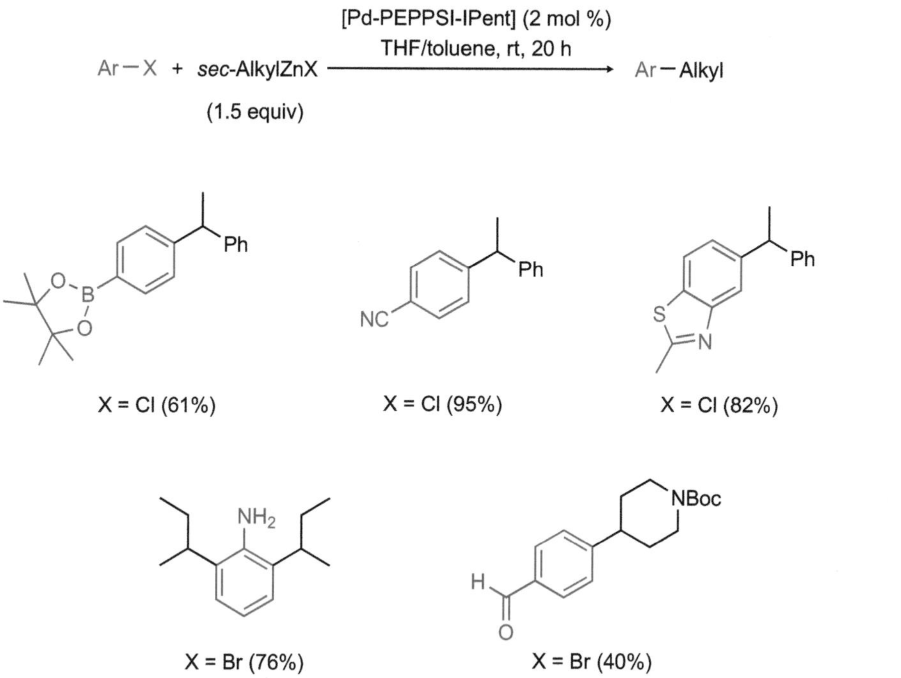

Fig. 1.37: Efficient cross-coupling of secondary alkylzincs with aryl halides by Organ et al. using Pd-PEPPSI-NHCs catalysts [34].

Particularly appealing procedures for alkyl-aryl cross-coupling reactions involving Buchwald-type ligands were also reported (Fig. 1.38). For example, excellent yields could be obtained with only 1 mol % of palladium associated with C-Phos ligand for the cross-coupling reaction of secondary alkylzincs with bromochromanes [64].

Among the sophistically tuned monophosphine ligands which display a strong efficiency in Pd-Negishi cross-coupling reactions, the extremely bulky and electron-rich Q-Phos ligand developed by Hartwig et al. (Fig. 1.11) allowed the cross-coupling of Reformatsky reagents of acetate enolates with unactivated chloroarenes (Fig. 1.39) [65]. Owing to the low stability of zinc acetate enolates, this coupling could not be achieved even with Fu catalyst $Pd(P(t\text{-}Bu)_3)_2$. It is also worth mentioning that the Q-Phos system can be applied to Reformatsky reagents synthesized *in situ* from the corresponding alkyl halide and Rieke's activated zinc dust [66]. For zinc enolates with a higher substitution degree, which are thermally more stable than their acetate analogues, the palladium(I) dimer $\{[P(t\text{-}Bu)_3]PdBr\}_2$ efficiently promoted the desired transformation (Fig. 1.40) [66].

A more recent example developed by the Jackson group highlights the high potential of Pd-catalyzed Negishi cross-couplings in late-stage functionalization processes in the field of amino-acid synthesis. A key-intermediate toward the synthesis of cyclic peptide OF4949-III is indeed achieved by an alkyl-aryl Negishi cross-coupling with a

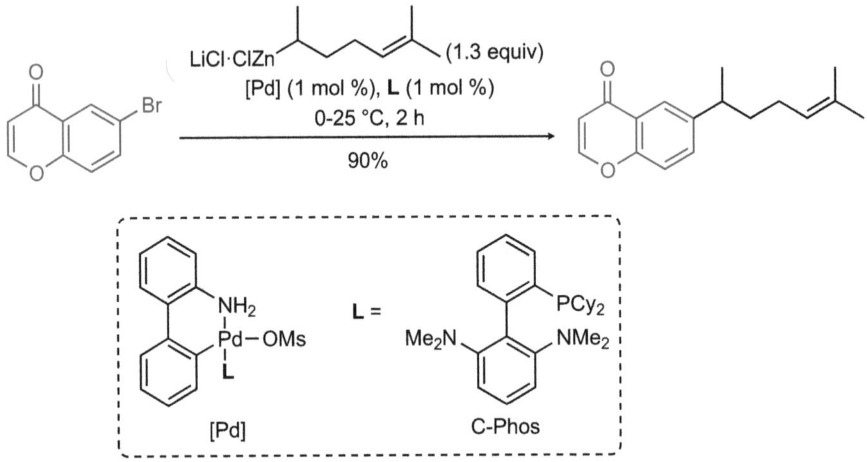

Fig. 1.38: Alkyl-aryl cross-coupling applied to functionalization of chromanes by Buchwald et al. [64].

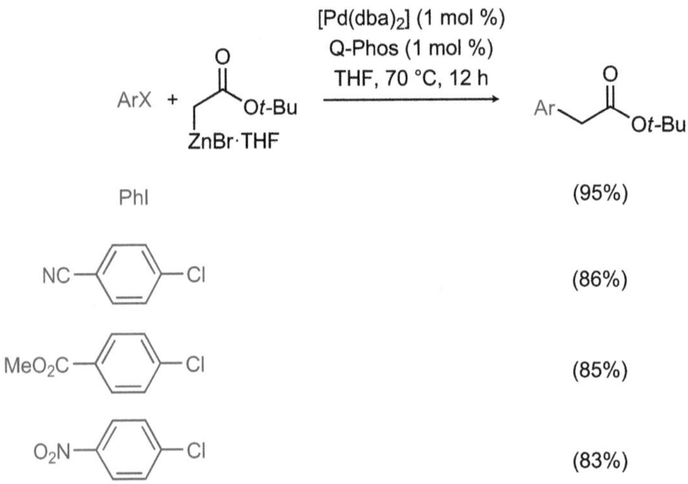

Fig. 1.39: Alkyl-aryl cross-coupling of Reformatsky enolates mediated by palladium(0) with electron-rich Q-Phos ligand [66].

particularly high functional group tolerance, including the presence of acidic NH bonds in the starting electrophile (Fig. 1.41) [67]. The application of Negishi cross-coupling reactions in the field of the synthesis of amino-acids has been recently covered by a review by Cobb and co-workers [68].

Several procedures for the cross-coupling reaction between alkylzincs and heteroaryl electrophiles were also described. Organ et al. demonstrated that secondary alkylzincs could be efficiently coupled with 2-bromopyridines using a modified Pd-PEPPSI-IPent catalyst, with an excellent 94% yield (Fig. 1.42) [69].

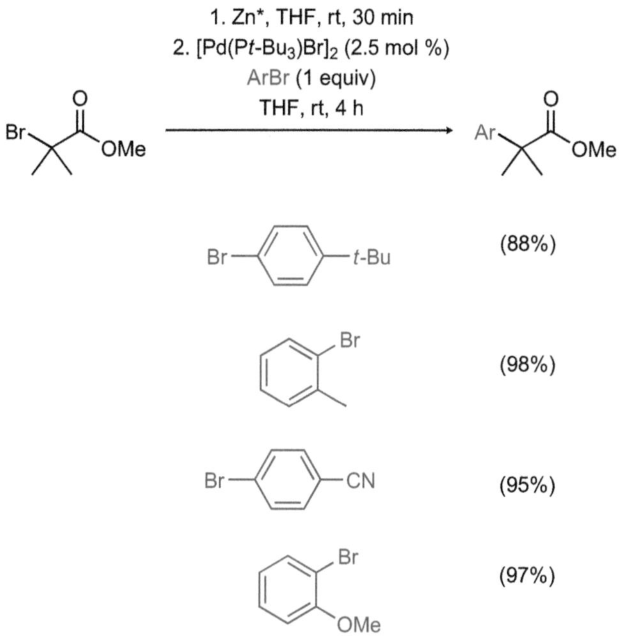

Fig. 1.40: Tandem formation of Reformatsky substituted enolates and alkyl-aryl cross-coupling reactions catalyzed by a palladium(I) precursor [66].

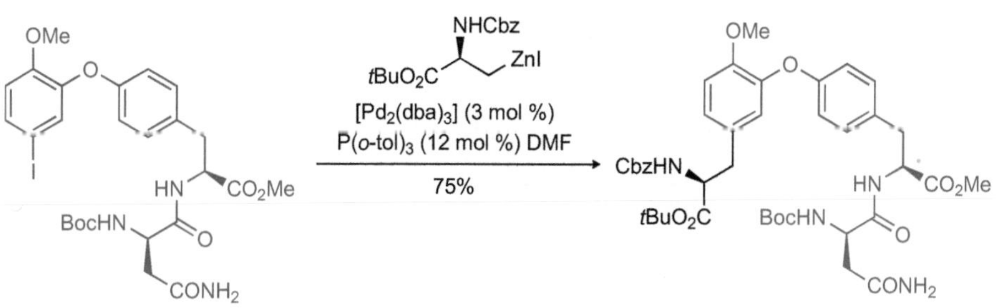

Fig. 1.41: Synthesis of cyclic peptide OF4949-III featuring a key alkyl-aryl Negishi cross-coupling by Jackson et al. [67].

Knochel et al. also reported a very efficient procedure for the cross-coupling of cycloalkylzincs with 2-iodopyridines (Fig. 1.43). A high diastereoselectivity is observed, and the palladium precatalyst was associated with a sterically demanding electron-rich phosphine ligand [70].

It is worth mentioning that alkyl-aryl Negishi-type cross-coupling reactions can also be efficiently promoted under Barbier conditions, starting from a mixture of alkyl and aryl (pseudo)halides. A representative example has been reported by Baudoin et al.,

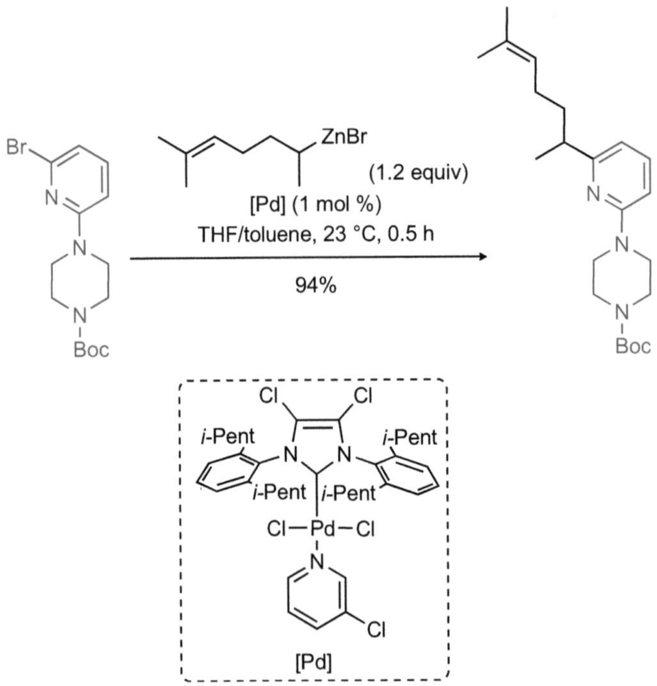

Fig. 1.42: Alkyl-heteroaryl coupling by Organ et al. with a modified Pd-PEPPSI-IPent catalyst [69].

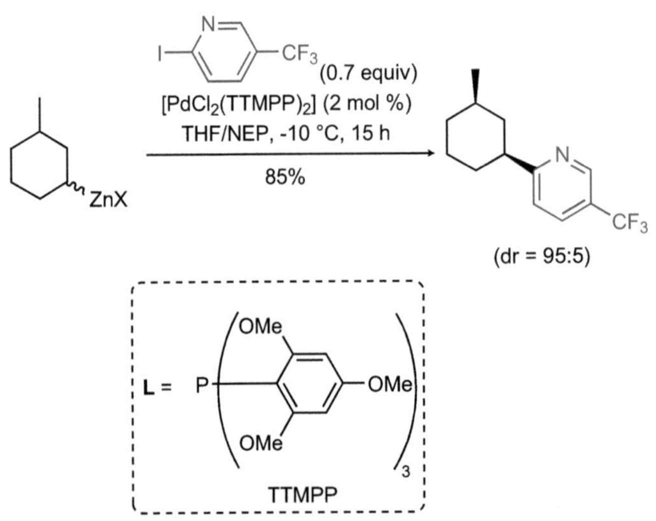

Fig. 1.43: Stereoconvergent alkyl-heteroaryl cross-coupling by Knochel et al. (NEP = *N*-ethyl-2-pyrrolidone) [70].

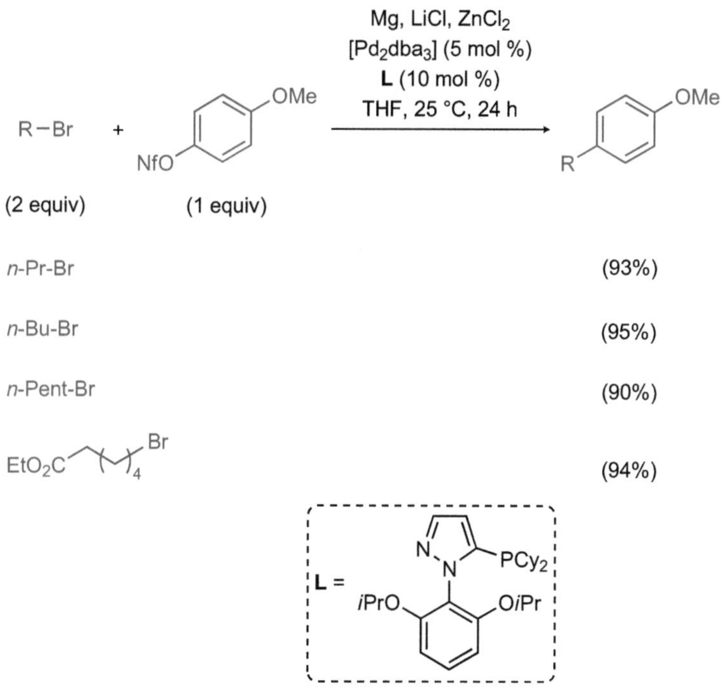

Fig. 1.44: Alkyl-aryl cross-coupling under Barbier conditions by Baudoin et al. [71].

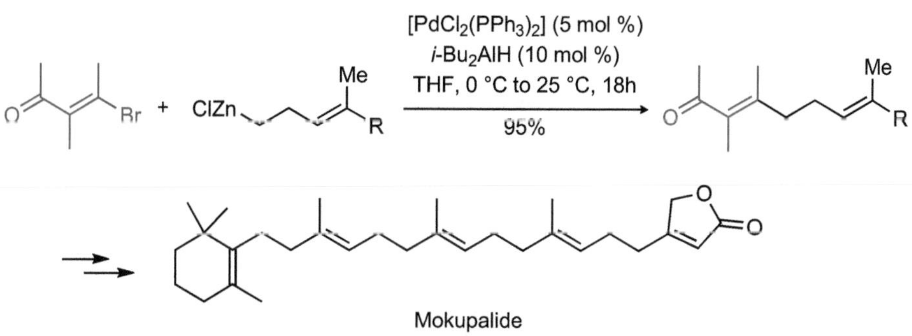

Fig. 1.45: Alkyl-vinyl Negishi coupling applied to mokupalide synthesis by Negishi et al. [72].

using a Buchwald-type ligand (Fig. 1.44) associated to Pd_2dba_3. In this case, selective insertion of Mg into an aliphatic halide followed by transmetallation with $ZnCl_2$ under Knochel conditions (use of LiCl additive) affords *in situ* an aliphatic nucleophile, which undergoes a cross-coupling reaction with either aryl triflates or nonaflates [71].

1.5.3.2 With vinyl electrophiles

Alkenylation of alkylzinc nucleophiles using a palladium catalyst is one of the most efficient Negishi cross-coupling which can be carried out. In line with his early reports, Negishi et al. demonstrated than homoallylzinc nucleophiles could efficiently be coupled with β-bromosubstituted α,β-unsaturated carbonyl derivatives, using either $Pd(PPh_3)_4$ or $PdCl_2(PPh_3)_2$ associated with i-Bu_2AlH as a reductant. One illustration of this cross-coupling reaction was the introduction of an alkyl-vinyl linkage of the lactone part of mokupalide (Fig. 1.45) [72].

The generation of stereochemically pure alkylzincs also proved to be a pivotal challenge in alkyl-vinyl cross-coupling reactions. Knochel et al. reported several efficient procedures for the generation of stereochemically pure alkylzinc compounds, which could be used successfully as coupling partners along with vinyl halides in diastereoselective processes (Fig. 1.46) [73].

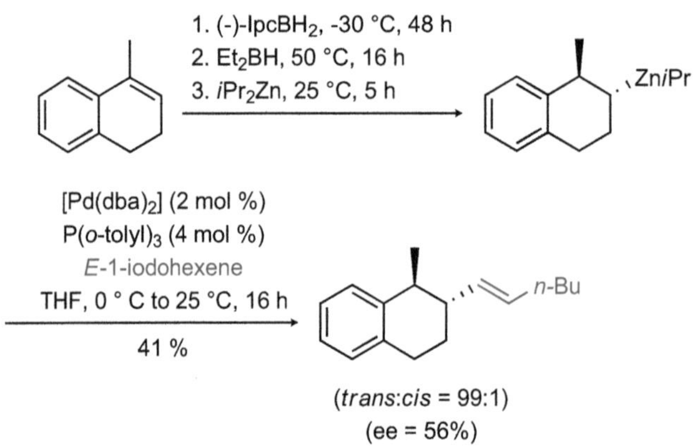

Fig. 1.46: Generation of stereochemically pure alkylzincs by Knochel et al. followed by stereospecific Negishi cross-coupling [73].

Alkyl-vinyl cross-coupling reactions also proceed with a remarkable chemoselectivity. A representative example of chemoselective alkyl-vinyl procedure has been reported by Negishi et al. in 2002, which perfectly tolerates the presence of alkyl electrophiles (Fig. 1.47). In that case, $Pd(dppf)Cl_2$ (Fig. 1.5) is used as a catalyst under very smooth conditions. The difference of reactivity between the alkyl and vinyl iodides has been explained by the presence of a double bond in the latter, which enhances by a 100-fold factor its reactivity toward the palladium(0) catalyst [74].

Numerous total syntheses of natural products were developed using a Negishi alkyl-vinyl cross-coupling as the key step. One can cite the example of the construction of the sp^3-sp^2 linkage in the course of the total synthesis of (-)-discodermolide (Fig. 1.48) [75, 76]. It is noteworthy that this efficient cross-coupling reaction could be

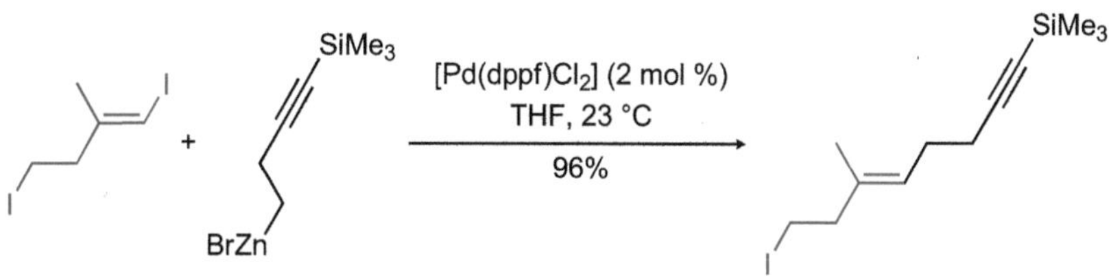

Fig. 1.47: Chemoselective alkyl-vinyl cross-coupling in the presence of alkyl electrophiles [74].

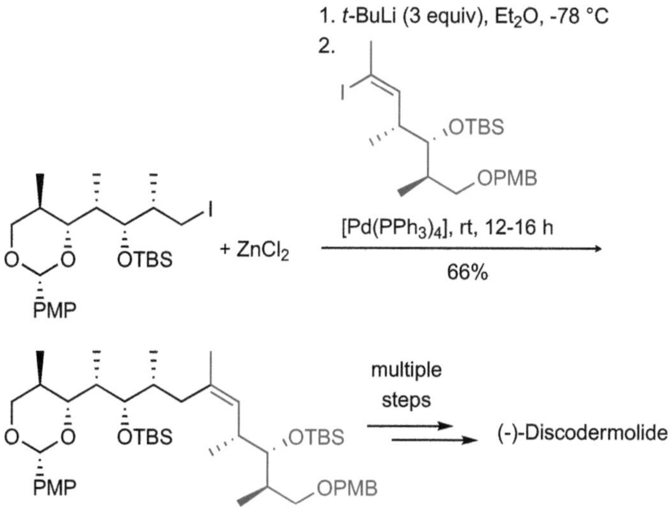

Fig. 1.48: Introduction of an alkyl-vinyl key linkage in the synthesis of (-)-discodermolide [75, 76].

carried out using the classic tetrakis $Pd(PPh_3)_4$ catalyst. It has been demonstrated, in this work, that the relative alkyl/Li/Zn ratio is of high importance for the success of the cross-coupling reaction. For instance, in the (-)-discodermolide synthesis, the best coupling performances were obtained by premixing *t*-BuLi (3 equiv) with the alkyl iodide reagent in the presence of $ZnCl_2$ prior to the addition of the vinyl electrophile.

A highly chemoselective cross-coupling reaction example has been reported by Cossy, Meyer et al. in 2001 for the synthesis of a derivative of 4*a*,5-dihydrostreptazolin, a potent antibiotic and antifungal compound (Fig. 1.49) [77]. The selective alkylation of a vinyl electrophile by an alkylzinc using $Pd(PPh_3)_4$ catalyst is also a key step in the synthesis of (*Z*)-γ-bisabolene reported by Negishi et al. (vinyl iodide) [78], as well as in synthesis of brevetoxin A realized by Nicolaou et al. (vinyl triflate) [79].

Moreover, similarly to what was discussed in Section 1.5.1.2 for aryl-vinyl couplings, the nature of the palladium ligands and external additives plays a crucial role in the stereochemical outcome of the alkyl-vinyl cross-coupling step. When $PdCl_2(Amphos)_2$ is used as the catalyst in the alkyl-vinyl coupling of (*Z*)-vinyl iodides in the

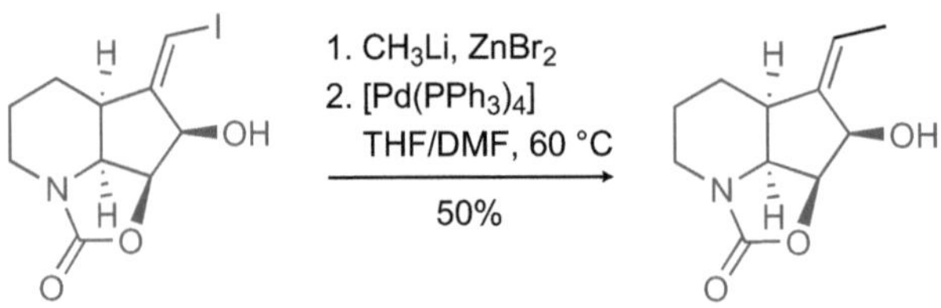

Fig. 1.49: Alkenylation of a methyl nucleophile used in the total synthesis of a *4a*,5-dihydrostreptazolin analogue by Cossy, Meyer et al. [77].

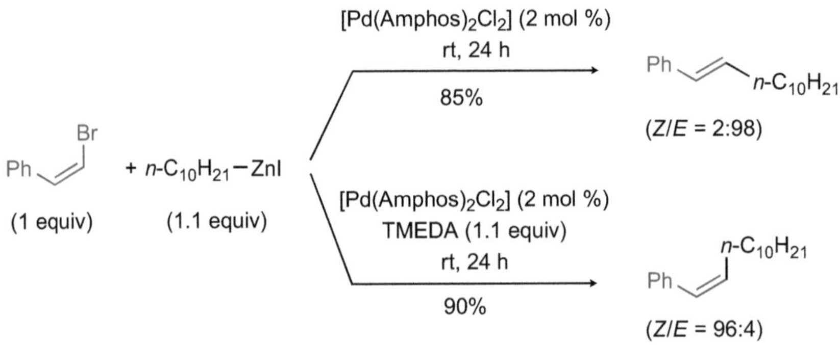

Fig. 1.50: Stereodivergence in an alkyl-vinyl cross-coupling governed by the use of TMEDA as additive by Lipshutz et al. [52].

absence of any additive, a total inversion of configuration is observed and the (*E*)-cross-coupling product is selectively obtained. When TMEDA is used as an additive, a total retention of the configuration is observed, the cross-coupling product (*Z*) being selectively formed (Fig. 1.50) [52].

1.5.3.3 With alkyl electrophiles

As discussed in Section 1.4 of this chapter, the alkyl-alkyl cross-coupling reaction is the most challenging cross-coupling to be achieved. Indeed, decomposition of active alkyl-palladium(II) species by β-hydride elimination pathways easily occurs, the alkyl chain coming from either the nucleophile or the electrophile. Therefore, the development of efficient coupling methods required the use of tailor-made ligands, classical precursors such as $Pd(PPh_3)_4$ failing at providing satisfactory yields. In 2003, Fu et al. reported that the electron-rich and sterically demanding trialkylphosphine $P(Cyp)_3$ (Fig. 1.5) could be associated with Pd_2dba_3 to achieve efficient alkyl-alkyl cross-coupling with a particularly good functional group tolerance (Fig. 1.51). This method can be applied efficiently to a variety of alkyl halides (I, Br, Cl) and pseudo-halides such as tosylates [55].

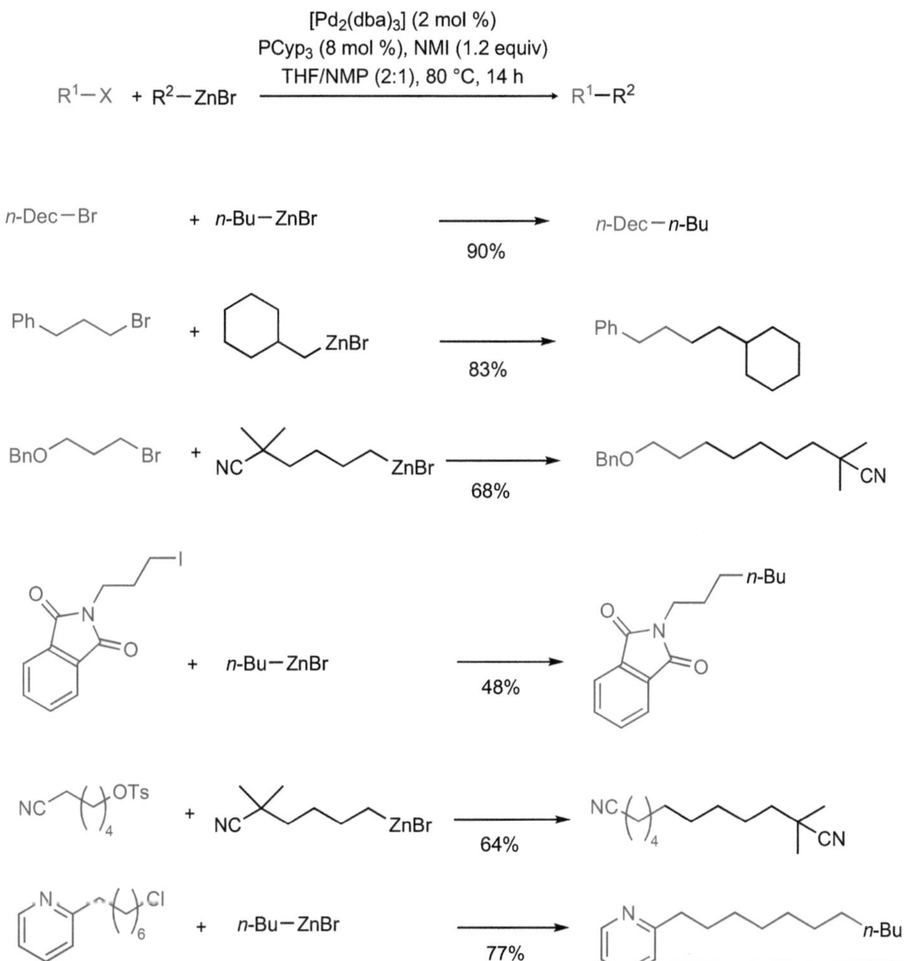

Fig. 1.51: Alkyl-alkyl coupling method relying on the use of an electron-rich phosphine ligand by Fu et al. [55].

Later on, considering the need of kinetically bypassing the β-elimination unwanted pathways, several groups developed bulky NHC ligands which enhanced the rate of the alkyl-alkyl reductive elimination (see Sections 1.3 and 1.4). Seminal alkyl-alkyl cross-coupling reaction using NHCs, as stabilizing platforms, was reported by Organ et al. (Fig. 1.13) [32]. In particular, it was demonstrated that SIPr provided satisfactory yields in the cross-coupling of functionalized alkylzincs with alkyl bromides (Fig. 1.52). A very good chemoselectivity was observed, since the C-Br bond of α,ω-bromochloroalkanes could be selectively transformed, opening the door for post-functionalization of the C-Cl bond by other cross-coupling reactions [80].

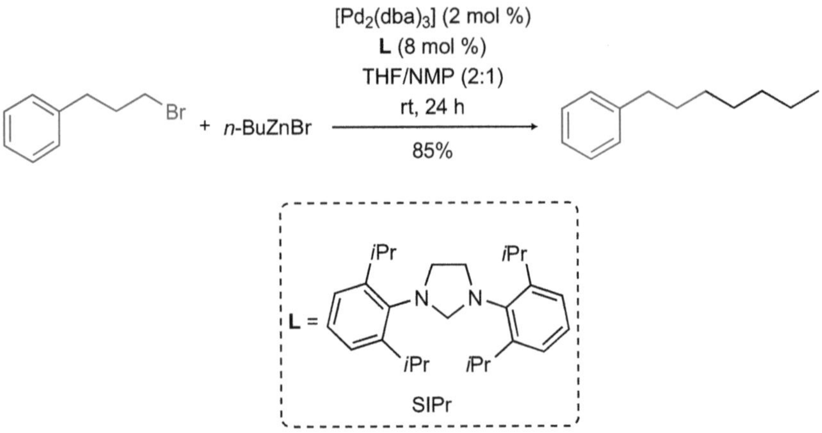

Fig. 1.52: Efficient alkyl-alkyl Negishi cross-coupling using NHC-stabilized palladium(0) catalysts by Organ et al. [80].

More recently, Organ et al. highlighted the particular efficiency of high-order zincates as transmetalating reagents in alkyl-alkyl cross-coupling reaction (see Section 1.3) [13]. Dianionic zincates such as $RZnBr_3^{2-}$ led to extremely selective cross-coupling products by reaction with alkyl bromides (Fig. 1.53) [13]. Tolerance of tertiary nitriles is also displayed by this system, catalyzed by the Pd-PEPPSI-IPent complex (Fig. 1.14). Those results echo the more classic use of halide salts as additives in various methods reported by Knochel et al., which, for example, enable the access to high-order cuprates in copper-catalyzed cross-couplings [81, 82].

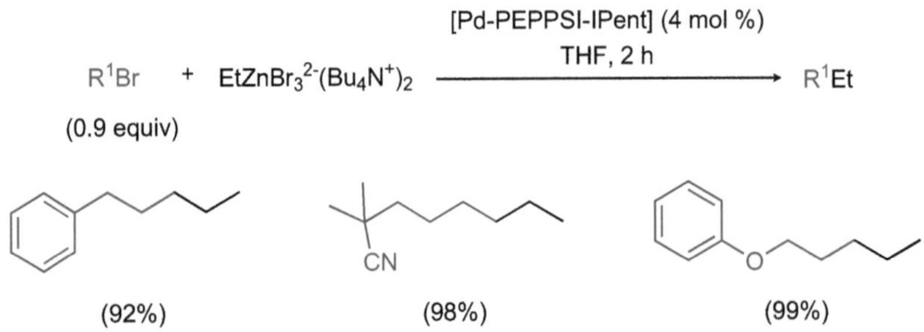

Fig. 1.53: High-order zincates as reactive species in alkyl-alkyl cross-couplings by Organ et al. [13].

1.5.4 Alkynylzinc

Palladium-mediated couplings involving alkynylzincs as nucleophilic partners were soon developed in this field. A review by Negishi et al. summarized the first pioneering examples of those transformations [83].

1.5.4.1 With aryl electrophiles

Early reports by Negishi et al. in 1978 highlighted the high efficiency of the cross-coupling reaction between alkynylzinc reagents and (hetero)aryl halides in the presence of $Pd(PPh_3)_4$ as a precatalyst (Fig. 1.54) [8]. Numerous examples of (hetero)aryl halides could be used as coupling partners, and some couplings also included the organozinc reagent derivatized from acetylene [84].

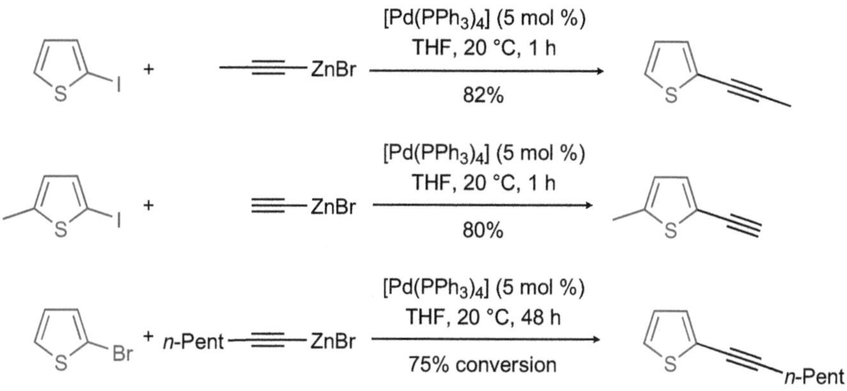

Fig. 1.54: Alkynyl-heteroaryl cross-couplings by Negishi et al. [8].

Those methods were soon extended to a variety of substrates, such as conjugated diynes, and proved to afford better yields than the classic copper-catalyzed Cadiot-Chodkiewicz procedure [85]. The nucleophilic alkynylzinc partner can be generated *in situ* from a vinyl chloride by treatment with *n*-BuLi followed by addition of $ZnBr_2$ (Fig. 1.55) [86].

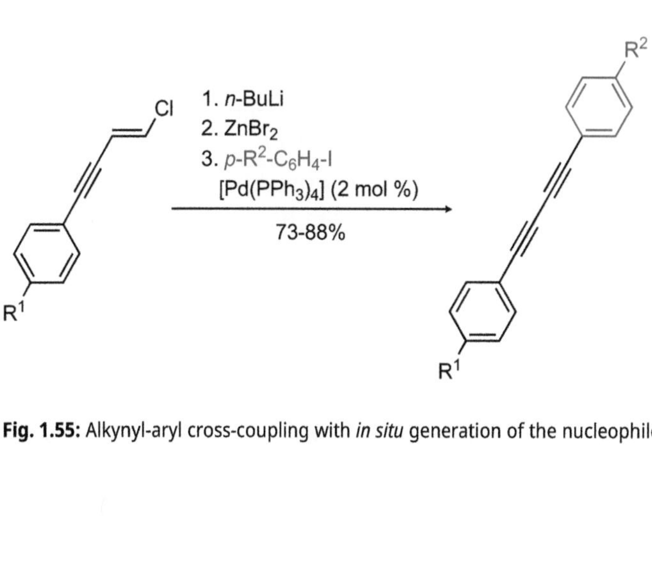

Fig. 1.55: Alkynyl-aryl cross-coupling with *in situ* generation of the nucleophile from a vinyl chloride [86].

Conjugated lithium acetylides trapped with $ZnCl_2$ can also afford suitable alkynylzinc partners for a Negishi cross-coupling using 1,1-dibromoolefins as a starting material (Fig. 1.56). In that case, a Fritsch-Buttenberg-Wieschell rearrangement allows the construction of the diyne fragment. Subsequent Negishi cross-coupling, mediated by $Pd(PPh_3)_4$, with aryl halides can then proceed with good yields to afford functionalized triynes [87].

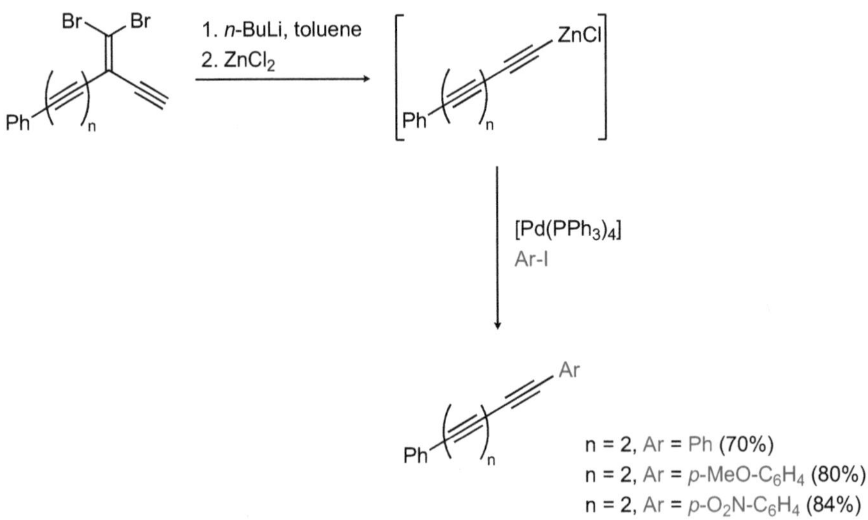

Fig. 1.56: Alkynyl-aryl cross-coupling with *in situ* generation of the nucleophile from a Fritsch-Buttenberg-Wieschell rearrangement [87].

A drawback of these methods is, however, the use of *n*-BuLi in the generation of the alkynylzinc partner, which prevents the transformation of functionalized starting materials. To circumvent this drawback, Negishi et al. developed a procedure relying on the use of LDA as a strong non-nucleophilic base, which is more tolerant to organic functions. Propynyl esters or ketones could thus be efficiently used as starting reagents to afford alkynylzinc derivatives, which could be further coupled with a variety of aryl iodides in the presence of $Pd(PPh_3)_4$ (Fig. 1.57) [88].

Moreover, stereo-conservative alkynyl-aryl cross-coupling reactions leading to the formation of 2,2′-dialkynyl-1,1′-binaphthyls have also been reported under microwave conditions, using the classical tetrakis $Pd(PPh_3)_4$ precatalyst (Fig. 1.58) [89].

1.5.4.2 With vinyl electrophiles

As outlined in Section 1.2 of this chapter, the alkynyl-vinyl cross-coupling using alkynylzincs as nucleophilic reagents was one of the very first efficient examples reported by

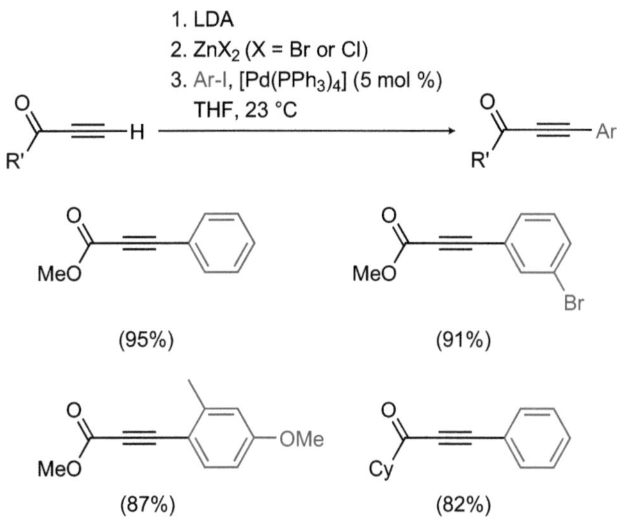

Fig. 1.57: Alkynyl-aryl cross-coupling with improved functional tolerance using LDA for the generation of the nucleophile [88].

the Negishi group (Fig. 1.2). The efficiency of this cross-coupling reaction often allows the use of simple catalytic systems, such as $Pd(PPh_3)_4$ or classic palladium(0) precursors stabilized with dibenzylideneacetone (dba), in the absence of external sophisticated ligand.

An early example of alkynyl-vinyl cross-coupling reaction in total synthesis was described by Negishi et al. in the course of the synthesis of xerulin. As discussed in Section 1.5.2.2, the synthesis of this complex polyenynyl not only involves vinyl-vinyl Negishi cross-coupling reactions, but also an alkynyl-vinyl selective cross-coupling reaction to prepare the key organozirconium dienyl synthon depicted in Fig. 1.30 (Fig. 1.59) [58].

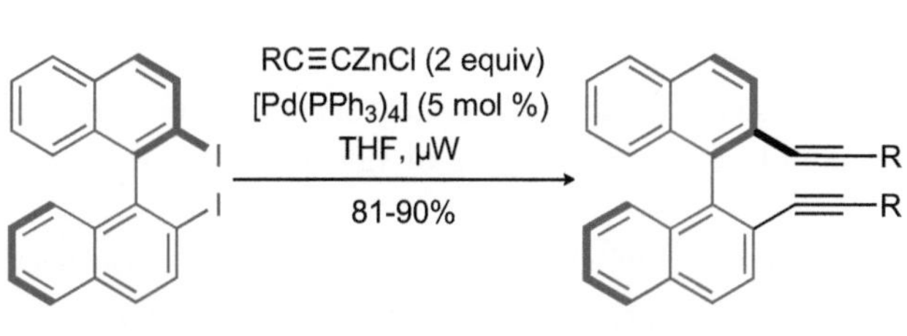

Fig. 1.58: Stereoconservative alkynyl-aryl cross-coupling on binaphthyl electrophiles [89].

A double chemoselective alkynyl-vinyl cross-coupling method was also reported by Negishi et al., which allows the step-by-step selective double functionalization of (*E*)-1-iodo 2-bromoethylene by two successive one-pot cross-coupling reactions with different alkynylzinc partners. Conjugated diynes can thus be obtained without using any

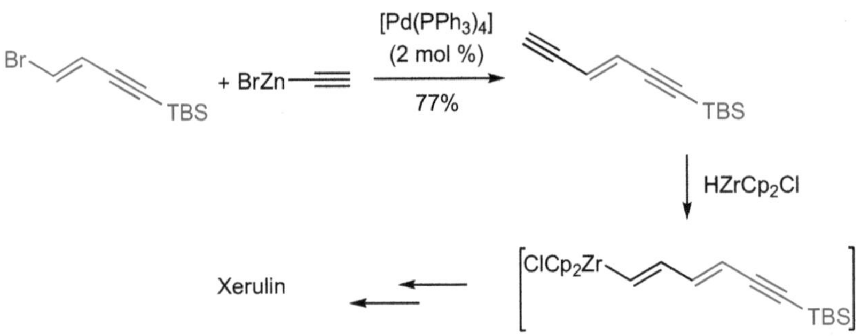

Fig. 1.59: Formation of a key synthon in the synthesis of xerulin by alkynyl-vinyl cross-coupling reactions [58].

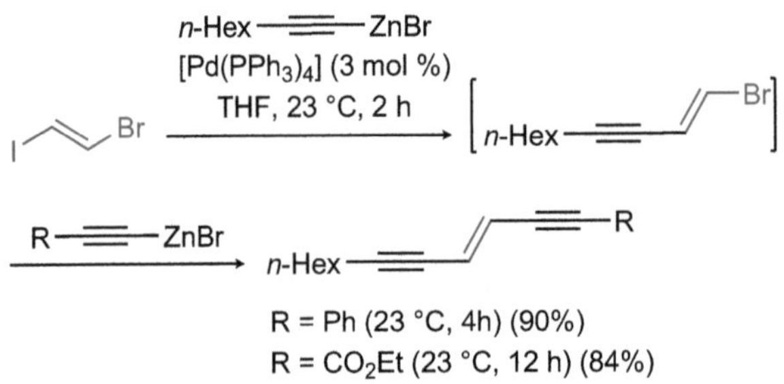

Fig. 1.60: Double one-pot functionalization of a dihalogeno-olefin by alkynyl-vinyl cross-coupling reactions [90].

alkynylation reagent in excess, which is a classic drawback of the Sonogashira alkynylation method (Fig. 1.60) [90].

Other methods have been reported ever since, some of them displaying interesting chemo- or stereoselective features. For example, an alkynyl-vinyl cross-coupling has been used as a key step in the total synthesis of the racemic tricholomenyn A (Fig. 1.61). This method moreover displays a high functional group tolerance, since the cross-coupling reaction can take place in the presence of silylethers, epoxides and conjugated ketones [91].

An efficient stereoselective alkynyl-vinyl cross-coupling reaction has also been used by Organ et al. in the synthesis of bupleurynol (Fig. 1.62). This cross-coupling is part of a queued cross-coupling Pd-mediated reactions allowing the synthesis of the complex target in a single step, making thus this new synthetic process considerably efficient compared to a previous nine-step procedure. A lithiation of a terminal alkyne followed by transmetallation with $ZnCl_2$ affords the nucleophilic organozinc partner, and $Pd(PPh_3)_4$ is used as the catalyst [92].

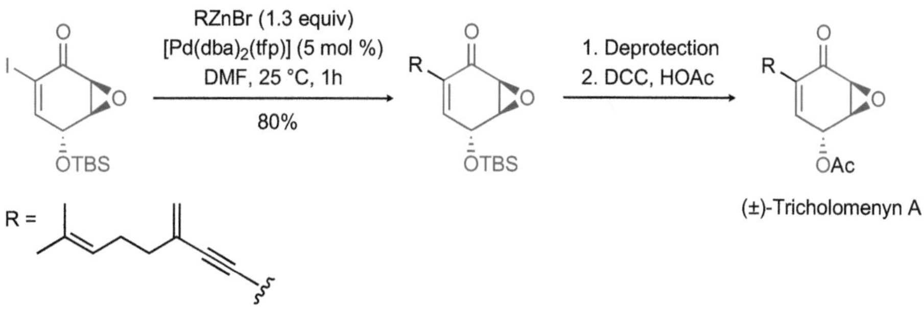

Fig. 1.61: Synthesis of racemic tricholomenyn A with a key alkynyl-vinyl Negishi cross-coupling (using the tfp ligand, Fig. 1.5) [91].

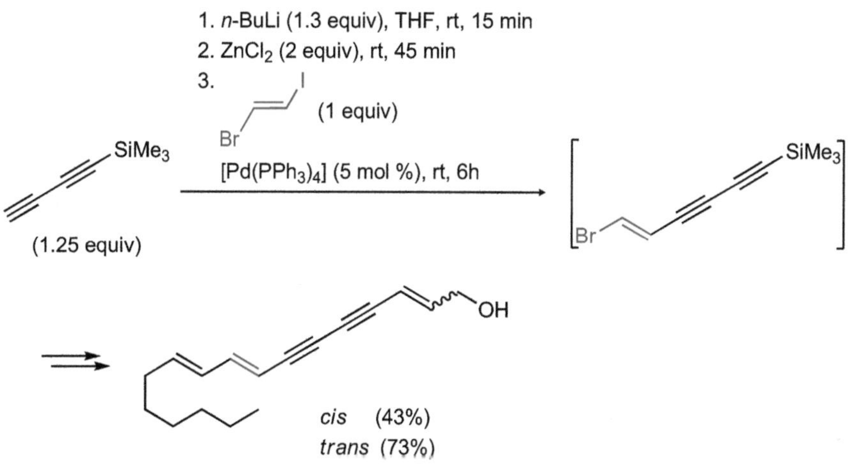

Fig. 1.62: Synthesis of bupleurynol by Organ et al. involving an alkynyl-vinyl Negishi cross-coupling [92].

1.5.4.3 With alkyl electrophiles

In line with the other cross-coupling reactions involving alkyl electrophiles discussed in this chapter, the alkynyl-alkyl Negishi cross-coupling is particularly challenging due to the slow reactivity of the alkyl electrophile with the palladium catalysts. It is worth mentioning that propargyl electrophiles can easily be coupled with alkynylzinc substrates using $Pd(PPh_3)_4$, but the allenyl product formed by conjugated addition is observed (Fig. 1.63) [93].

More recently, an interesting alkynyl-alkyl strategy relying on an oxidative homocoupling of organozinc reagents was reported by Lei et al. The first nucleophilic organozinc partner can be used as an *in situ* metalating reagent for the formation of the second alkynylzinc nucleophile obtained from the corresponding alkyne (Fig. 1.64) [94]. To do so, an excess of organozinc is used (3 equiv). This procedure is extremely appealing from a

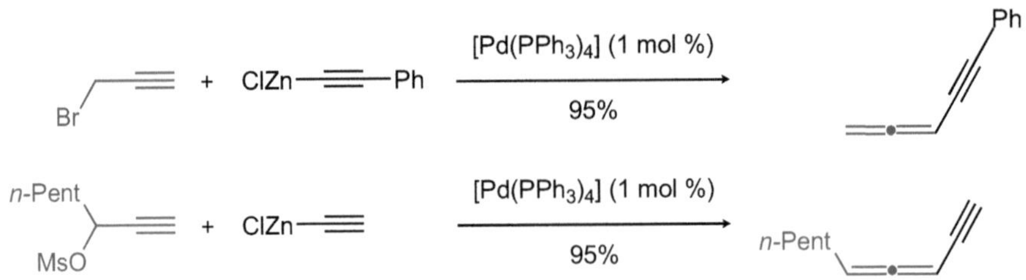

Fig. 1.63: Synthesis of allenyl compounds by addition of alkynylzinc on propargyl electrophiles [93].

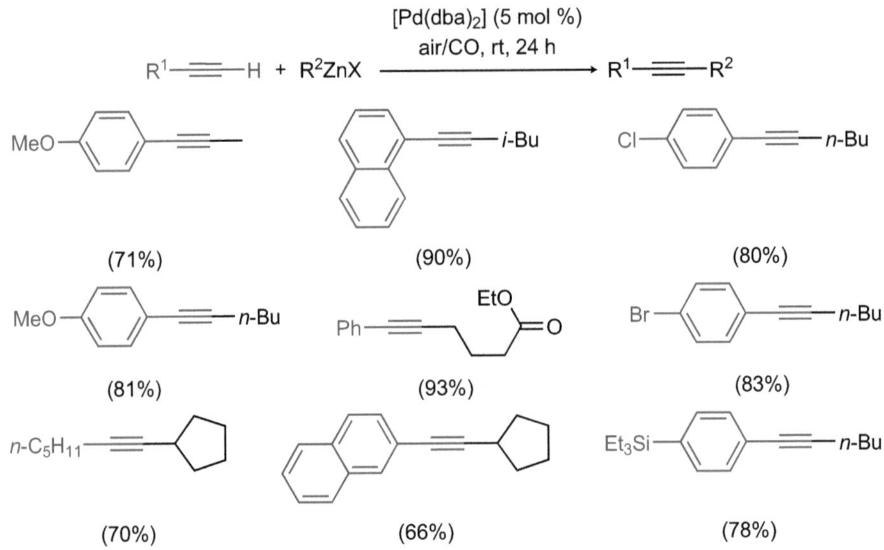

Fig. 1.64: Alkynyl-alkyl oxidative homocoupling using O_2 from air as a sacrificial oxidant [94].

synthetic standpoint as O_2 (dry air) is used as a sacrificial oxidant, unlike other similar oxidative homocoupling procedures which required the use of toxic organic oxidants such as desyl chloride PhCHClC(O)Ph [95]. A remarkable scope of internal alkynes can be obtained in the present case.

From a mechanistic standpoint, the role of CO atmosphere proved to be critical due to the π-accepting properties of CO when it acts as a ligand to palladium intermediates. Formation of Pd-CO intermediates seems to enhance the reductive elimination step, which otherwise proceeds too slowly to obtain efficient cross-coupling yields.

1.5.4.4 With alkynyl electrophiles

One of the most efficient methods to perform a sp-sp cross-coupling reaction, leading to the formation of conjugated diynes, relies so far on copper catalysis in the absence of palladium (Cadiot-Chodkiewicz coupling). It involves a terminal alkyne and a 1-haloalkyne as cross-coupling partners, and can be applied to a variety of substrates [85].

Alternative versions using alkynylzinc reagents and 1-haloalkynes have also been reported, but usually the reactions proceed with unsatisfactory yields, statistical mixtures of the desired cross-coupling products with homocoupling compounds being observed. In some rare cases, several specific procedures have been reported, which afforded satisfactory yields of the expected cross-coupling product (Fig. 1.65) [83].

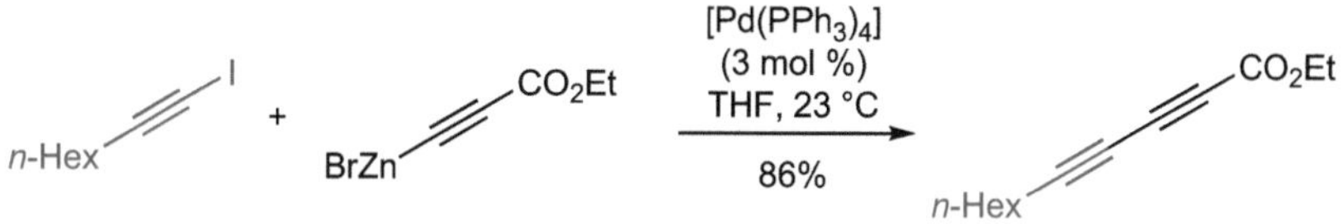

Fig. 1.65: Example of alkynyl-alkynyl cross-coupling involving alkynyl halides by Negishi et al. [83].

1.6 Development of high-scale cross-coupling processes in modern organic synthesis

1.6.1 Addressing the question of palladium contamination

The development of new and appealing synthetic strategies in organic synthesis must also take into account several paradigms in order to be applicable at large industrial scales, especially in the synthesis of pharmaceutical compounds. In the particular field of cross-coupling reactions, mediated by palladium catalysis, the purification of the final product is of high importance. Indeed, an extensive removal of palladium traces from the product must be performed, and palladium contents, usually less than 2 ppm, must be reached. Classical methods for palladium removal have been reported, an efficient one being the use of trimercaptotriazine as a chelating reagent [96].

However, a classic issue in heterocyclic synthesis is that nitrogen-containing heterocycles can act as efficient ligands to palladium or nickel, making the metal removal step more difficult. Hence, no general method for palladium traces removal can be described. In 2003, Manley and Acemoglu et al. reported a high-scale synthesis of a Phase I batch of PDE472, an inhibitor of phosphodiesterase which relies on a key aryl-aryl Negishi cross-coupling (Fig. 1.66). An efficient separation of the palladium traces from the batch could be performed *via* crystallization of the product with the hemimaleate salt, leading to a palladium content less than 0.5 ppm [97].

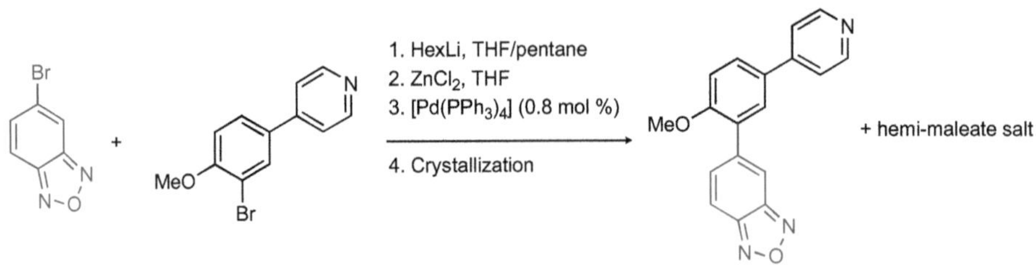

Fig. 1.66: Gram-scale synthesis of a biologically active compound by aryl-aryl Negishi cross-coupling [97].

1.6.2 Toward easily scalable processes: Cross-couplings and flow chemistry

The recent technical development of continuous flow chemistry also represents a major milestone in high-scale industrial synthesis. This technique allows a fine control of the temperature of the reaction medium as well as of the residence time of the reagents in the reactor. This is particularly useful in the development of high-scale cross-coupling transformations. Indeed, transition-metal cross-coupling reactions are usually highly exothermic transformations, which can induce a difficult control of the thermal exchanges between the reaction medium and the other parts of the system (e.g. cooling bath) and, hence, to altered selectivities.

Alcazàr et al. reported the first alkyl-aryl Negishi cross-coupling using commercially available palladium catalysts immobilized onto silica gel (SiliaCat DPP-Pd, Fig. 1.67) [98]. The cross-coupling reaction proceeds in smooth conditions, with a broad functional group tolerance, and the obtained product displays a remarkably low 0.027 ppm palladium content [99].

Similarly, the group of Organ recently reported efficient selective alkyl-(hetero)aryl cross-coupling reaction featuring aliphatic zinc nucleophiles with a silica-supported Pd-PEPPSI-IPr catalyst under continuous flow conditions (Fig. 1.68). This supported highly functionalized catalyst moreover proved to be active under continuous flow conditions for more than 15 h, making it particularly suitable for high-scale synthesis purposes. Leaching of the palladium from the supported catalyst is, however, observed at long reaction times (30% of metal loss after 5 h under flow conditions, 40% loss after 15 h) [100].

Such methods display a double advantage. At first, they allow the recovery of the catalyst thanks to its heterogenization on functionalized silica; in second, it is theoretically possible to scale-up a continuous flow process by simply carrying the transformation on a longer period, with no modification of the experimental setup, in stark contrast with the batch chemistry.

A challenging cross-coupling reaction between Reformatsky zinc enolates and aryl bromides or activated (hetero)aryl chlorides under flow conditions, using a soluble Pd-Zn bimetallic intermediate activated under photochemical conditions, has also

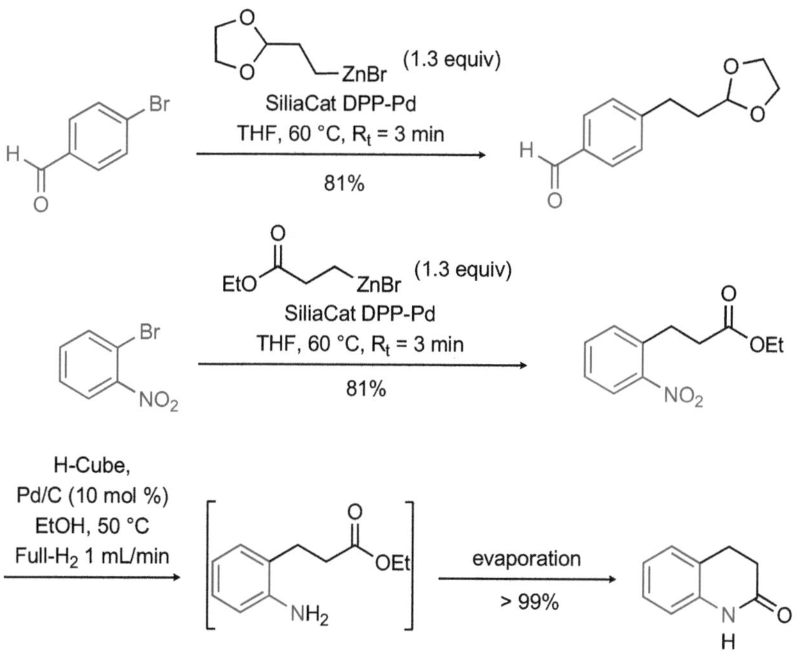

Fig. 1.67: Alkyl-aryl cross-coupling reaction in flow conditions using an immobilized Pd catalyst [99].

been reported by Alcazàr et al. It is believed that the photochemical activation triggers the oxidative addition step between the palladium(0) active oxidation state (stabilized by a Buchwald-type ligand) and the organic halide (Fig. 1.69) [101].

1.6.3 Green and sustainable Negishi cross-coupling in aqueous solvents

A significant challenge in sustainable organometallic catalysis is the development of synthetic methods which require the lower possible amount of organic solvents and, ideally, which can be carried out in water. Water-compatible methods involving main group organometallics such as organozincs are far from being trivial, given the basic Brønsted character of these latter species. A significant breakthrough has also been achieved by Lipshutz et al. who developed conditions allowing palladium-catalyzed Negishi cross-coupling reactions in aqueous solvents in the presence of TMEDA, under micellar conditions [102–104]. The key of the success of this water-compatible Negishi cross-coupling is the use of a surfactant, which seems to provide the micellar hydrophobic pocket in which the water-sensitive organozinc reagents are formed and react *in situ* (Fig. 1.70). From a mechanistic standpoint, it has been suggested that TMEDA ligates to the zinc(II) center after oxidative addition of zinc(0) onto the alkyl halide. The

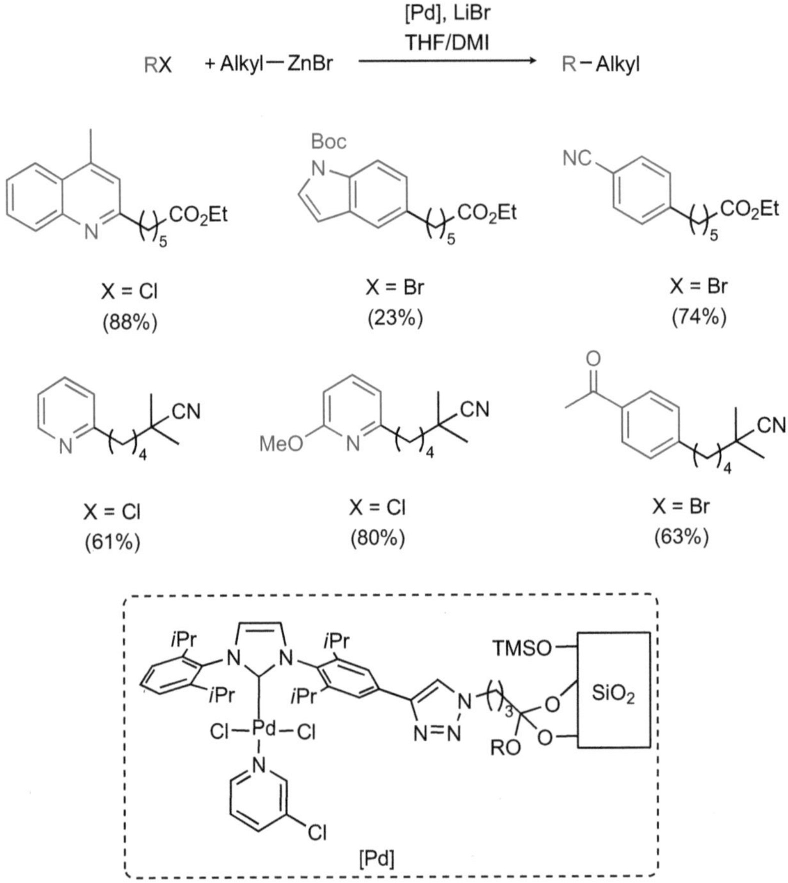

Fig. 1.68: Alkyl-(hetero)aryl coupling by Organ et al. using a supported Pd-PEPPSI-NHC catalyst [100].

key of the stability of the alkylzinc moiety in water is due to the ionization of the Zn-I bond, leading to the corresponding cationic species which prevents protonolysis of the Zn-C bond by water, hence allowing the subsequent coupling step to occur (Fig. 1.70). These studies were sustained by mass spectroscopy experiments (using Infrared multiple photon dissociation – IRMPD) along with DFT calculations [105].

Application of Lipshutz micellar-Negishi cross-coupling methods in total synthesis was also recently reported. Synthesis of biologically active aspergilides was performed using a key alkyl-vinyl Negishi cross-coupling under Barbier conditions, using $PdCl_2(Amphos)_2$ (Fig. 1.11) in a micellar medium (Fig. 1.71). It is worth mentioning that this total synthesis represents a very attractive example of sustainability since all the carbon atoms present in the final molecule originate from biomass-derived chemicals such as ethanol, levulinic acid and 5-hydroxymethylfurfural [106].

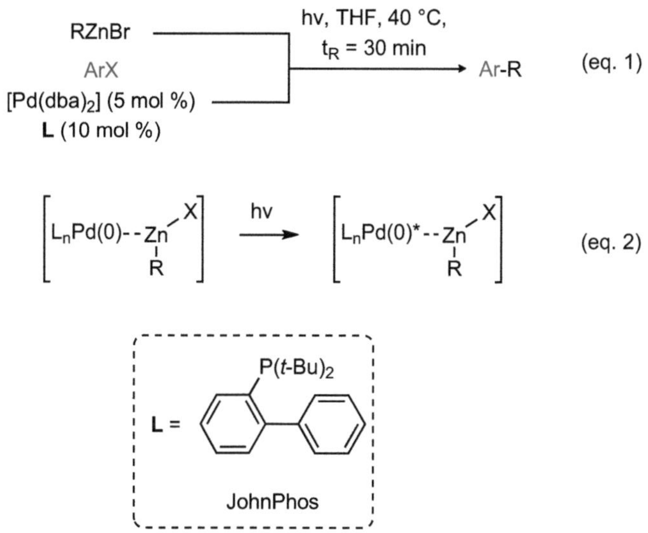

Fig. 1.69: Flow conditions for coupling of aryl bromides or activated (hetero)aryl chlorides with zinc enolates by Alcazàr et al. (eq. 1) and enhancement of the oxidative addition rate by photochemical activation of a palladium(0) complex (eq. 2) [101].

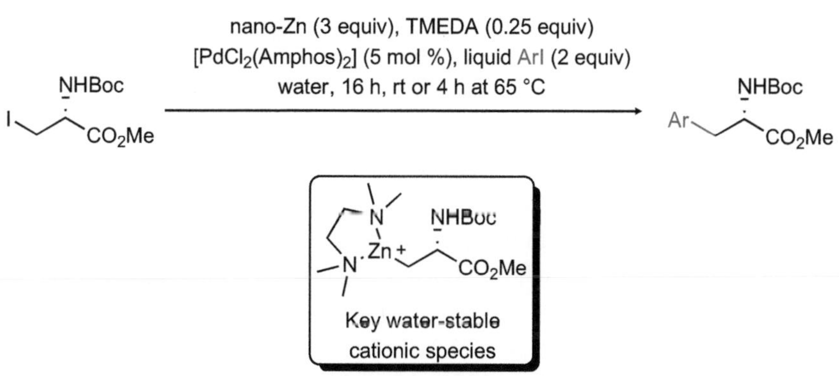

Fig. 1.70: Formation of a key cationic zinc(II) water-stable species [105].

1.7 Synthesis of complex targets

1.7.1 Other applications of Pd-catalyzed Negishi couplings in total synthesis

Given the broad scope of hybridization patterns that can be encompassed in palladium-mediated Negishi cross-coupling reactions, those reactions have emerged as a

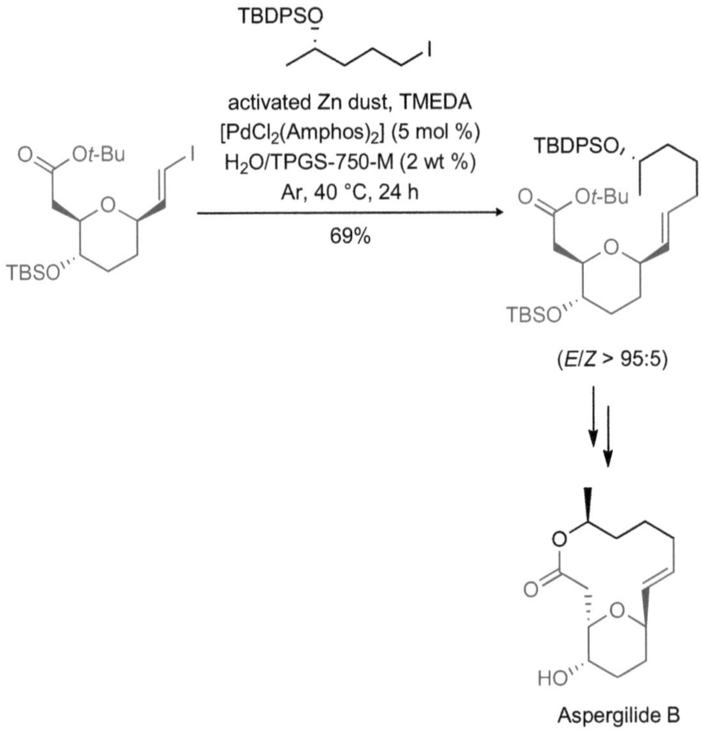

Fig. 1.71: Application of Lipshutz micellar-Negishi coupling to total synthesis of aspergilides [106].

classic, reliable, synthetic tool for the synthesis of complex targets of biological interest. Numerous total syntheses, including several examples discussed in Section 1.5, were reported in the past; we will discuss in this Section several classic targets of pharmaceutical interest which can be obtained using a key Negishi cross-coupling.

As a showpiece for highlighting the high potential of Negishi cross-coupling for the synthesis of classic pharmaceutical targets, Goossen and Gessner et al. reported that ibuprofen could easily be synthesized using two alkyl-aryl Negishi cross-couplings mediated by a palladium source associated with ylide-functionalized phosphine ligands (Fig. 1.72) [107]. A first chemoselective coupling on the bromine functionality followed by a second coupling of the chlorine or triflate functionality with a Reformatsky enolate leads, after hydrolysis, to ibuprofen with an excellent yield.

The total synthesis of (-)-callystatin A, a potent antiproliferative molecule with promising cytotoxic properties, has also stimulated numerous research groups. Given the presence of vinyl-vinyl linkages in the structure of this challenging compound, Panek et al. suggested a route in which one of those linkages is introduced by a stereocontrolled Negishi cross-coupling, the vinylzinc nucleophile being obtained by hydrozirconation of a terminal alkyne (Fig. 1.73) [108].

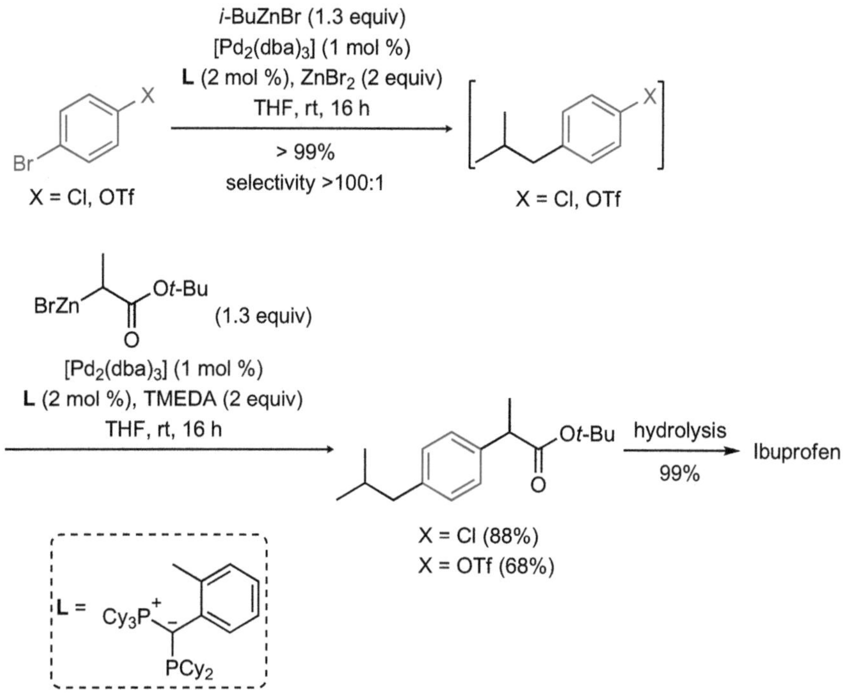

Fig. 1.72: Synthesis of ibuprofen by Goossen and Gessner et al. using a double Negishi cross-coupling mediated by a palladium source in the presence of functionalized ylide phosphines [107].

As outlined in Section 1.4 of this chapter, the use of dialkylbiarylphosphines (Buchwald's ligands) opened the door to new challenging Negishi cross-coupling patterns. Such catalysts have been efficiently used in late-stage functionalization processes, making them particularly attractive for applications in total synthesis. Representative examples include an elegant late-stage alkyl-aryl cross-coupling between functionalized decalin and isoindolinone subunits, mediated by (S-Phos)palladium(0) catalyst (Fig. 1.5) and reported by the Magauer group (Fig. 1.74) [109]. This Negishi procedure is used at a late stage of the total synthesis of the antiviral meroterpenoid, (+)-stachyflin.

1.7.2 Toward functionalization of highly sensitive targets

In Sections 1.5 and 1.6 of this chapter, numerous chosen examples highlighted the high versatility of the Pd-mediated Negishi cross-coupling methods and especially their tolerance to a large variety of organic functions.

The tolerance of Negishi cross-coupling methods for boron-based functionalities has also been exploited in total synthesis. Cho et al. also reported an efficient alkyl-(hetero)aryl Negishi cross-coupling of difunctionalized borylmethylzinc(II) reagents,

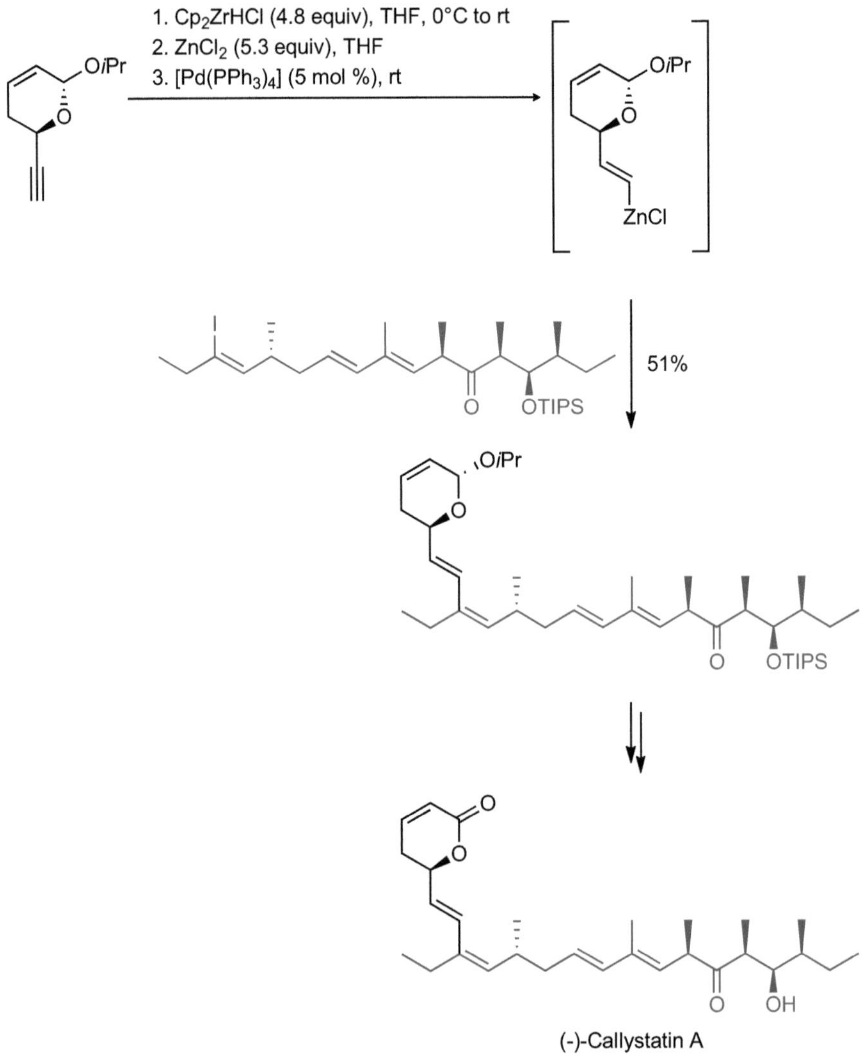

Fig. 1.73: Synthesis of (-)-callystatin A by Panek et al., featuring a vinyl-vinyl linkage introduction by Negishi cross-coupling [108].

which has been efficiently used at a high scale (5 mmol) in the late-stage functionalization of δ-tocopherol triflate. This example demonstrates the high interest of such cross-coupling methods encompassing a large scope of functionalities, which can afford substrates used in post-functionalization transformations (Fig. 1.75) [110].

In a similar way, it was also demonstrated that sensitive azaborine derivatives could be functionalized by a Negishi cross-coupling, and a large variety of substituents (alkyl, (hetero)aryl, vinyl groups) could be installed thanks to this strategy (Fig. 1.76) [111].

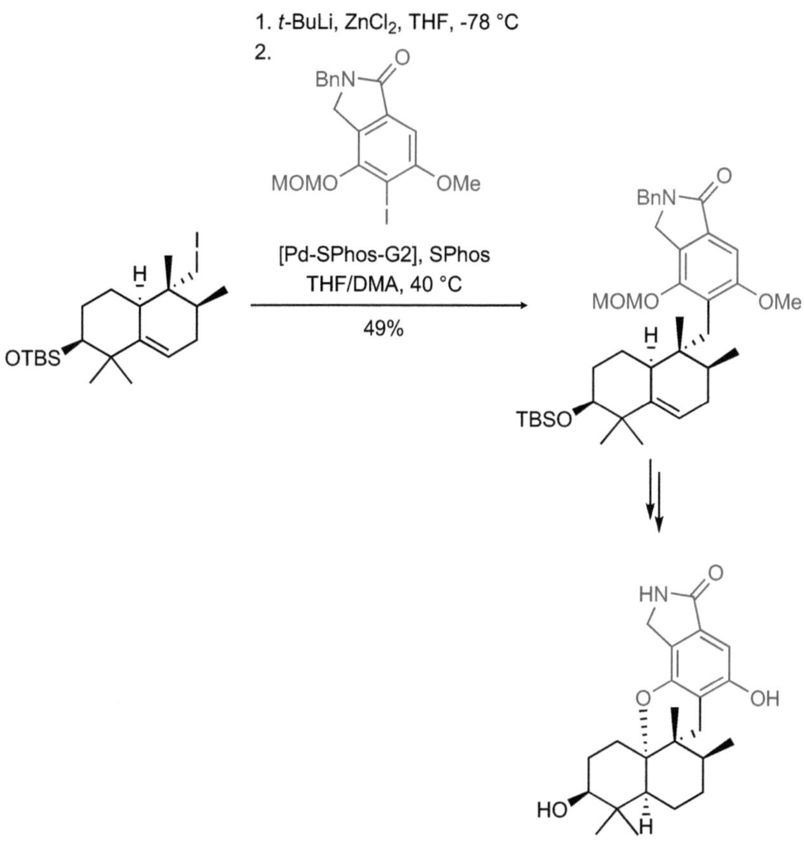

Fig. 1.74: Late-stage Negishi cross-coupling in the synthesis of (+)-stachyflin by Magauer et al. [109].

The development of cross-coupling methods featuring difficulty related to nucleophiles, such as fluorinated species with a very poor nucleophilic character, still draws the attention of synthetic groups. For example, a challenging aryl-alkyl cross-coupling reaction involving a CF_2H-based electrophilic partner could recently be achieved (Fig. 1.77) [112]. In this case, the use of a diphosphine ligand with a large bite angle such as XantPhos triggers the reductive elimination step, which is particularly difficult given the strong electron-withdrawing properties of the CF_2H group.

1.7.3 Applications in radiolabeling chemistry

Among the rare examples of aryl-alkyl cross-coupling reactions reported in literature (see Section 1.5.1.3), an interesting system allowing the ^{11}C-radiolabeling of (hetero)

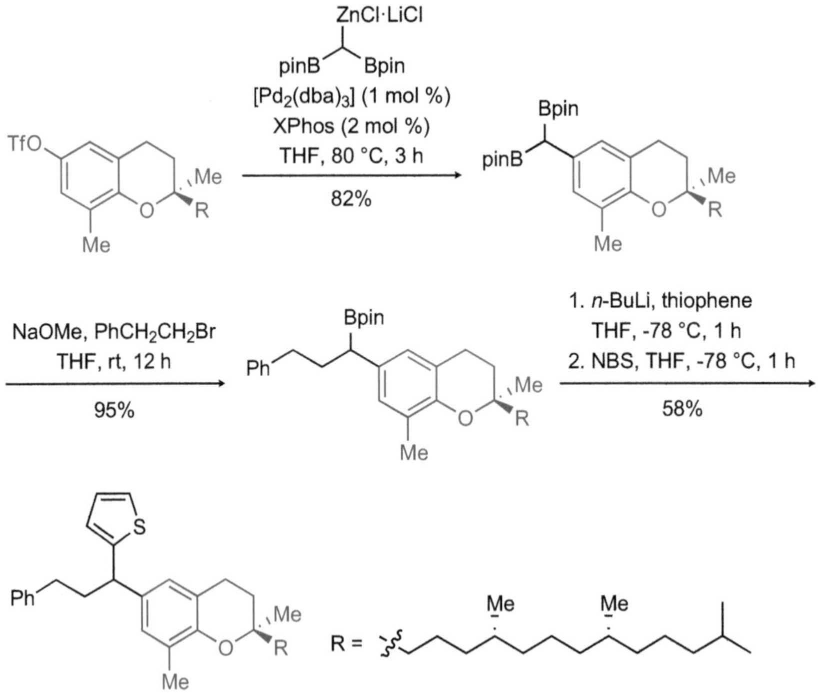

Fig. 1.75: Negishi cross-coupling using a functionalized borylmethylzinc nucleophile by Cho et al. (XPhos: see Fig. 1.24) [110].

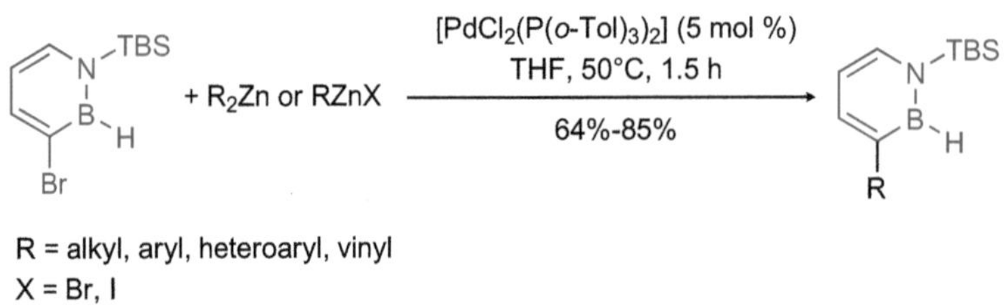

Fig. 1.76: Functionalization of azaborine derivatives by a Negishi cross-coupling [111].

aryl nucleophiles has been described by Huiban et al. [113]. ^{11}C-Labeled radiocompounds are of high interest in positron emission tomography (PET) since they enable the monitoring of the fate of biologically active compounds in *in vivo* conditions. Using $PdCl_2(PPh_3)_2$ as a catalyst and $[^{11}C]H_3I$ as an electrophile, a variety of (hetero) arylzinc nucleophiles could be obtained with very good RadioChemical Yields (RCY) (Fig. 1.78) [114].

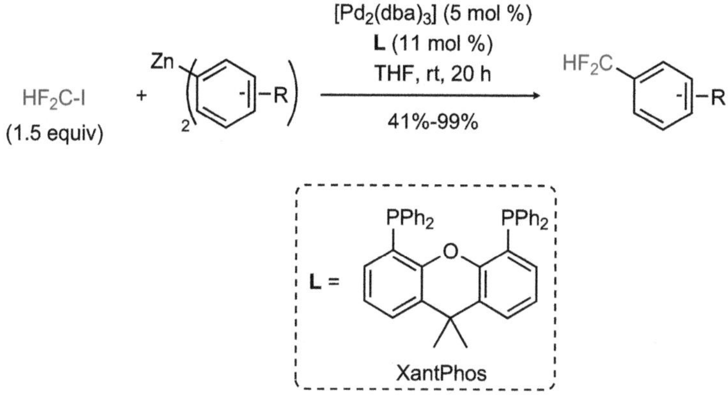

Fig. 1.77: Challenging Aryl-CF_2H Negishi cross-coupling by Mikami et al. [112].

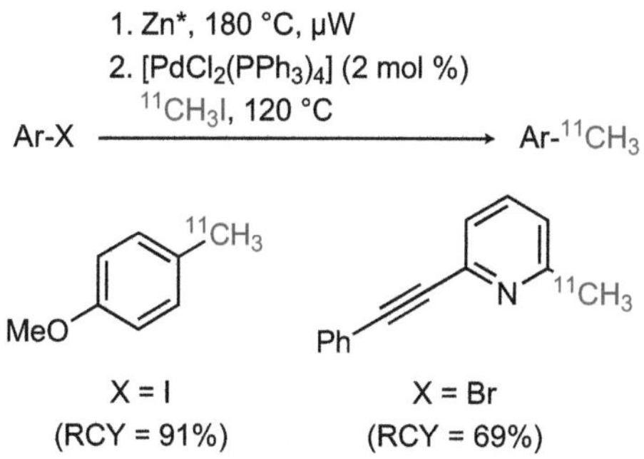

Fig. 1.78: Formation of ^{11}C-labeled (hetero)aromatic compounds by aryl-alkyl Negishi cross-coupling [113].

1.8 Summary and conclusion

This chapter summarizes the main advantages and the application scopes of palladium-mediated Negishi cross-couplings. It is worth noting that an important variety of coupling patterns can be carried out using simple, commercially available palladium sources and ligands (e.g. $Pd_2(dba)_3$, $Pd(PPh_3)_4$). The Negishi cross-coupling strongly benefited from the important progresses made in comprehensive palladium coordination chemistry during the last decades. New tailor-made ligands were developed and used efficiently to unlock coupling reactivity patterns which were, so far, difficult to carry out. Remarkable examples are the development of electron-rich alkylphosphines enabling the activation of strong electrophiles such as chloroarenes, as well as the development of sterically crowded ligands enhancing the reductive elimination rates, which are particularly useful for the alkyl-alkyl coupling patterns.

Moreover, given the moderate reactivity of organozinc nucleophiles, lower than that of their lithium or magnesium analogues, the Negishi cross-coupling reactions soon proved to display an appreciable functional group tolerance, allowing late-stage functionalization of complex targets. Hence, it has been used in a large number of total syntheses as a key C-C bond formation step.

Because of the aforementioned advantages (large application scope, broad functional group tolerance), the palladium-mediated Negishi cross-coupling appears as an appealing synthetic tool for the formation of C-C bonds in high-scale industrial syntheses. To do so, this research field also took advantage of technical improvements of continuous flow chemistry, and several cross-coupling reactions were successfully performed under those conditions, some of them using promising heterogeneous silica-supported palladium catalysts.

References

[1] Haas D, Hammann JM, Greiner R, Knochel P, ACS Catal. 2016, 6, 1540–1552.
[2] Phapale VB, Cárdenas DJ, Chem. Soc. Rev. 2009, 38, 1598–1607.
[3] Johansson Seechurn CC, Kitching MO, Colacot TJ, Snieckus V, Angew. Chem. Int. Ed. 2012, 51, 5062–5085.
[4] Negishi E, Baba S, J. Chem. Soc. Chem. Commun. 1976, 1995, 596b–597b.
[5] Negishi E, King AO, Okukado N, J. Org. Chem. 1977, 42, 1821–1823.
[6] King AO, Okukado N, Negishi E, J. Chem. Soc. Chem. Commun. 1977, 42, 683–684.
[7] Fauvarque JF, Jutand A, J. Organomet. Chem. 1977, 132, C17–C19.
[8] King AO, Negishi E, Villani FJ, Silveira A, J. Org. Chem. 1978, 43, 358–360.
[9] Piber M, Jensen AE, Rottländer M, Knochel P, Org. Lett. 1999, 1, 1323–1326.
[10] Böck K, Feil JE, Karaghiosoff K, Koszinowski K, Chem. Eur. J. 2015, 21, 5548–5560.
[11] González-Pérez AB, Álvarez R, Faza ON, de Lera AR, Aurrecoechea JM, Organometallics 2012, 31, 2053–2058.
[12] Chass GA, O'Brien CJ, Hadei N, Kantchev EAB, Mu W, Fang D, Hopkinson AC, Csizmadia IG, Organ MG, Chem. Eur. J. 2009, 15, 4281–4288.
[13] McCann LC, Hunter HN, Clyburne JAC, Organ MG, Angew. Chem. Int. Ed. 2012, 51, 7024–7027.
[14] Achonduh GT, Hadei N, Valente C, Avola S, O'Brien CJ, Organ MG, Chem. Commun. 2010, 46, 4109–4111.
[15] McCann LC, Organ MG, Angew. Chem. Int. Ed. 2014, 53, 4386–4389.
[16] Polynski MV, Pidko EA, Catal. Sci. Technol. 2019, 9, 4561–4572.
[17] Fuentes B, Garcìa-Melchor M, Lledós A, Maseras F, Casares JA, Ujaque G, Espinet P, Chem. Eur. J. 2010, 16, 8596–8599.
[18] Del Pozo J, Salas G, Álvarez R, Casares JA, Espinet P, Organometallics 2016, 35, 3604–3611.
[19] van Asselt R, Elsevier CJ, Organometallics 1994, 13, 1972–1980.
[20] Casares JA, Espinet P, Fuentes B, Salas G, J. Am. Chem. Soc. 2007, 129, 3508–3509.
[21] Gioria E, Martínez-Ilarduya JM, Espinet P, Organometallics 2014, 33, 4394–4400.
[22] Liu Q, Lan Y, Liu J, Li G, Wu Y, Lei A, J. Am. Chem. Soc. 2009, 131, 10201–10210.
[23] Lundgren RJ, Stradiotto M, Chem. Eur. J. 2012, 18, 9758–9769.
[24] Dai C, Fu GC, J. Am. Chem. Soc. 2001, 123, 2719–2724.

[25] Fu GC, Acc. Chem. Res. 2008, 41, 1555–1564.
[26] Zapf A, Ehrentraut A, Beller M, Angew Chem. Int. Ed. 2000, 39, 4153–4155.
[27] Kataoka N, Shelby Q, Stambuli JP, Hartwig JF, J. Org. Chem. 2002, 67, 5553–5566.
[28] Guram AS, King AO, Allen JG, Wang X, Schenkel LB, Chan J, Bunel EE, Faul MM, Larsen RD, Martinelli MJ, Reider PJ, Org. Lett. 2006, 8, 1787–1789.
[29] Krasovskiy A, Duplais C, Lipshutz BH, Org. Lett. 2010, 12, 4742–4744.
[30] Old DW, Wolfe JP, Buchwald SL, J. Am. Chem. Soc. 1998, 120, 9722–9723.
[31] Hayashi T, Konishi M, Kobori Y, Kumada M, Higuchi T, Hirotsu K, J. Am. Chem. Soc. 1984, 106, 158–163.
[32] Hadei N, Kantchev EAB, O'Brie CJ, Organ MG, Org. Lett. 2005, 7, 3805–3807.
[33] Hartwig JF, Inorg. Chem. 2007, 46, 1936–1947.
[34] Çalimsiz S, Organ MG, Chem. Comm. 2011, 47, 5181–5183.
[35] Valente C, Belowich ME, Hadei N, Organ MG, Eur. J. Org. Chem. 2010, 2010, 4343–4354.
[36] Tucker CE, Majid TN, Knochel P, J. Am. Chem. Soc. 1992, 114, 3983–3985.
[37] Rottländer M, Palmer N, Knochel P, Syntlett 1996, 1996, 573–575.
[38] Blanksby SJ, Ellison GB, Acc. Chem. Res. 2003, 36, 255–263.
[39] Tsuji H, Mitsui C, Ilies L, Sato Y, Nakamura E, J. Am. Chem. Soc. 2007, 129, 11902–11903.
[40] Çalimsiz S, Sayah M, Mallik D, Organ MG, Angew. Chem. Int. Ed. 2010, 49, 2014–2017.
[41] Ponce-de-León J, Espinet P, Chem. Commun. 2021, 57, 10875–10878.
[42] Alimardanov A, Schmieder-van de vondervoort L, De vries AHM, Adv. Synth. Catal. 2004, 346, 1812–1817.
[43] Shen J, Wong B, Gu C, Zhang H, Org. Lett. 2015, 17, 4678–4681.
[44] Morii K, Yasuda Y, Morikawa D, Mori A, Okano K, J. Org. Chem. 2021, 86, 13388–13401.
[45] Colombe JR, Bernhardt S, Stathakis C, Buchwald SL, Knochel P, Org. Lett. 2013, 15, 5754–5757.
[46] Sase S, Jaric M, Metzger A, Malakhov V, Knochel P, J. Org. Chem. 2008, 73, 7380–7382.
[47] Luzung MR, Patel JS, Yin J, J. Org. Chem. 2010, 75, 8330–8332.
[48] Jensen AE, Dohle W, Sapountzis I, Lindsay DM, Vu VA, Knochel P, Synthesis 2002, 4, 565–569.
[49] Potter GA, McCague R, J. Org. Chem. 1990, 55, 6184–6187.
[50] Sutherland A, Gallagher T, Sharples CGV, Wonnacott S, J. Org. Chem. 2003, 68, 2475–2478.
[51] Clososki GC, Rohbogner CJ, Knochel P, Angew. Chem. Int. Ed. 2007, 46, 7681–7684.
[52] Krasovskiy A, Lipshutz BH, Org. Lett. 2011, 13, 3818–3821.
[53] Hansen AL, Ebran J, Gøgsig TM, Skrydstrup T, J. Org. Chem. 2007, 72, 6464–6472.
[54] Negishi E, Hu Q, Huang Z, Qian M, Wang G, Aldrichimica Acta 2005, 38, 71–87.
[55] Zhou J, Fu GC, J. Am. Chem. Soc. 2003, 125, 12527–12530.
[56] Schaus JV, Panek JS, Org. Lett. 2000, 2, 469–471.
[57] Vincent P, Beaucourt J, Pichat L, Tetrahedron Lett. 1982, 23, 63–64.
[58] Negishi E, Alimardanov A, Xu C, Org. Lett. 2000, 2, 65–67.
[59] Sorg A, Brückner R, Angew Chem. Int. Ed. 2004, 43, 4523–4526.
[60] Duffault J, Einhorn J, Alexakis A, Tetrahedron Lett. 1991, 32, 3701–3704.
[61] Hung DT, Nerenberg JB, Schreiber SL, J. Am. Chem. Soc. 1996, 118, 11054–11080.
[62] Zeng X, Hu Q, Qian M, Negishi E, J. Am. Chem. Soc. 2003, 125, 13636–13637.
[63] Luo X, Zhang H, Duan H, Liu Q, Zhu L, Zhang T, Lei A, Org. Lett. 2007, 9, 4571–4574.
[64] Yang Y, Niedermann K, Han C, Buchwald SL, Org. Lett. 2014, 16, 4638–4641.
[65] Hama T, Hartwig JF, Org. Lett. 2008, 10, 1549–1552.
[66] Hama T, Ge S, Hartwig JF, J. Org. Chem. 2013, 78, 8250–8266.
[67] Nolasco L, Perez Gonzalez M, Caggiano L, Jackson RFW, J. Org. Chem. 2009, 74, 8280–8289.
[68] Brittain WDG, Cobb SL, Org. Biomol. Chem. 2018, 16, 10–20.
[69] Pompeo M, Froese RDJ, Hadei N, Organ MG, Angew. Chem. Int. Ed. 2012, 51, 11354–11357.

[70] Thaler T, Haag B, Gavryushin A, Schober K, Hartmann E, Gschwind RM, Zipse H, Mayer P, Knochel P, Nat. Chem. 2010, 2, 125–130.
[71] Zhang K, Christoffel F, Baudoin O, Angew. Chem. Int. Ed. 2018, 57, 1982–1986.
[72] Kobayashi M, Negishi E, J. Org. Chem. 1980, 45, 5223–5225.
[73] Boudier A, Darcel C, Flachsmann F, Micouin L, Oestreich M, Knochel P, Chem. Eur. J. 2000, 6, 2748–2761.
[74] Negishi E, Liou S, Xu C, Huo S, Org. Lett. 2002, 4, 261–264.
[75] Smith III AB, Qiu Y, Jones DR, Kobayashi K, J. Am. Chem. Soc. 1995, 117, 12011–12012.
[76] Smith III AB, Beauchamp TJ, LaMarche MJ, Kaufman MD, Qiu Y, Arimoto H, Jones DR, Kobayashi K, J. Am. Chem. Soc. 2000, 122, 8654–8664.
[77] Cossy J, Pévet I, Meyer C, Eur. J. Org. Chem. 2001, 2001, 2841–2850.
[78] Anastasia L, Dumond YR, Negishi E, Eur. J. Org. Chem. 2001, 3039–3043.
[79] Nicolaou KC, Shi G, Gunzner JL, Gärtner P, Wallace PA, Ouellette MA, Shi S, Bunnage ME, Agrios KA, Veale CA, Hwang C, Hutchinson J, Prasad CVC, Ogilvie WW, Yang Z, Chem. Eur. J. 1999, 5, 628–645.
[80] Hadei N, Kantchev EAB, O'Brien CJ, Organ MG, J. Org. Chem. 2005, 70, 8503–8507.
[81] Knochel P, Gavryushin A, Malakhov V, Krasovskiy A, DE 102006015378A1 20071004, 2007.
[82] Piller FM, Metzger A, Schade MA, Haag BA, Gavryushin A, Knochel P, Chem. Eur. J. 2009, 15, 7192–7202.
[83] Negishi E, Anastasia L, Chem. Rev. 2003, 103, 1979–2018.
[84] Negishi E, Xu C, Tan Z, Kotora M, Heterocycles 1997, 46, 209–214.
[85] Sindhu KS, Thankachan AP, Sajitha PS, Anilkumar G, Org. Biomol. Chem. 2015, 13, 6891–6905.
[86] Negishi E, Hata M, Xu C, Org. Lett. 2000, 2, 3687–3689.
[87] Luu T, Morisaki Y, Cunningham N, Tykwinski RR, J. Org. Chem. 2007, 72, 9622–9629.
[88] Anastasia L, Negishi E, Org. Lett. 2001, 3, 3111–3113.
[89] Krascsenicsová K, Walla P, Kasák P, Uray G, Kappe CO, Putala M, Chem. Commun. 2004, 2606–2607.
[90] Negishi E, Qian M, Zeng F, Anastasia L, Babinski D, Org. Lett. 2003, 5, 1597–1600.
[91] Negishi E, Tan Z, Liou SY, Liao B, Tetrahedron 2000, 56, 10197–10207.
[92] Ghasemi H, Antunes LM, Organ MG, Org. Lett. 2004, 6, 2913–2916.
[93] Ruitenberg K, Kleijn H, Westmijze H, Meijer J, Vermeer P, Recl. Trav. Chim. Pays-Bas 1982, 101, 405–409.
[94] Chen M, Zheng X, Li W, He J, Lei A, J. Am. Chem. Soc. 2010, 132, 4101–4103.
[95] Xin J, Zhang G, Deng Y, Zhang H, Lei A, Dalton Trans. 2015, 44, 19777–19781.
[96] Rosso WV, Lust DA, Bernot PJ, Grosso JA, Modi SP, Rusowicz A, Sedergran TC, Simpson JH, Srivastava SK, Humora MJ, Anderson NG, Org. Process Res. Dev. 1997, 1, 311–314.
[97] Manley PW, Acemoglu M, Marterer W, Pachinger W, Org. Proc. Res. Dev. 2003, 7, 436–445.
[98] Lemay M, Pandarus V, Simard M, Marion O, Tremblay L, Béland F, Top. Catal. 2010, 53, 1059–1062.
[99] Egle B, de Muñoz JM, Alonso N, De Borggraeve WM, de la Hoz A, Díaz-Ortiz A, Alcázar J, J. Flow Chem. 2014, 4, 22–25.
[100] Price GA, Bogdan AR, Aguirre AL, Iwai T, Djuric SW, Organ MG, Catal. Sci. Technol. 2016, 6, 4733–4742.
[101] Abdiaj I, Huck L, Mateo JM, de la Hoz A, Gomez MV, Díaz-Ortiz A, Alcazar J, Angew. Chem. Int. Ed. 2018, 57, 13231–13236.
[102] Krasovskiy A, Duplais C, Lipshutz BH, Am J, Chem. Soc. 2009, 131, 15592–15593.
[103] Duplais C, Krasovskiy A, Wattenberg A, Lipshutz BH, Chem. Commun. 2010, 46, 562–564.
[104] Lipshutz BH, Abela AR, Bošković ZV, Nishikata T, Duplais C, Krasovskiy A, Top. Catal. 2010, 53, 985–990.
[105] Ross AJ, Dreiocker F, Schäfer M, Oomens J, Meijer AJHM, Pickup BT, Jackson RFW, J. Org. Chem. 2011, 76, 1727–1734.
[106] Koh P, Loh T, Green Chem. 2015, 17, 3746–3750.

[107] Hu Z, Wei X, Handelmann J, Seitz A, Rodstein I, Gessner VH, Goossen LJ, Angew. Chem. Int. Ed. 2021, 60, 6778–6783.
[108] Langille NF, Panek JS, Org. Lett. 2004, 6, 3203–3206.
[109] Haut F, Speck K, Wildermuth R, Möller K, Mayer P, Magauer T, Tetrahedron 2018, 74, 3348–3357.
[110] Lee H, Lee Y, Cho SW, Org. Lett. 2019, 21, 5912–5916.
[111] Brown AN, Li B, Liu SH, Tetrahedron 2019, 75, 580–583.
[112] Nitta J, Motohashi H, Aikawa K, Mikami K, Asian J. Org. Chem. 2019, 8, 698–701.
[113] Kealey S, Passchier J, Huiban M, Chem. Commun. 2013, 49, 11326–11328.
[114] Rejc L, Gómez-Vallejo V, Alcázar J, Alonso N, Andrés JI, Arrieta A, Cossío FP, Llop J, Chem. Commun. 2018, 54, 4398–4401.

Olivier Piva

2 Organozinc reagents and nickel

2.1 Introduction

In a historical point of view, the direct reaction between an organic halide (or any other electrophilic species) and nucleophilic organometallic compounds such as Grignard reagents represents a straightforward approach for the formation of C-C bonds. Unfortunately, these reagents or other alkali are so basic that the presence of electrophilic centers on the substrates is rarely suitable. Moreover, this type of coupling reaction was in a recent past, mainly limited to Csp^3 electrophiles. The discovery that palladium and, in some cases, nickel(0) could favor the coupling of Csp^2 electrophilic species when opposed to suitable organometallic reagents led to a real breakthrough in organic synthesis. Many efforts have been made to enhance the regioselectivity and the efficiency of the process. Beside the reactions conditions (solvent, temperature, ligands), the role of the organometallic derivatives was crucial. To the Kumada-Corriu reaction, involving magnesium derivatives, the Negishi-type reaction with organozinc reagents or the Suzuki-Miyaura reaction involving boron reagents are nowdays, preferred transformations. The mild conditions allow direct application in the field of natural product synthesis [1]. At this point, it is important to note that mechanisms for coupling reactions, catalyzed by palladium or nickel, can strongly differ especially with alkyl derivatives [2]. The classical pathway which is applicable for Csp^2-Csp^2 bond formation includes three key steps: oxidative addition/transmetallation/reductive elimination. In the case of alkyl groups, the coupling Csp^3-Csp^3 under nickel catalysis could follow a radical pathway, while the reaction under palladium catalysis led mainly to a competitive β-H elimination [2–4].

In this chapter, we will focus only on the reactivity of preformed zinc derivatives and also electrophilic species in the presence of nickel complexes, a process discovered nearly 45 years ago [5–7]. While most of the cross-coupling reactions have been initially designed using air-sensitive organozinc halides [8], a considerable gap of efficiency has been reached by using Knochel's organozinc pivalates [9, 10]. Related results, recently described in the literature, will be therefore emphasized [11]. A special focus will be also made on the formation and control of stereogenic or quaternary centers. Other processes including nickel-catalyzed reductive cross-coupling in which zinc(0) promotes solely the reduction of nickel(III) intermediates will not be treated in this chapter [12–15] as well as the catalytic arylation of C-H bonds in heteroaryl derivatives [16] and nickel-catalyzed dicarbofunctionalization of alkenes [17].

Olivier Piva, Université Claude Bernard LYON 1

https://doi.org/10.1515/9783110728859-002

2.2 Cross-coupling reactions of arylzinc reagents

2.2.1 With aryl derivatives

2.2.1.1 With aryl chlorides and bromides

Beside the well-established cross-coupling reactions of aryl halides with zinc derivatives, in the presence of palladium catalysts, the process involving cheaper nickel derivatives is really appealing and has been extensively considered. Conditions discovered by Knochel et al. constitute a real breakthrough. Even with substrates possessing N-H or O-H acidic protons, the cross-coupling occurred with arylZnCl.LiCl complex when performed with low charge (2 mol %) of $Ni(acac)_2$ and 2,2'-bipyridine (Fig. 2.1) [18]. However, to be successful, the reaction requires the slow addition of the zinc derivative in slight excess to the aryl halide via a syringe pump over 90 min.

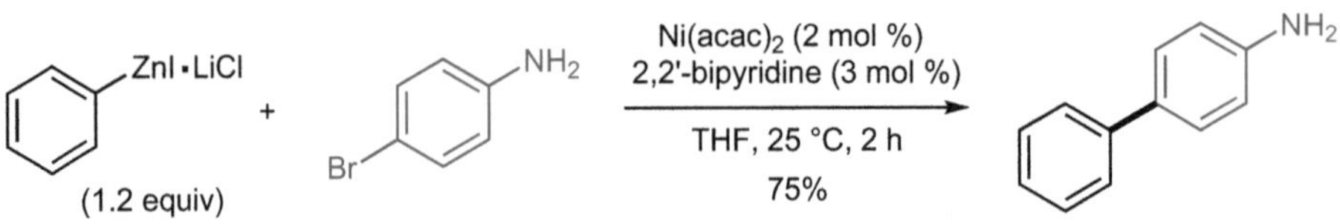

Fig. 2.1: Cross-coupling reaction between phenylzinc chloride and *p*-bromoaniline.

It is worth mentioning that lower charge loading has been reached when the reaction was performed with binuclear **NHC-Ni** complexes (Fig. 2.2) [19].

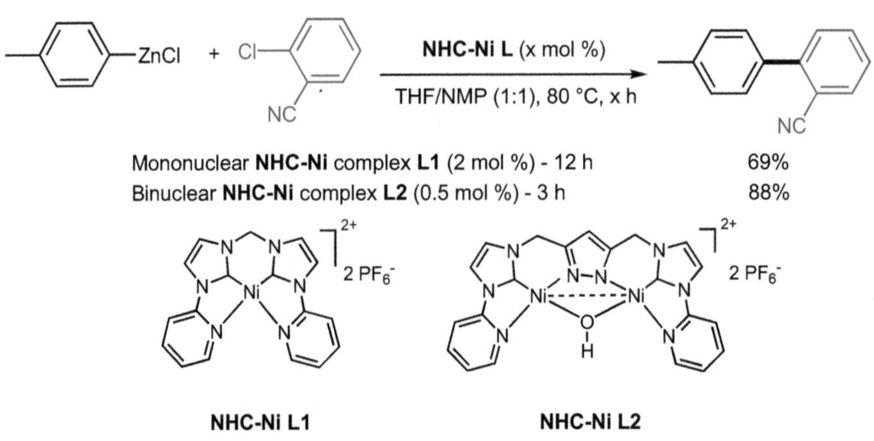

Fig. 2.2: Cross-coupling reactions catalyzed by **NHC-Ni** complexes.

Trifluoromethyl groups are relevant in agrochemicals and pharmaceutical compounds. In this context, an indirect functionalization of bis-chloroarenes has been achieved using trifluoroaryl zinc derivatives (Fig. 2.3). However, the high temperature required led to a bis-functionalization of the substrate [20].

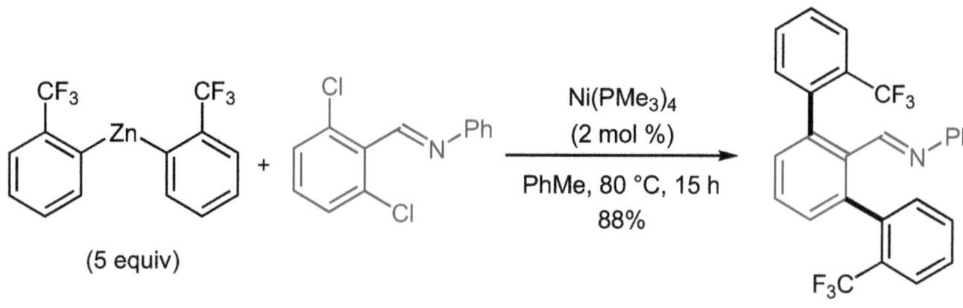

Fig. 2.3: Bis-functionalization of dichloroarenes.

2.2.1.2 With aryl fluorides

Polyfluoroaromatics constitute promising substrates in drug design and to access new materials. However, the cleavage of only one C-F bond is still a challenge. To be successful, three major problems should be overcome. The robustness of the C-F bond does not deserve itself the oxidative addition step. Moreover, the presence of at least one additional electron-withdrawing group strongly reduces the rate of the reductive elimination. In addition, monosubstitution appears to be dramatic due to the capacity of nickel catalysts to achieve ring walking and activation of the second C-F bond. Fortunately, when the reaction was performed in the presence of ligands such as alkoxydiphosphine ligand, the reaction occurred smoothly and furnished monarylated adducts in moderate to good yields (Fig. 2.4) [21].

The presence of suitable directing groups is also of great interest to promote selective C-C bond formation. A phenacyl group has been considered for this purpose and gives high selectivities with various arylzinc derivatives, when used in the presence of commercially available $Ni(PCy_3)_2Cl_2$ (Fig. 2.5) [22]. A high level of selectivity was also noticed with benzylimines. In that case, a smooth hydrolysis performed during treatment led to the formyl derivatives [23].

A preliminary study by Wang et al. demonstrated that nickel complex is necessary to promote the cross-coupling excluding a direct nucleophilic substitution [22]. The reaction was not affected by the presence of 1,1-diphenylethylene excluding the formation of radical species. Therefore, the reaction is considered to take place according to the well-known catalytic cycle (Fig. 2.6) including the classical tryptic: oxidative addition/transmetallation with organozinc/reductive elimination, the nickel(II) catalyst being reduced by an excess of zinc reagents.

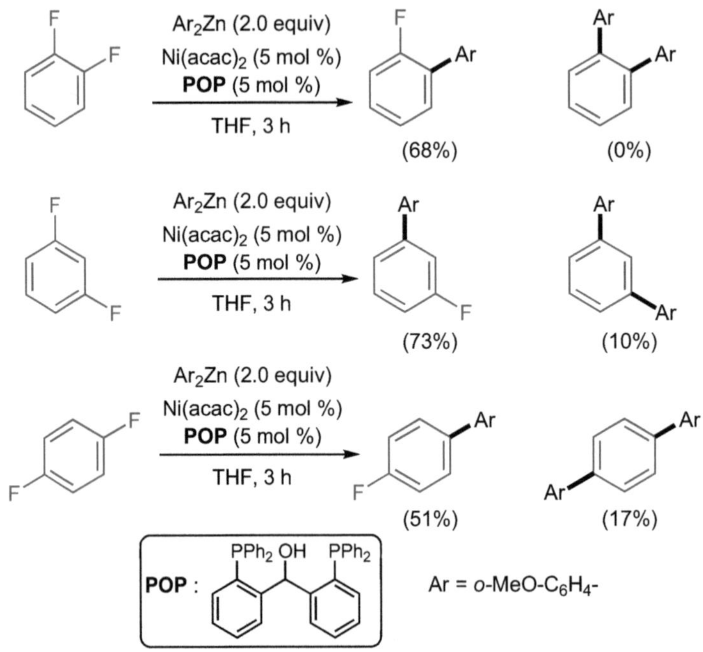

Fig. 2.4: Monoarylation of difluorobenzenes.

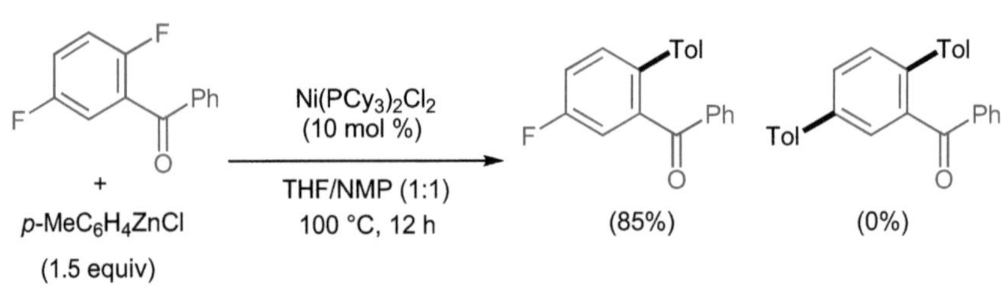

Fig. 2.5: Regioselective monoarylation of difluorobenzene derivatives.

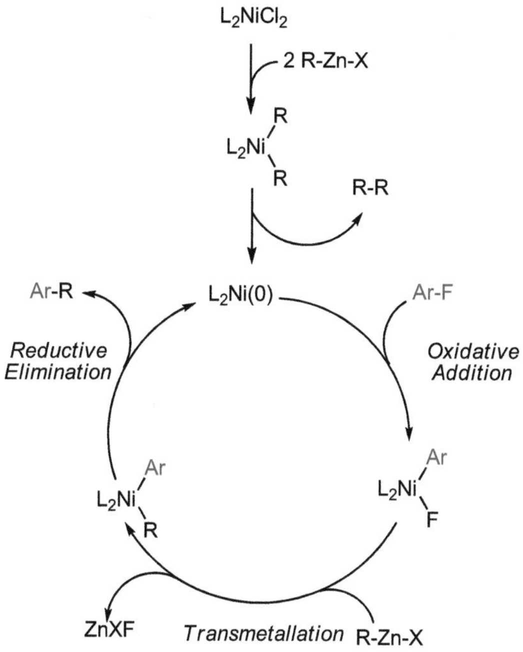

Fig. 2.6: Catalytic cycle involving arylfluorines and organozinc species.

2.2.1.3 With phenol esters

2.2.1.3.1 With aryl pivalates

Naphthyl and phenyl pivalates, easily prepared from the corresponding phenols, react with phenylzinc chloride in a 1:1 mixture of THF/DMA and moderate temperature to deliver the biaryl compounds in high yields (Fig. 2.7). The reaction was catalyzed with

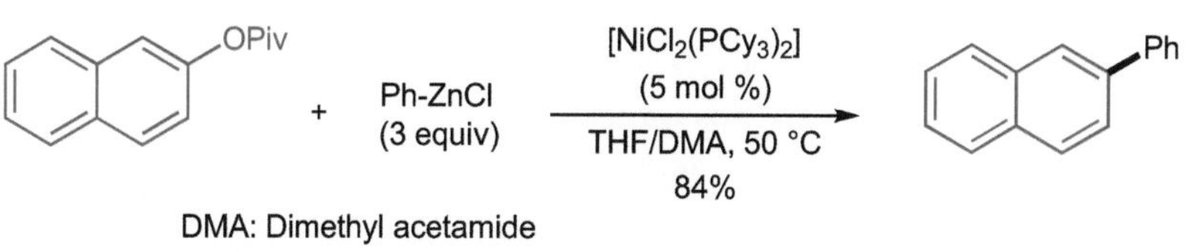

Fig. 2.7: Coupling between a naphthylpivalate and phenylzinc reagent.

Ni(II) salts which were *in situ* reduced thanks to an excess of the organozinc chloride. Interestingly, the same reaction with less hindered esters (acetate or isobutyrate) led to disappointing results presumably due to the direct attack of the zinc reagent on the electrophilic center. On the other hand, reaction with highly sterically hindered arylzinc reagents (as mesitylzinc chloride) was totally inefficient, affecting probably the transmetallation process. One other major drawback is the necessity to perform the reaction under strict inert atmosphere [24].

2.2.1.3.2 With aryl triflates

Among all organozinc reagents, arylzinc pivalates have been shown to promote highly efficiently cross-coupling reactions [25]. They are easily prepared from Grignard reagents and further transmetallation with $Zn(OPiv)_2.LiCl$. Interestingly, they are highly stable even when exposed to air. In the presence of $NiCl_2(PPh_3)_2$, anisylzinc pivalate reacted with different aryl- and heteroaryltriflates or nonaflates to give diaryl compounds in excellent yields (66–95%) (Fig. 2.8) [26].

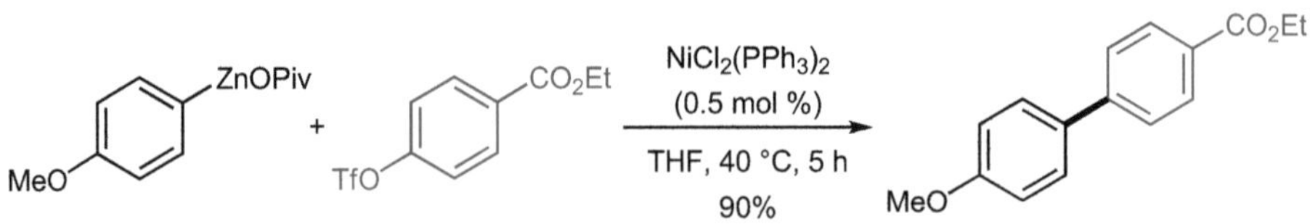

Fig. 2.8: Cross-coupling reaction between an aryltriflate and an arylzinc pivalate.

2.2.1.4 With aryl ethers

Aryl methyl ethers were recently considered to be reluctant toward C(aryl)-O bond cleavage. However, recent studies have demonstrated that dianionic zincates such as $PhZnMe_3Li_2$ (only these species) can promote the expected cross-coupling reaction [27]. The reaction suffered the presence of enolizable groups but was initially limited to naphthyl derivatives (Fig. 2.9) [28].

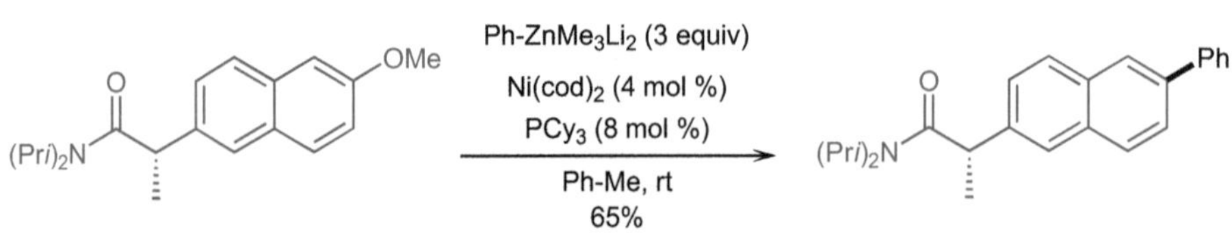

Fig. 2.9: Negishi-type coupling from naphthyl methyl ethers.

This limitation has been recently overcome by using aryl 2-pyridyl ethers. The reaction smoothly proceeds in the presence of Ni(II) catalyst which is reduced by a slight excess of arylzinc chloride reagent. A wide number of functionalities including electron-donating and electron-withdrawing groups both on the substrate and the arylzinc reagent are tolerated (Fig. 2.10). In a synthetic point of view, the reaction was performed on gram scale demonstrating the robustness of the process [29].

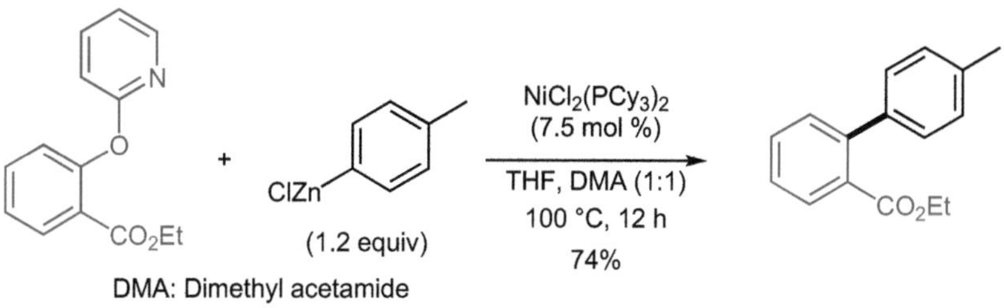

Fig. 2.10: Reaction of aryl 2-pyridyl ether with arylzinc reagents.

A catalytic cycle disclosed in Fig. 2.11 features a speculate N,O-chelate intermediate which could dramatically weaken the C(Ar)-O bond. The transmetallation which occurs after the coordination of the organozinc reagent to the nitrogen and oxygen atoms should be the rate-determining step.

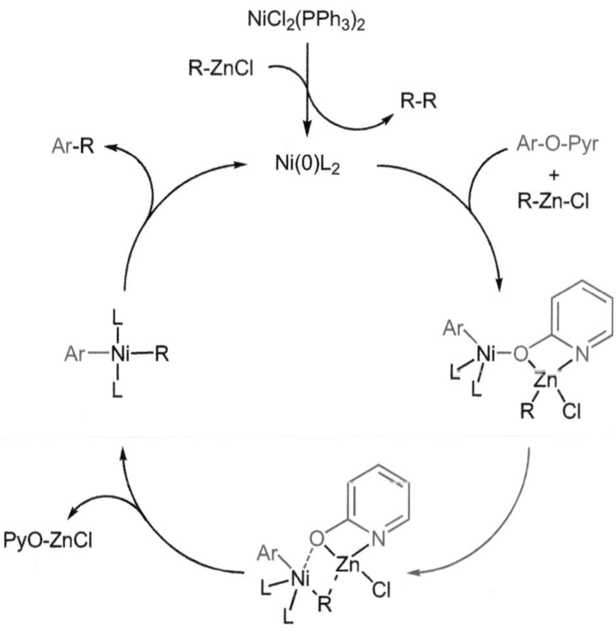

Fig. 2.11: Catalytic cycle for 2-pyridyl ethers as electrophilic substrates.

2.2.1.5 With aryltrimethylammonium iodides and triflates

Anilines are versatile derivatives in organic synthesis. Wang et al. reported that aryl trimethylammonium iodides can be opposed to arylzinc reagents to deliver biaryl adducts with impressive yields (Fig. 2.12). The reaction suffered the presence of numerous functionalities on both the substrate and the arylzinc reagent, even at the *ortho* positions [30].

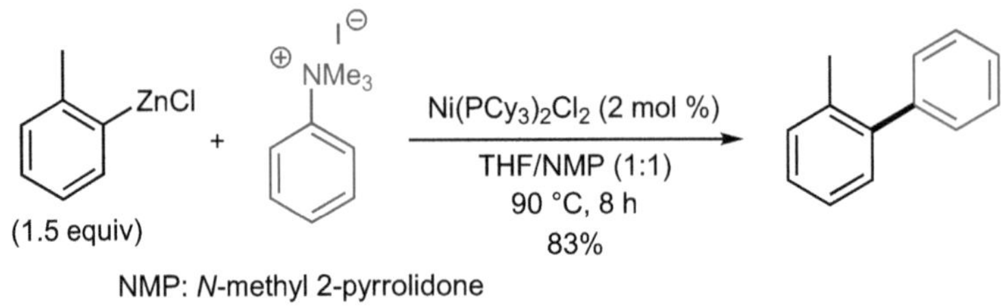

Fig. 2.12: Formation of biphenyls from phenyltrimethylammonium iodides.

Interestingly, the charge in catalyst is low, only 2 mol %, while the co-solvent N-methyl 2-pyrrolidone (NMP) stabilizes the nickel salts. However, attempts to perform the reaction with arylammonium triflates instead of the iodides, under the same conditions, were unsuccessful. These results are consistent with the reactivity of the same arylammonium iodides with organoboranes in the parent Suzuki cross-coupling reaction [31]. More recently, the lack of reactivity of aryltriflates was overcome, and the group of Wang reported a very interesting results using a *P,N,N*-pincer nickel complex **L3** in only 0.01 mol % (Fig. 2.13) [32]. Huang and coworkers have developed a sequential process in which the ammonium triflate is prepared *in situ* from an aniline and

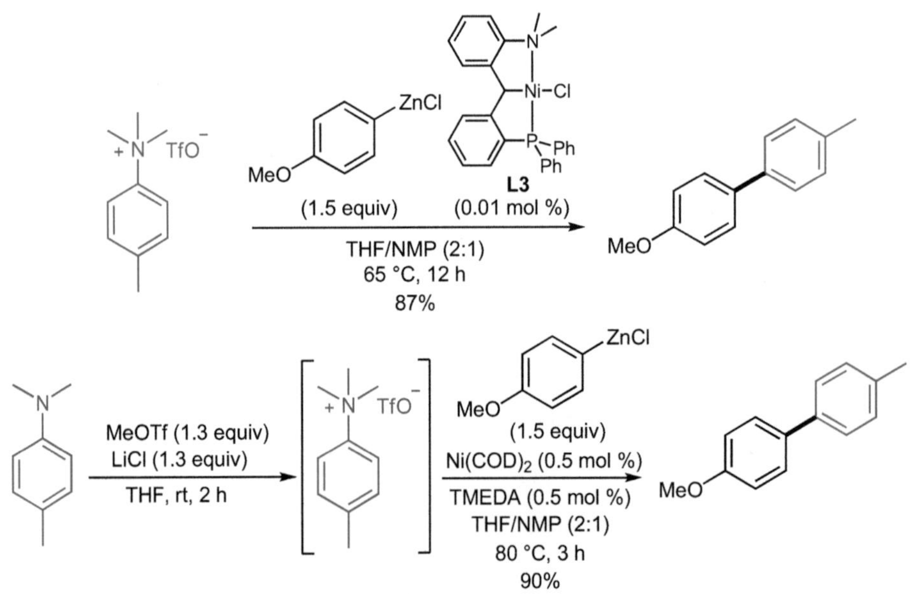

Fig. 2.13: Access to biphenyl from trimethylammonium triflates.

methyltriflate. In that case, the charge of nickel catalyst was much more significant (0.5 mol %) [33].

2.2.1.6 With arylthioethers

Thiomethylated *N*-heterocycles can be readily functionalized through a Csp^2-Csp^2 cross-coupling reaction catalyzed by nickel(0) derivatives performed at room temperature. Among the precatalysts tested, $NiCl_2$.DME and $Ni(acac)_2$ associated with the DPE-Phos ligand gave the best results (Fig. 2.14) [34].

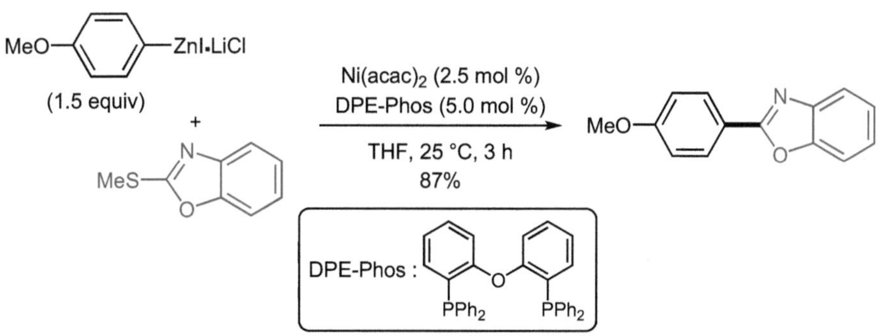

Fig. 2.14: Nickel-catalyzed formation of Csp^2-Csp^2 bonds from arylthioethers.

2.2.1.7 With arylsulfonium salts

An interrupted Pummerer/nickel-catalyzed cross-coupling reaction has been designed by the group of Procter [35]. This elegant process was promoted by treatment of electron-rich arene in the presence of the commercially available sulfoxide **S1**, triflic anhydride and potassium carbonate. The sulfonium intermediate was then exposed to anisylzinc chloride to deliver the biphenyl derivative in 62% overall yield (Fig. 2.15).

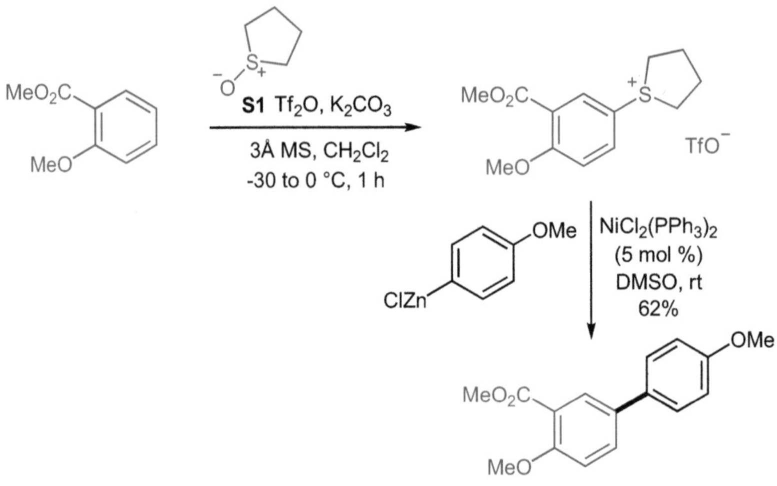

Fig. 2.15: Synthesis of biphenyls from aryl sulfonium salts.

2.2.1.8 With arylmethylsulfoxides

The coupling of aryl zinc reagents and aryl methyl sulfoxides has been successfully carried out by the group of Yorimitsu [36]. A methylsulfenate anion is supposed to be generated during the course of the reaction. This intermediate is in equilibrium with potential oxidizing reagents which could be detrimental to the efficiency of the reaction. A careful choice of the zinc reagent (in slight excess) was crucial to consume the sulfenate anion. By this way, impressive yields in the cross-coupling products were obtained (Figs. 2.16 and 2.17).

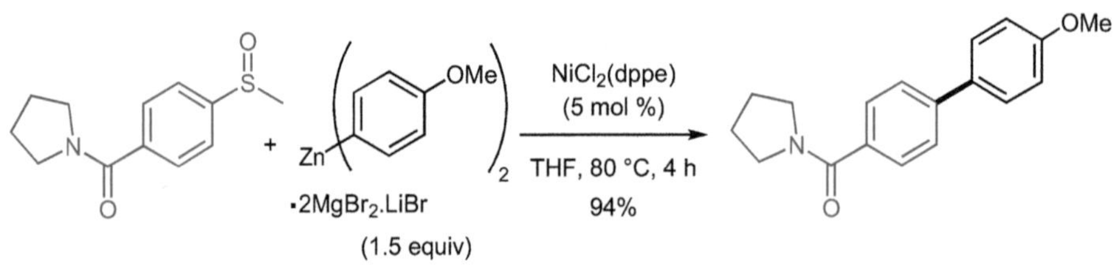

Fig. 2.16: Cross-coupling reaction from aryl sulfoxides – Synthesis of biphenyls.

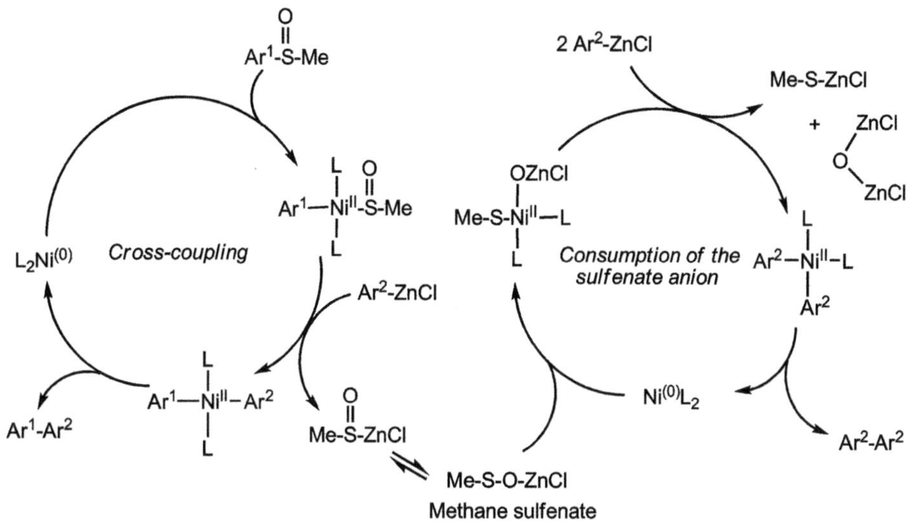

Fig. 2.17: Catalytic cycle for coupling between arylzinc reagent and arylmethylsulfoxydes.

2.2.2 With alkenyl derivatives

2.2.2.1 With enol ethers

2.2.2.1.1 With vinyl pivalates

As noticed before (Section 2.1.3.1), phenylzinc chloride can react with enol pivalates to furnish the corresponding adducts in appreciable yields (Fig. 2.18) [24].

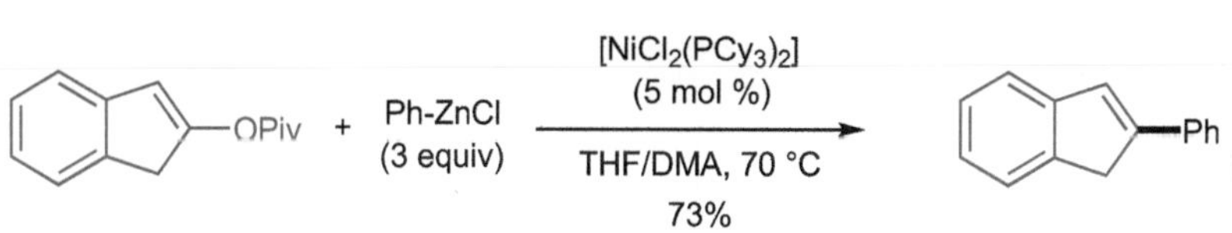

Fig. 2.18: Nickel-catalyzed arylation of enol esters.

2.2.2.1.2 With vinyl triflates

The formation of C-C bond from enol triflates (or nonaflates) with organozinc reagents has been recently reviewed [37]. The reaction can be achieved on numerous substrates. Interestingly the stereochemistry of the electrophilic species can be maintained during the coupling with zinc pivalates, while with organozinc chloride, an erosion of the stereochemistry was observed and the formation of the *E*-isomer was mainly observed (Fig. 2.19) [26].

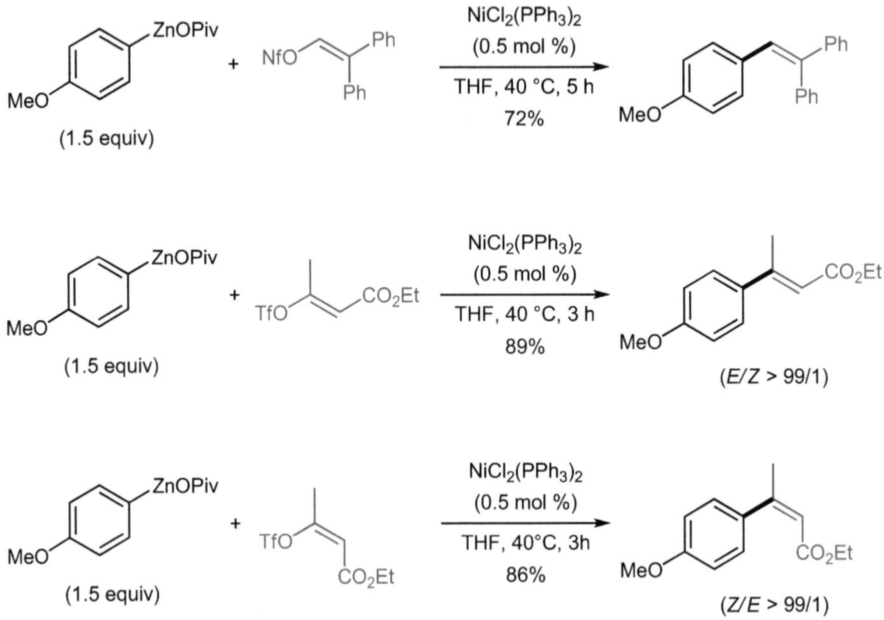

Fig. 2.19: Stereoselective cross-coupling reaction between arylzinc pivalates and enol triflates.

2.2.2.1.3 With vinyl phosphates

4-Diethylphosphonyloxycoumarin was implicated into a nickel Negishi-type cross-coupling with aromatic and heteroaromatic zinc bromides or iodides (Fig. 2.20). The resulting products were isolated in high yields (75–90%) [38].

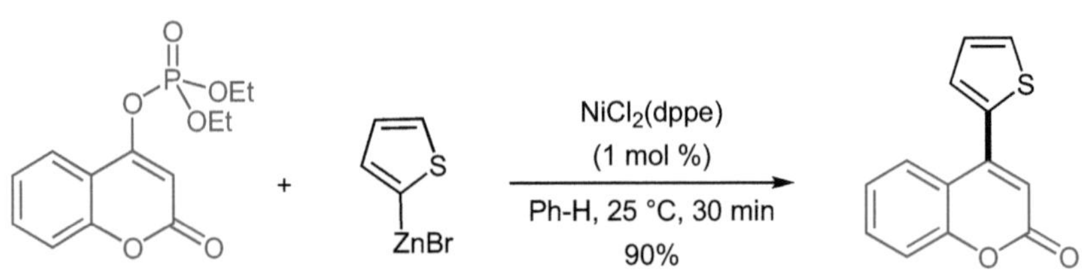

Fig. 2.20: Synthesis of 4-heteroarylcoumarins.

A similar transformation was carried out on acyclic derivatives (Fig. 2.21). Interestingly, the process appeared chemo- and highly stereoselective [39].

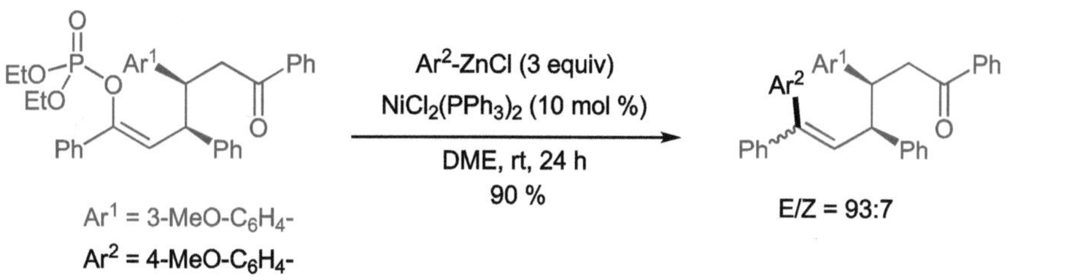

Fig. 2.21: Formation of a Csp^2-Csp^2 bond from a vinylphosphate.

2.2.2.2 With sulfonium salts

The synthesis of stilbene derivatives has been reported from styrenes (Fig. 2.22). The sequence required the *in situ* preparation of vinyl sulfonium salts by an interrupted Pummerer rearrangement associated with a cross-coupling reaction using an arylzinc chloride reagent and a nickel(II) salt catalyst, which is reduced by the slight excess of zinc reagent [35].

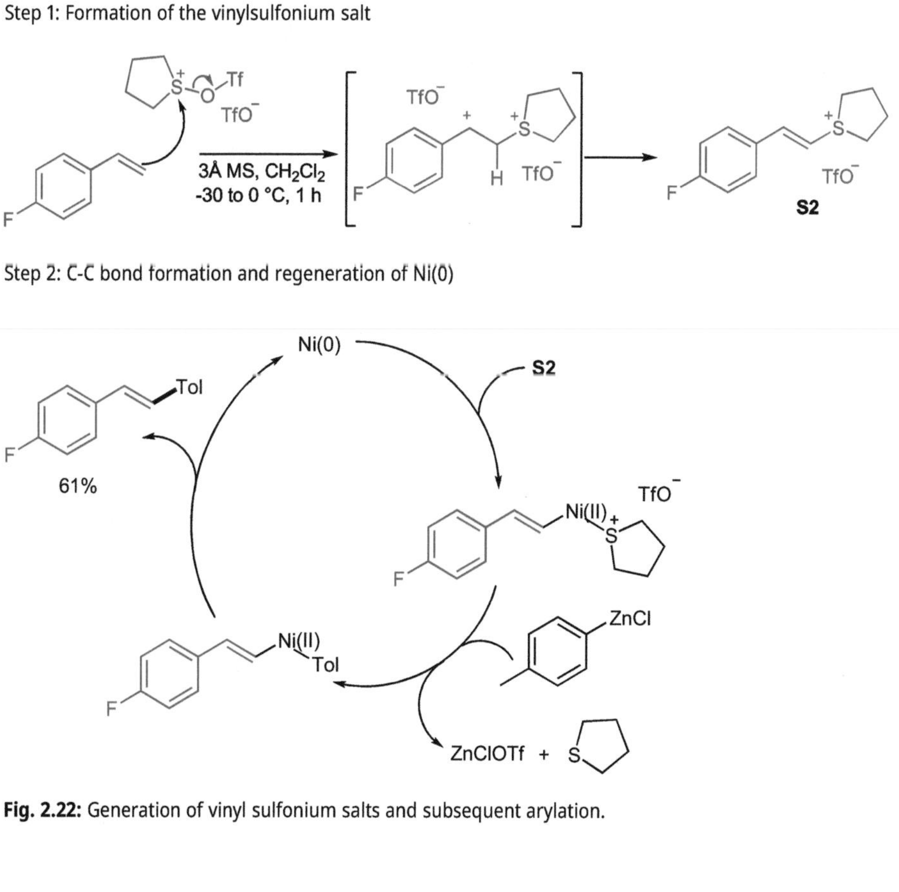

Fig. 2.22: Generation of vinyl sulfonium salts and subsequent arylation.

The coupling reaction follows the classic catalytic cycle including an oxidative addition, a transmetallation with an arylzinc reagent and a final reductive elimination. An unsuspected drawback is probably the release of stinking sulfur derivatives.

Alternatively, hindered and cyclic alkenyl sulfonium salts were prepared from homopropargylic arenes (Fig. 2.23). A tandem Pummerer reaction/Friedel-Crafts cyclization led to intermediate **S3** which was next implicated in a nickel Negishi-type cross-coupling reaction.

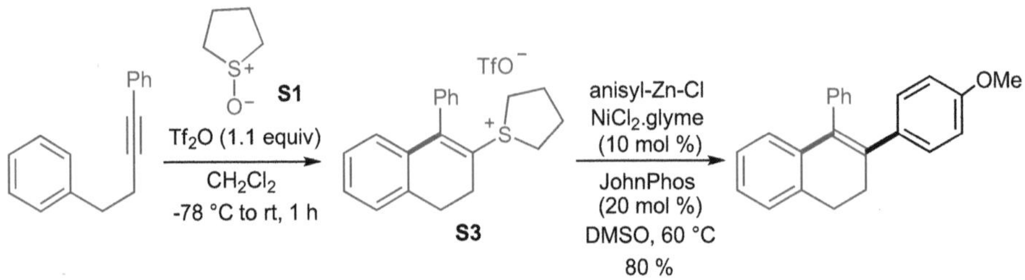

Fig. 2.23: Tandem Pummerer reaction/Friedel-Crafts cyclization and Negishi reaction.

2.2.3 With alkynyl derivatives

The reaction developed by Procter et al. was also used to prepare bis-arylacetylene. An interrupted Pummerer rearrangement led to an alkynyl sulfonium salt which was immediately implicated into a nickel cross-coupling reaction (Fig. 2.24). The yield for this single example is moderate – only 32% – compared to the well-established Sonogashira reaction [35].

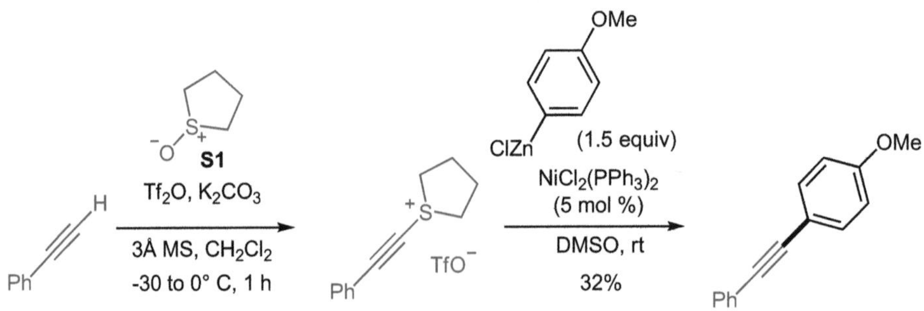

Fig. 2.24: Tandem interrupted Pummerer rearrangement and cross-coupling reaction.

2.2.4 With alkyl derivatives

One real advantage using organozinc reagents in the presence of nickel salts is the possibility, compared to palladium catalysis, to build C-C bonds from aliphatic electrophiles even possessing one or two β-hydrogen atoms. While palladium is reluctant to this issue, the Negishi cross-coupling reaction can be nowadays applied to the formation of Csp^2-Csp^3 and also Csp^3-Csp^3 bonds. In this section, the functionalization of alkyl groups with aryl derivatives will be described depending on the electrophilic species used. Furthermore, thanks to chiral ligands introduced in the media, high stereocontrol can be achieved. This aspect will be described in more detail in Section 2.5. Contributions by the groups of Knochel and Fu were at the origin of this important breakthrough.

2.2.4.1 With secondary alkyl halides

2.2.4.1.1 With CF_3-substituted secondary alkyl halides

Secondary trifluoroalkyl halides can react with arylzinc chlorides in the presence of nickel(II) salts and **L4** as chiral bis(oxazoline) ligand (Fig. 2.25). The cross-coupling product is highly enantioselective [40].

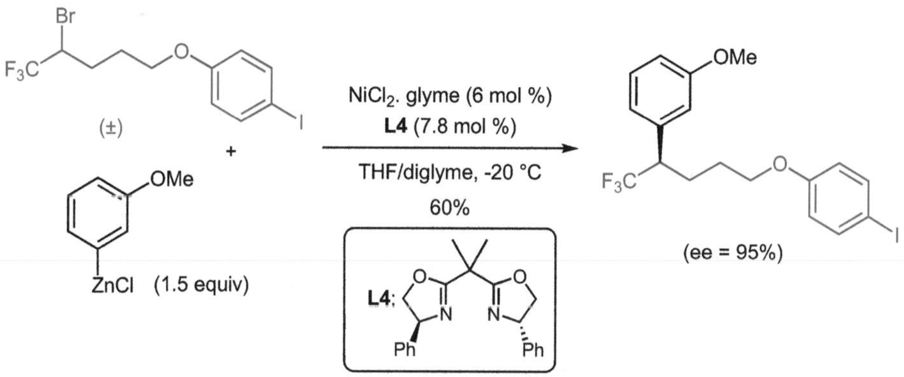

Fig. 2.25: Enantioselective synthesis of trifluoromethyl derivatives by Negishi-type reaction.

2.2.4.1.2 With α-haloamides

α-Aryl difluoroamides and their difluoro analogues constitute important frameworks in numerous bioactive compounds in the pharmaceutical and agrochemical domains [41]. Ando et al. developed an efficient synthesis starting from arylzinc derivatives and secondary or tertiary bromodifluoroacetamides [42]. The reaction was conducted with an excess of the arylzinc reagent (3 equiv) while TMEDA and bisoxazoline **L4** have been introduced to respectively stabilize the zinc and nickel species (Fig. 2.26).

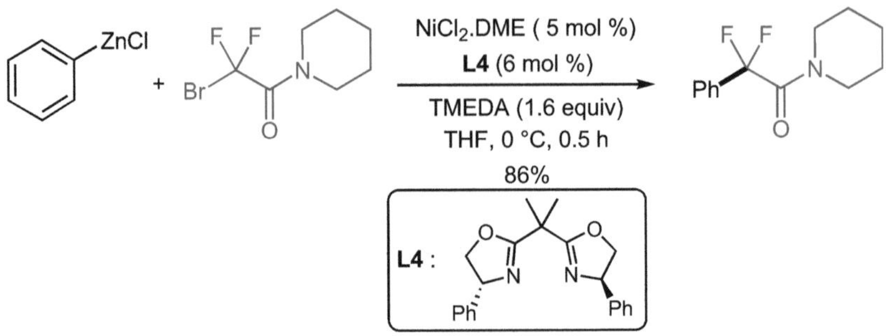

Fig. 2.26: Synthesis of α,α-difluoroamides.

2.2.4.1.3 With α-bromo alkylnitriles

A coupling reaction between α-bromonitriles and arylzinc reagents has been achieved by the Fu's group [43]. The reaction took place with high *ee* despite the presence of a highly enolizable center (Fig. 2.27). The use of diarylzinc reagent and a bisoxazoline ligand, associated with $NiCl_2$.glyme as well as low temperatures, was required for this process.

Among the different ligands tested, bisoxazoline **L5** gave the highest *ee*; in contrast, PyBox **L7** was inefficient in terms of yields and enantioselectivities.

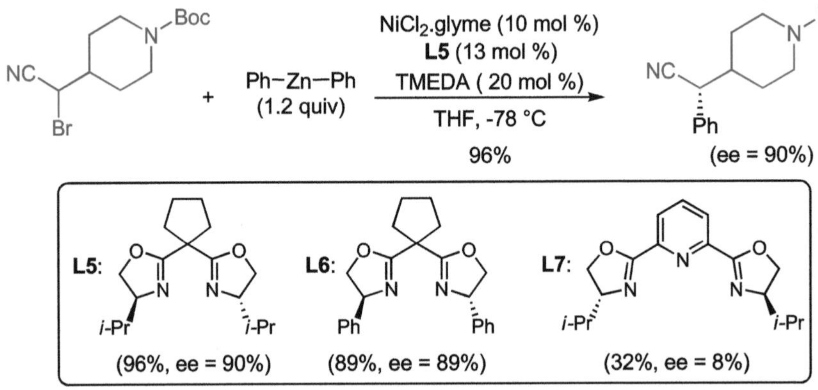

Fig. 2.27: Enantioselective functionalization of α-bromonitriles.

2.2.4.2 With secondary benzyl mesylates

In relation with the synthesis of 1,1-diarylalkanes, an important framework of numerous drugs, Fu and coworkers have investigated the asymmetric cross-coupling reaction of readily available benzylic mesylates and arylzinc iodides [44]. The ligand of choice to reach the highest selectivities was the bis(oxazoline) **L8** while PyBox ligands were once again not efficient in terms of yields and selectivities (Fig. 2.28).

Maintaining the temperature at −45 °C was crucial; otherwise, the selectivities decreased to low values. Addition of an excess of LiI was also primordial but was not simply relevant with the *in situ* formation of the parent benzylic iodide. As pointed out by Knochel, coordination of iodide to the alkylzinc iodide would deliver a more reactive zincate species [45]. The process was carried out on gram scale and applied to the synthesis of a precursor of Zoloft™ with a high level of enantioselectivity.

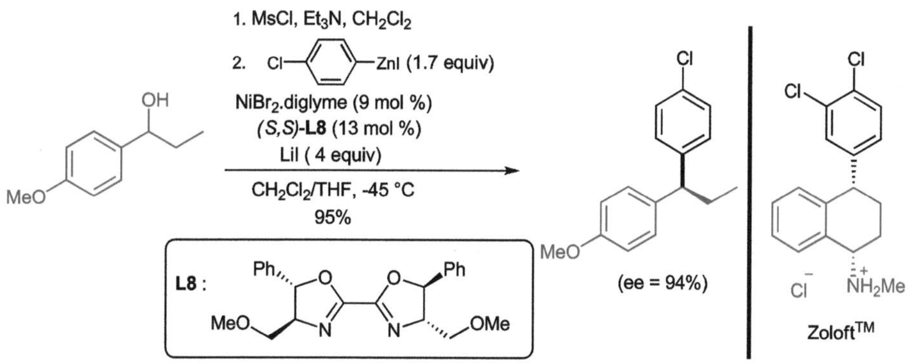

Fig. 2.28: Stereoselective synthesis of 1,1-diarylmethane derivatives from benzylic alcohols.

2.2.4.3 With allyl methyl ethers

In the presence of well-designed *P,N,N*-pincer nickel complexes as **L9**, the coupling of arylzinc chlorides with different benzylallyl ethers led to the exclusive formation of 1,3-diarylpropenes with appreciable yields (Figs. 2.29 and 2.30) [46]. This regioselectivity could be due to a minimization of steric hindrance and favorable conjugation of the double bond with the aryl group already present.

When the aryl group is replaced by an alkyl group, the regioselectivity is affected, leading mainly to the linear adduct (Fig. 2.31). These results are clearly in accordance with the formation of a π-allyl nickel complex intermediate.

The reaction has been also carried out in a sequential manner starting from an allylether possessing an additional reactive site. After a first functionalization, catalyzed by **L9**, a second coupling reaction was achieved in the presence of a more active nickel catalyst (Fig. 2.32).

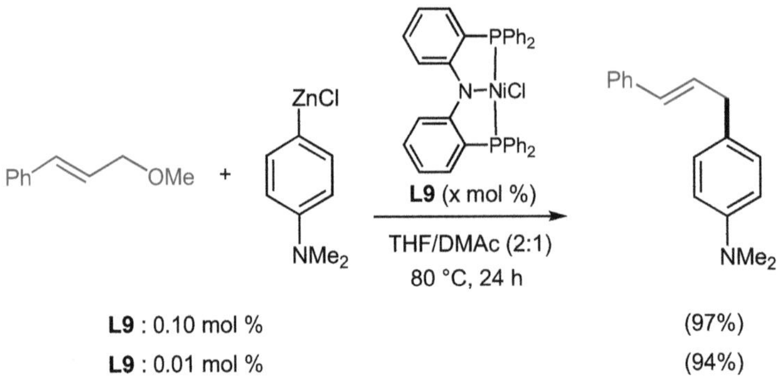

Fig. 2.29: Negishi-type reaction from allyl methyl ethers.

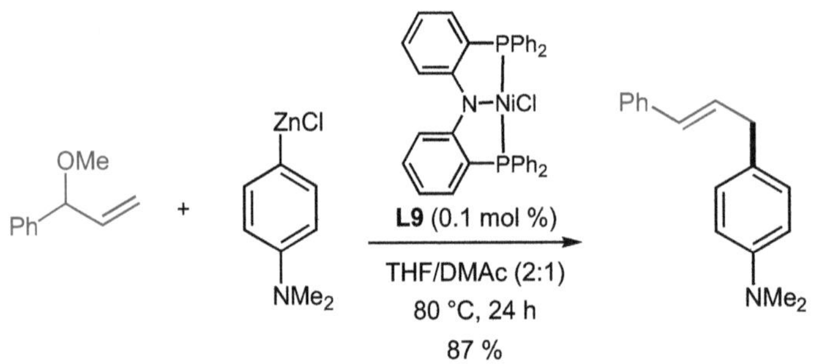

Fig. 2.30: Regioselective Negishi-type reaction from allyl methyl ethers.

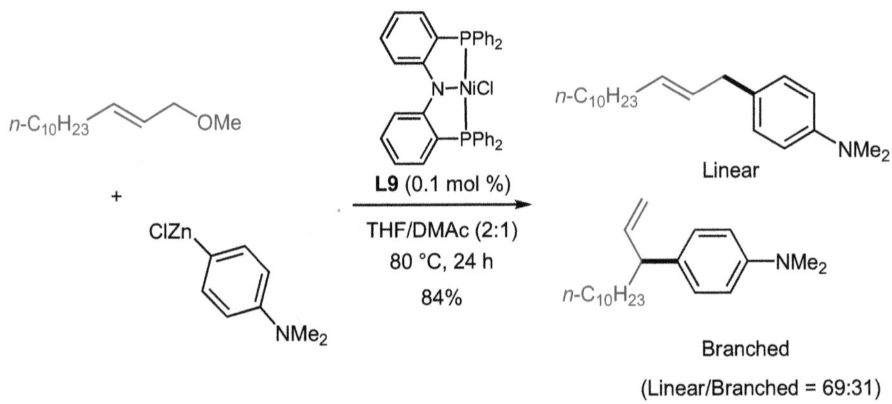

Fig. 2.31: Negishi-type reaction from allyl methyl ethers.

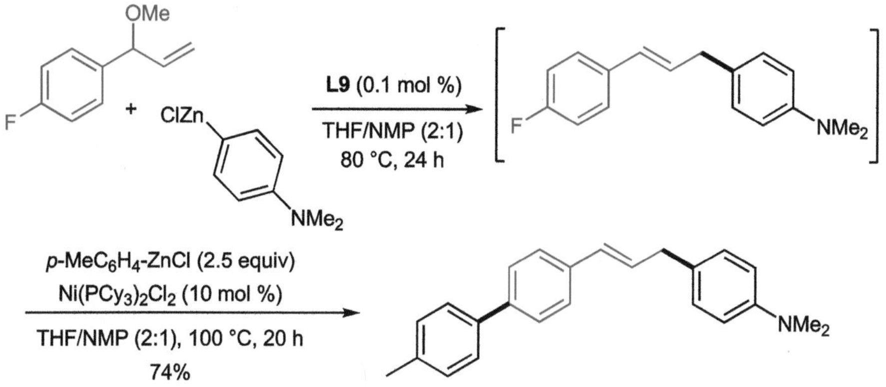

Fig. 2.32: Tandem Negishi-type reaction with an allyl methyl ether and an aryl fluoride.

2.2.4.4 With propargylic carbonates

Fu and coworkers have reported a stereoconvergent enantioselective cross-coupling reaction between racemic propargylic carbonates and organozinc reagents in the presence of **L10**, a chiral C2-symmetry nickel/PyBox catalyst derived from 1-aminoindan-2-ol (Fig. 2.33) [47].

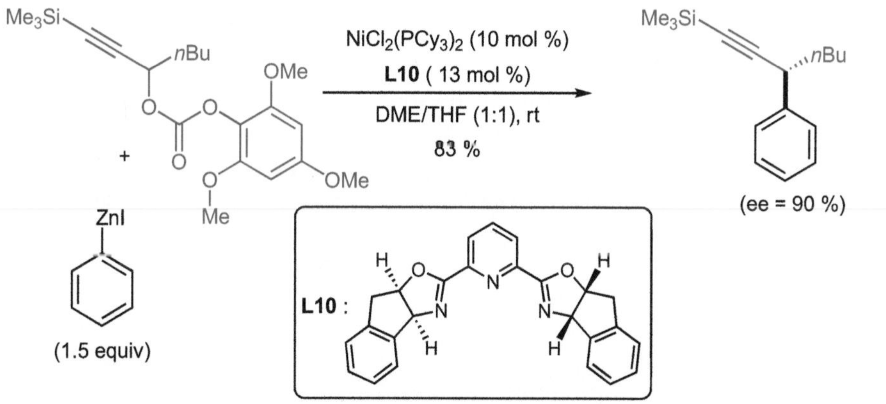

Fig. 2.33: Stereoconvergent cross-coupling reaction between a propargyl carbonate and an arylzinc reagent.

The reaction has been optimized, and 2,4,6-trimethoxyphenol carbonates led to the highest *ee*. Using the same experimental conditions, the same level of yields and selectivities were obtained for propargyl bromides and chlorides. The mechanism has also been investigated. Thanks to the synthesis of an isolated arylnickel(II) complex, a radical chain pathway has been proposed. It is also supported by the inhibition of the process when the reaction is carried out in the presence of TEMPO as a radical scavenger [48].

2.2.4.5 With thiocarbamates

Rousseaux et al. have achieved the α-arylation of esters starting from α-hydroxy esters readily converted into their thiocarbamates [49]. The reaction took place nicely when performed with a large excess of arylzinc in the presence of nickel(II) acetate and the achiral PyBox ligand **L11** (Fig. 2.34). The reaction was general, rapid (1 h) and the yields culminate to 81% with electron-enriched aryl groups. Starting from *(S)*-lactate ethyl ester, racemization of the α-center was observed during the process. Further experiments demonstrated that radicals are involved in the reaction.

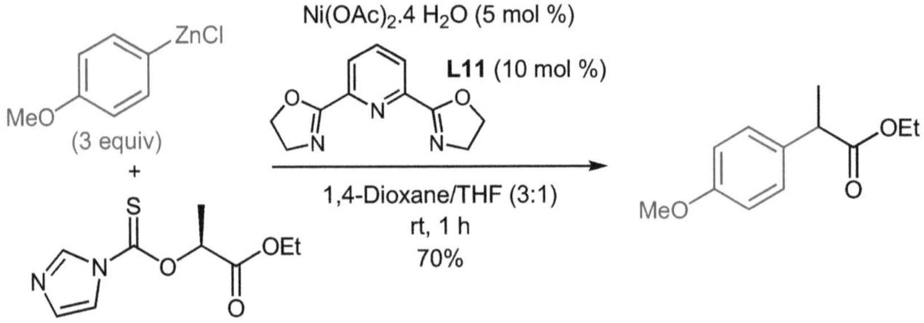

Fig. 2.34: Access to α-aryl esters from α-thiocarbamates.

2.2.4.6 With trialkylsulfonium salts

Trialkylsulfonium salts, generated by methylation of sulfides with methyl triflate, are able to react with arylzincates prepared themselves from aryl Grignard in the presence of $ZnCl_2$ and LiCl (Fig. 2.35) [50]. Mechanistically, the C-S bond cleavage could take place according to a Single Electron Transfer (SET) from low-valent nickel species to give the most carbon-centered radical. This is also supported by a rearrangement

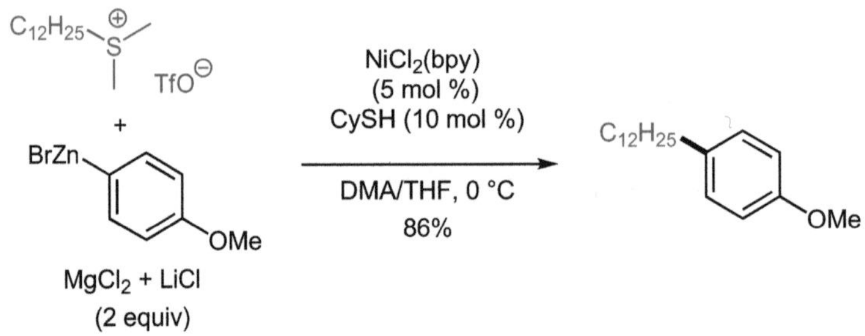

Fig. 2.35: Cross-coupling reaction between a trialkylsulfonium salt and an arylzinc reagent.

observed from the cyclopropylmethyl derivatives (Fig. 2.36). The yields were also considerably improved when the reaction was conducted in the presence of cyclohexylthiol which could favor the C-S cleavage.

2.2.4.7 With aliphatic carboxylic acids

Recently, direct decarboxylation of acids in the presence of photoredox catalysts, nickel salts and aryl halides has emerged as a new and attractive process to create new Csp^3-Csp^2 bonds [51]. During this Ir/Ni dual catalytic system, the radicals generated during the photoredox decarboxylation catalytic cycle interact with organonickel(II) intermediates obtained during a concomitant nickel catalytic cycle. The resulting high valent nickel(III) complex led, after reductive elimination, to the expected compound [52, 53]. While carboxylic acids are abundant, inexpensive and really stable reagents on the bench, other procedures have been investigated. Thus, the coupling of redox active *N*-hydroxyphthalimide (NHP) esters with arylzinc reagents catalyzed by nickel salts has been first described by Baran et al. [54, 55]. In the presence of di-*t*-Bubipy ligand **L12**, to stabilize the nickel complex, the reaction occurred under very mild conditions at rt (Fig. 2.37).

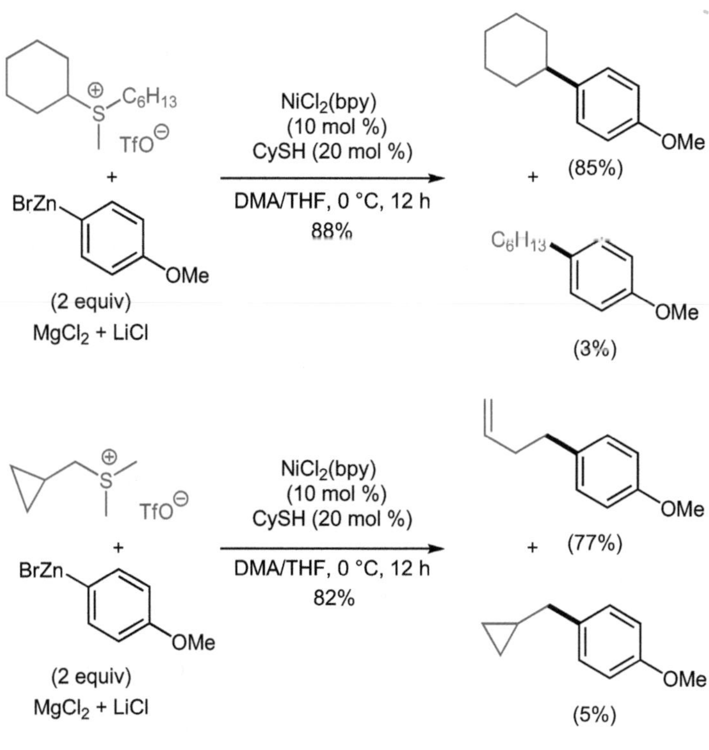

Fig. 2.36: Selectivity and rearrangement observed with various alkylsulfonium salts.

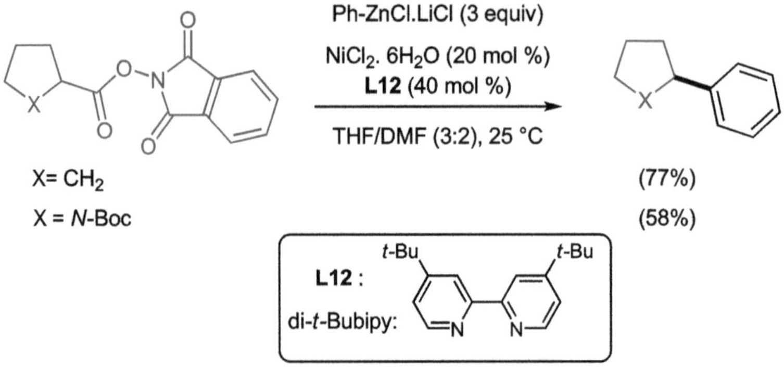

Fig. 2.37: Decarboxylative Negishi-type reaction from redox active NHP esters.

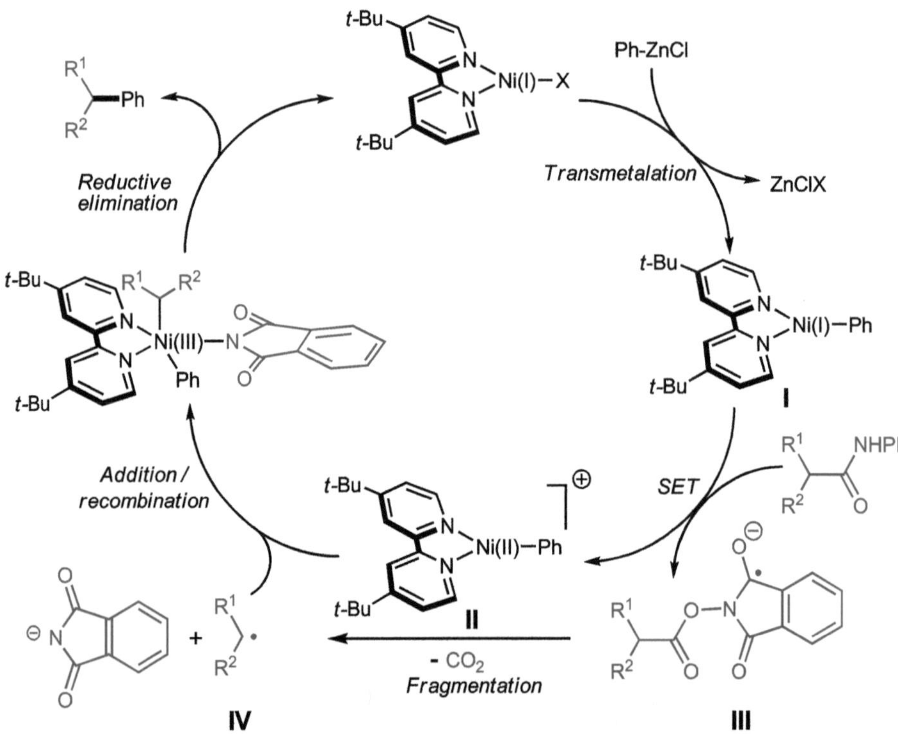

Fig. 2.38: Radical process using redox active NHP esters.

The scope of the reaction is broad, and the process can be applied to a large number of aryl and heteroaryl reagents and also to secondary, tertiary carboxylic acids even if they are sterically hindered. The reaction was also performed on gram-scale.

A radical mechanism has been proposed in which aryl-Ni(I) complex **I** was oxidized by a SET to a new Ni(II) species **II** while the radical anion **III** was further fragmented with CO_2 expulsion. The radical obtained can be recombined with **II** to finally undergo a reductive elimination leading to the expected Csp^2-Csp^3 bond (Fig. 2.38).

Based on the same process, the group of Baran designed the formation of benzylic quaternary centers from *N*-hydroxytetrachlorophthalimido (TCNHPI) esters [56]. In that case, the diamine ligand was not suitable and was switched to dicarbonyl-based ligand as dipivaloylmethane (dpm). Addition of $ZnBr_2$ as catalyst [57], as previously mentioned by Molander et al. has a slight, but effective, impact on the yields (Fig. 2.39). This method was directly applied to the synthesis of intermediates of important pharmaceutical drugs on gram scale.

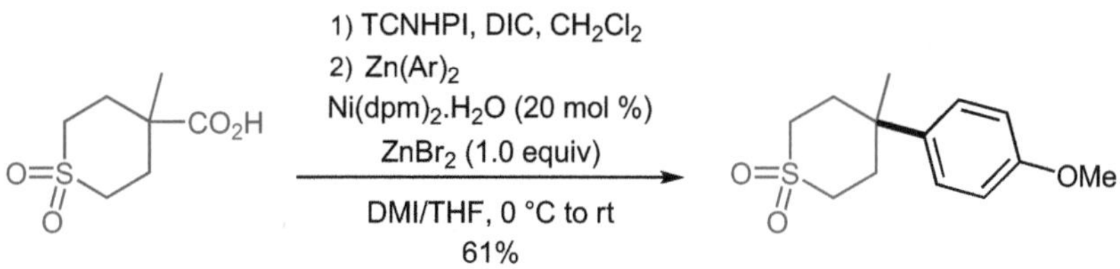

Ar = *p*-MeO-C_6H_4-
DIC : di*iso*propylcarbodiimide
dpm : dipivaloylmethane or 2,2,6,6-tetramethylheptane-3,5-dione
TCNHPI: *N*-hydroxytetrachlorophthalimide

Fig. 2.39: Formation of quaternary benzylic centers from carboxylic acids.

Starting from tertiary *N*-hydroxyphthalimide (NHPI) redox esters, a three-component reaction has been reported by the Baran's group. After decarboxylation, the tertiary radical formed could react with an acrylate according to a Giese-type reaction. The new stabilized radical species can further react with a Knochel's reagent (PhZnCl.LiCl) to create a new Csp^3-Csp^2 bond. This straightforward method allows a rapid access to α-aryl carboxylic derivatives (Fig. 2.40) [58].

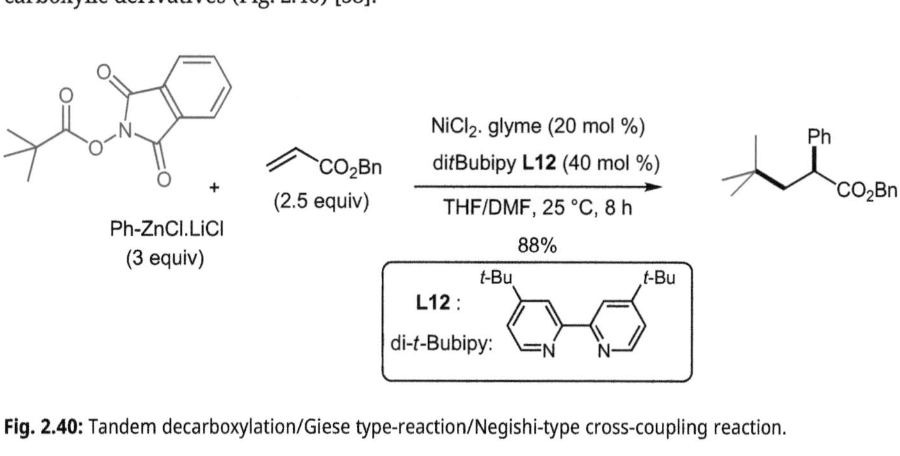

Fig. 2.40: Tandem decarboxylation/Giese type-reaction/Negishi-type cross-coupling reaction.

2.2.4.8 With alkylsulfones

Following a desulfonylative radical process, the formation of Csp^2-Csp^3 has been carried out from alkylsulfones and in the presence of a slight excess of arylzinc reagent [59, 60]. The *N*-phenyl tetrazolyl was the group of choice to maximize the yields (Fig. 2.41).

The process was also applied to the synthesis of mono-, di- and trifluorobenzyl derivatives (Fig. 2.42).

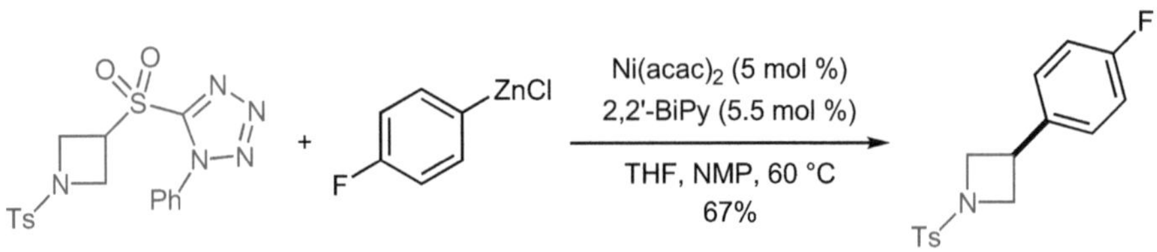

Fig. 2.41: Access to 3-arylazetidines.

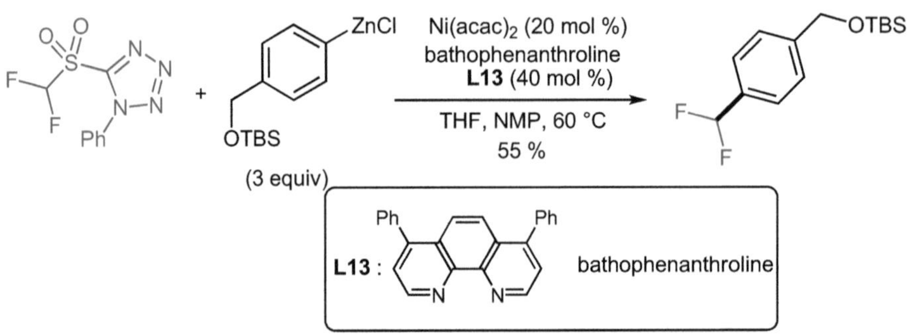

Fig. 2.42: Synthesis of difluorobenzyl derivatives.

2.3 Coupling reactions of alkenylzinc reagents

For the preparation of functionalized alkenyl reagents, the method developed by Knochel et al. is, without any doubt, the milder method reported to date. Alkenyl bromides bearing even highly sensitive carbonyl groups can be converted into lithium alkenylzincates (Fig. 2.43) [61].

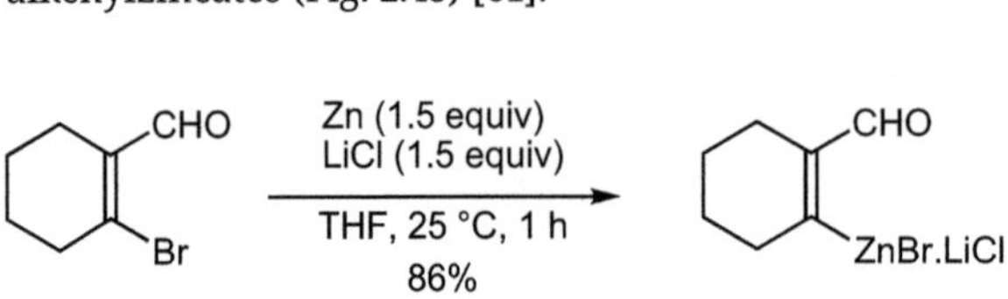

Fig. 2.43: Formation of lithium alkenylzincates from alkenyl bromides.

2.3.1 With alkyl derivatives

2.3.1.1 With α-bromonitriles

An enantioselective alkenylation of α-bromonitriles has been reported by Fu and co-workers. Bis(oxazoline) ligand **L5** was the best choice to reach high *ee* (range 80–92%) whatever the structure of the nucleophile (Fig. 2.44) [43]. No racemization or reconjugation occurred during the treatment.

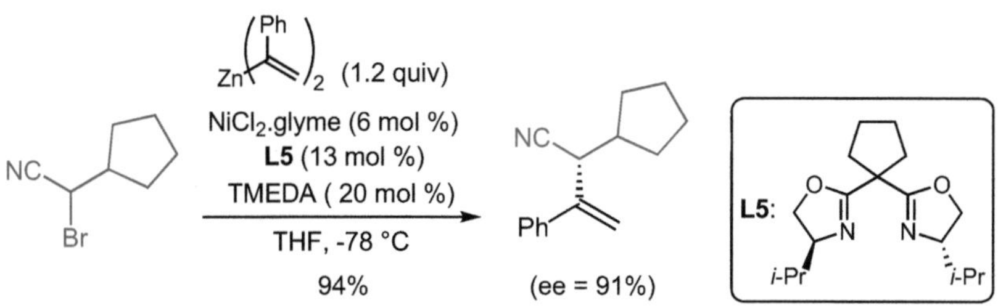

Fig. 2.44: Enantioselective functionalization of alkyl bromonitriles.

2.3.1.2 With activated esters

Baran et al. have investigated the decarboxylative alkenylation of activated esters [62]. They demonstrated that TCNHPI esters when submitted to an excess of zincate derivatives and in the presence of $Ni(acac)_2$ and bipyridine as catalysts led to the cross-coupling product (Fig. 2.45). The reaction was performed on numerous substrates.

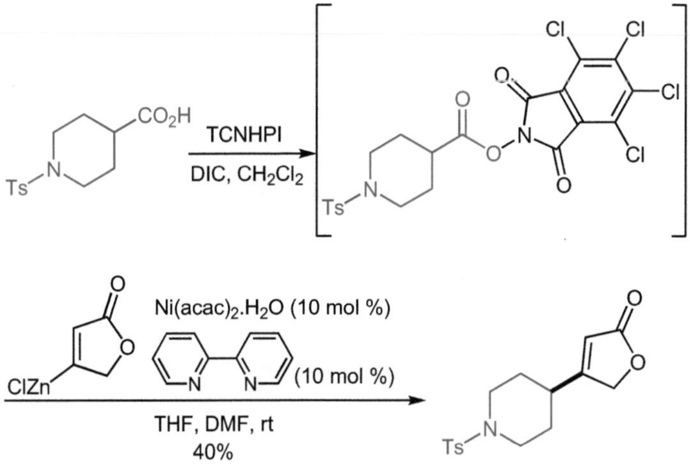

Fig. 2.45: Decarboxylative alkenylation of active TCNHPI esters.

Starting from a tricyclic carboxylic acid **1**, the reaction led to the norbornene derivative **2** after rearrangement and release of the strain; this result is in favor of a radical process (Fig. 2.46).

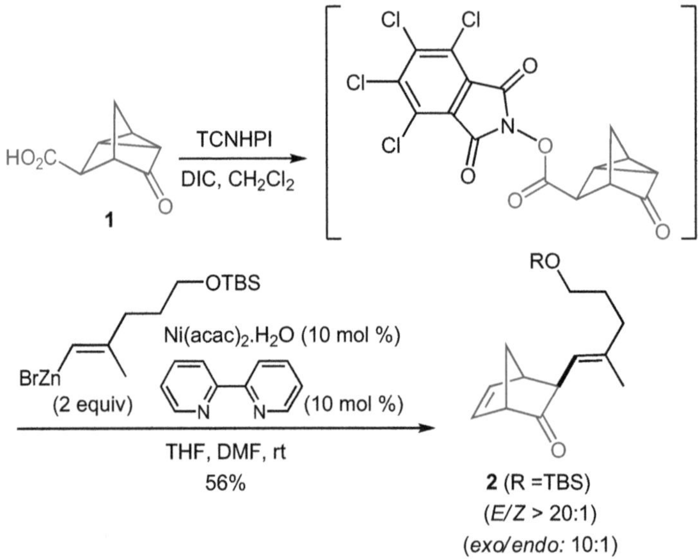

Fig. 2.46: Rearrangement and decarboxylative alkenylation of active TCNHPI esters.

2.4 Coupling reactions of alkynylzinc reagents

The reaction of choice to prepare alkynes is with no doubt the Sonogashira coupling reaction catalyzed by palladium(0). However, the Negishi cross-coupling reaction can also be used for such a purpose. The group of Knochel described a general method to prepare alkynylzinc reagents based on a deprotonation of terminal alkynes by the complex TMPZnOPiv.LiCl generated as depicted in Fig. 2.47. Interestingly, these solid derivatives are air stable for several hours without significant decomposition [63].

2.4.1 With aryl derivatives

The coupling of alkynylzinc pivalates with aryl halides has been studied by the group of Knochel. These processes were mainly reported in the presence of palladium catalysts such as $Pd(dba)_2$ in the presence of DavePhos ligand [63].

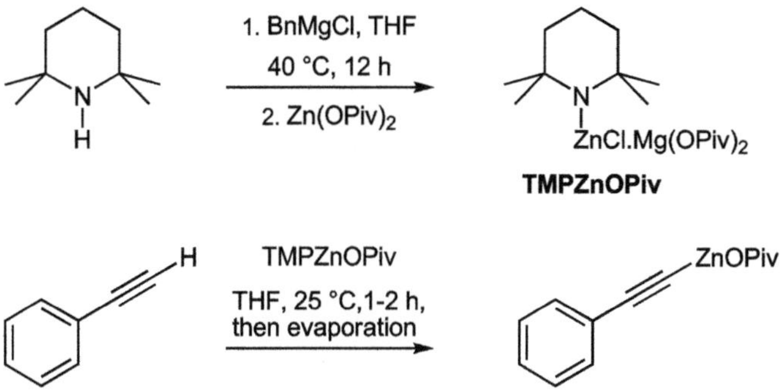

Fig. 2.47: Preparation of alkynylzinc reagents.

2.4.2 With alkenyl derivatives

2.4.2.1 With alkenyl triflates

The reagents described above were mainly cross-coupled with different electrophilic species in the presence of palladium. Nickel catalysis was only efficient with aryl and heteroaryl triflates (Fig. 2.48) [26].

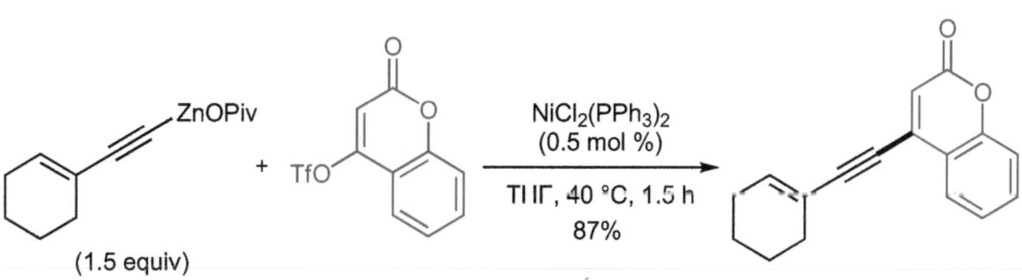

Fig. 2.48: Cross-coupling reaction between a vinyltriflate and an alkynylzinc reagent.

2.4.2.2 With alkenyl sulfonium salts

Alkenyl sulfonium salts generated from styryl derivatives are able to promote a Csp^2-Csp cross-coupling reactions from alkynylzinc chlorides in the presence of nickel(0) (Fig. 2.49) [35].

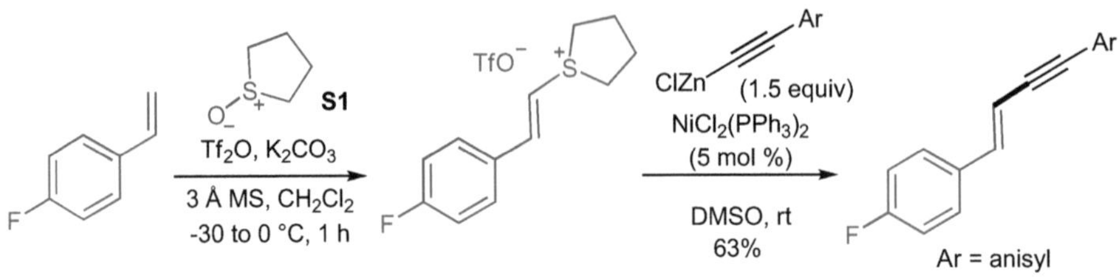

Fig. 2.49: Tandem Pummerer rearrangement and cross-coupling reaction with an alkynylzinc reagent.

2.4.3 With alkyl derivatives

2.4.3.1 With *N*-hydroxyphthalimide aliphatic esters

Decarboxylative alkynylation of carboxylic acids can be directly performed according to a two-step sequence (Fig. 2.50) [64]. After coupling with *N*-hydroxytetrachlorophthalimide under di*iso*propylcarbodiimide (DIC) activation, the resulting redox active ester was then treated with ethynylzinc chloride, prepared from the corresponding Grignard and $ZnCl_2$. After optimization, ligand **L14** and $NiCl_2$ were essential to observe the formation of Csp^3-Csp bonds in acceptable yields. The reaction has been performed on a large panel of substrates possessing numerous functionalities. Alternatively, the same reaction was also tested with substituted alkynyl Grignard reagents combined with Fe(II) salts at lower temperature.

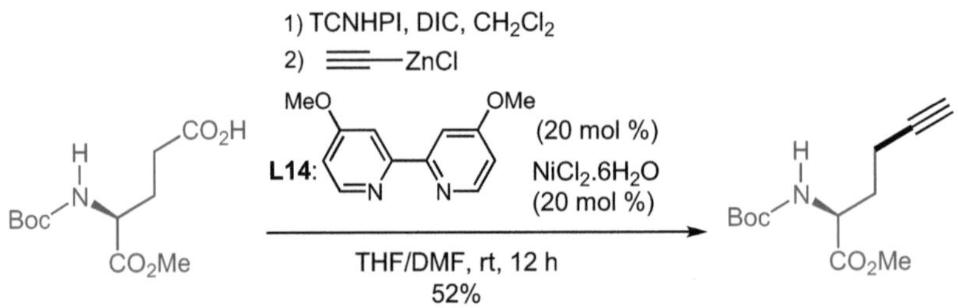

Fig. 2.50: Decarboxylative alkynylation of a terminal carboxylic acid.

2.5 Coupling reactions of alkylzinc reagents

2.5.1 With aryl derivatives

2.5.1.1 With aryl halides

In their seminal article, Negishi and coworkers reported the successful Csp^3-Csp^2 cross-coupling reaction between benzylzinc chloride and methyl *p*-bromobenzoate (Fig. 2.51) [5]. In contrast, the same reaction using benzylmagnesium chloride and the same nickel catalyst was poorly efficient and delivered the expected compound in only 12 % yield.

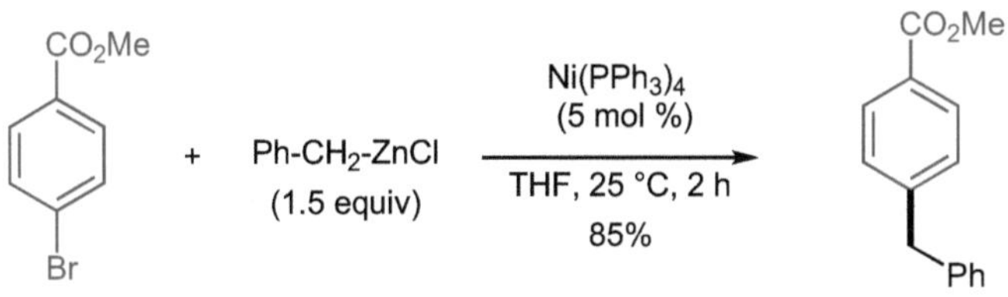

Fig. 2.51: Cross-coupling reaction between an aryl bromide and a benzylzinc reagent.

Arenylmethylzinc can be generated *in situ* from aryl iodide and bis(iodozincio)methane (Fig. 2.52). The reaction is catalyzed by nickel salts (especially with pyridine derivatives). Subsequent transmetallation with CuCN and addition of a suitable activated electrophile lead to the cross-coupling product [65].

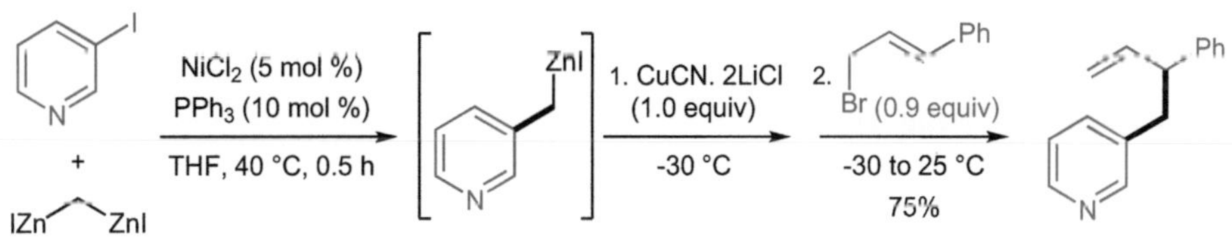

Fig. 2.52: Sequential cross-coupling reactions from bis(iodozincio)methane.

2.5.1.2 With aryl fluorides

Monofunctionalization of aryldifluorides possessing a directing group (imino or carbonyl group) can be achieved with *in situ* generated alkylzinc reagents. The reaction occurred solely at the *ortho* position (Fig. 2.53). The reaction is general and tolerates a wide range of functionalities on the alkylzinc reagent (Fig. 2.54) [22, 23].

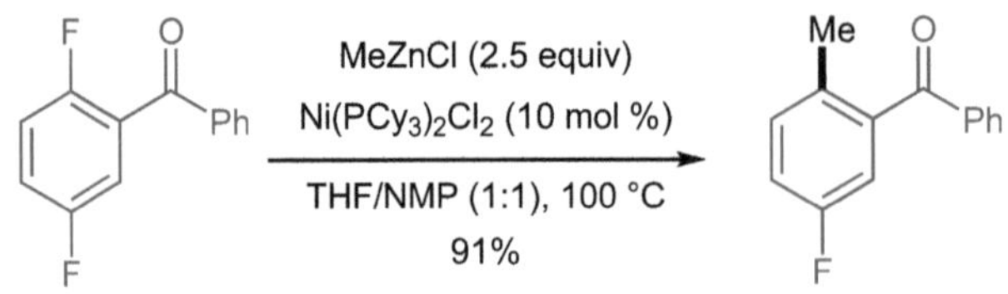

Fig. 2.53: Regioselective C-C bond formation from difluoroarenes.

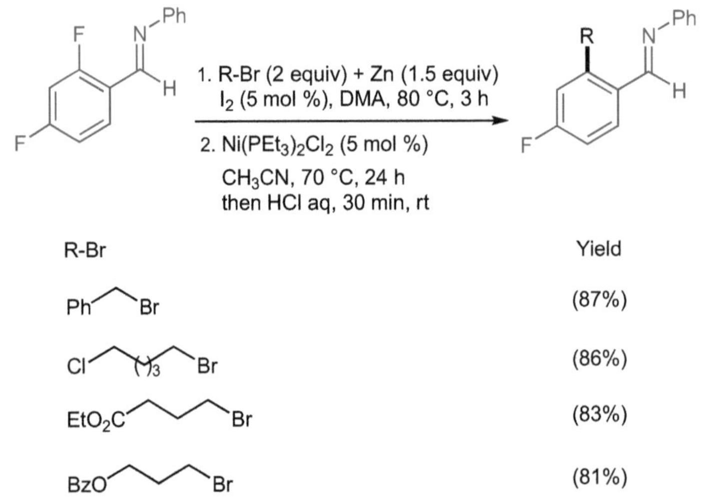

Fig. 2.54: Mono and regioselective functionalization of imino derivatives.

2.5.1.3 With aryl triflates

The presence of a nitrogen atom can be problematic during the formation of C-C bonds using organometallic catalysts. To solve this problem, the amino group is usually deactivated and transformed into an amide or a sulfonamide. Knochel et al. observed the coupling reaction between organozinc derivatives bearing an aminoalkyl group and aryl bromides or aryl triflates (Fig. 2.55) [66]. The reaction, achieved at room temperature, was really efficient in the presence of Ni(acac)$_2$ and DPE-Phos.

2.5.1.4 With phenol esters

Both aryl pivalates and phenol esters can undergo a decarboxylation/alkylation sequence when heated respectively at 70 °C or 150 °C in the presence of alkyl zincates, a nickel(0) catalyst and an appropriate diamine ligand. Among the different ligands tested,

1,2-bis(dicyclohexylphosphino)ethane (dcype) gave the best results [67]. This method has been advantageously applied to the synthesis of naproxen analogues (Fig. 2.56).

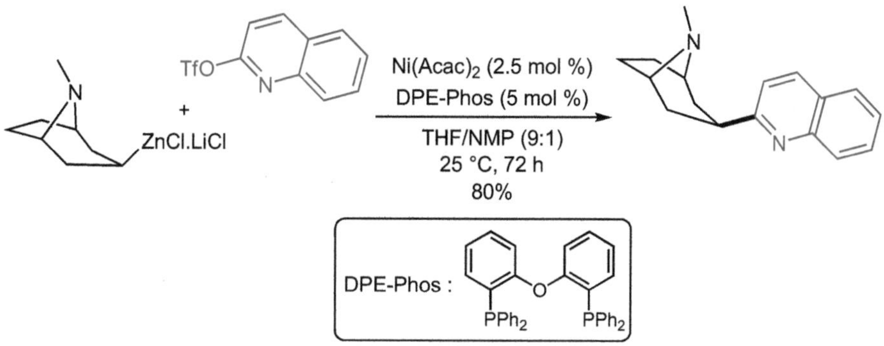

Fig. 2.55: Cross-coupling reaction between an aryl triflate and a bicyclic alkylzinc derivative.

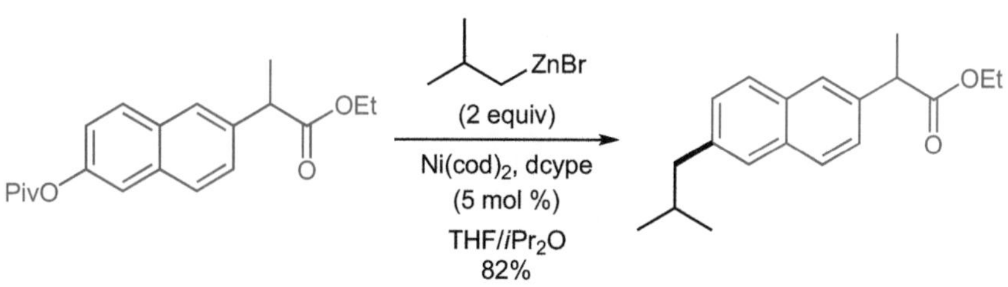

Fig. 2.56: Functionalization of an aryl pivalate and an alkylzinc bromide.

2.5.1.5 With aryl ethers

As already disclosed (Section 2.1.4), aryl 2-pyridyl ethers are able to react with alkyl zinc chlorides to form a new C-C bond. The reaction was efficient with methyl and benzyl derivatives (yields 71–87%) (Fig. 2.57). However, alkyl zinc reagents containing a β-hydrogen atom such as EtZnCl led to a mixture of cross-coupling and reduced compounds.

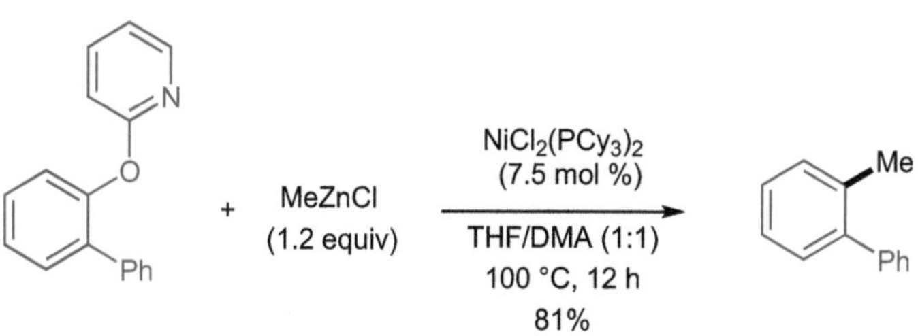

Fig. 2.57: C-C bond formation from an aryl 2-pyridyl ether and methylzinc chloride.

2.5.1.6 With aryl thiomethylethers

Knochel et al. reported the chemoselective functionalization of thiomethyl *N*-heterocycles by using benzyl- and alkyl-organozinc chlorides in the presence of $Ni(acac)_2$ as the precatalyst associated with DPE-Phos ligand at rt [34, 68]. Butylthiooxazole has been also functionalized in the presence of $NiCl_2(PPh_3)_2$ catalyst but at higher temperature (Fig. 2.58) [69, 70].

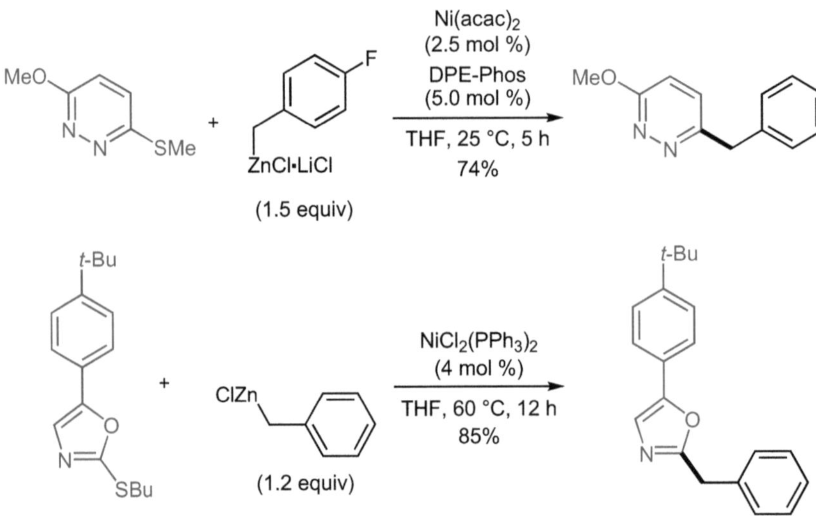

Fig. 2.58: Cross-coupling reactions between aryl thioalkylethers and benzylzinc reagents.

2.5.1.7 With aryl trimethylammonium iodides

Only primary alkylzinc reagents (methyl and benzyl zinc chlorides) smoothly underwent a C-C coupling with aryltrimethylammonium iodide salts. The reaction gave better yields when these reagents possessed an electron-withdrawing group (Fig. 2.59) [30].

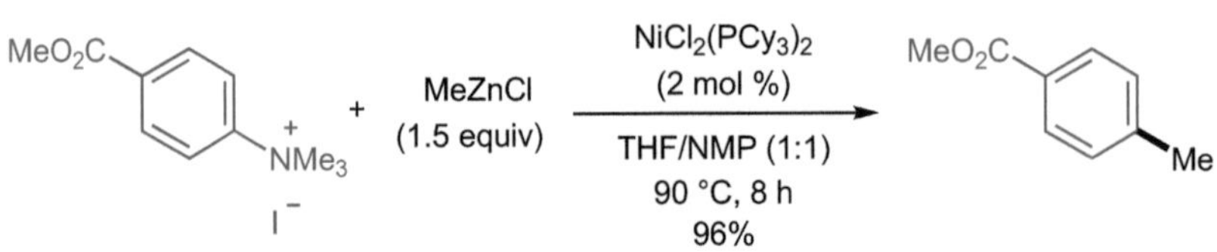

Fig. 2.59: Coupling reaction between a trimethylaryl iodide and methylzinc chloride.

2.5.2 With alkenyl derivatives

2.5.2.1 With α-oxy-vinylsulfones

The group of Zhang and Niu reported the synthesis of *Z*-enol ethers from α-oxy-vinyl sulfones and alkylzinc reagents. After optimization, the use of $Ni(OAc)_2$ as the precatalyst and 5-methylbipyridine (5-Mebipy) **L15** as the ligand appeared to be the best conditions to reach the highest yields (Fig. 2.60) [71].

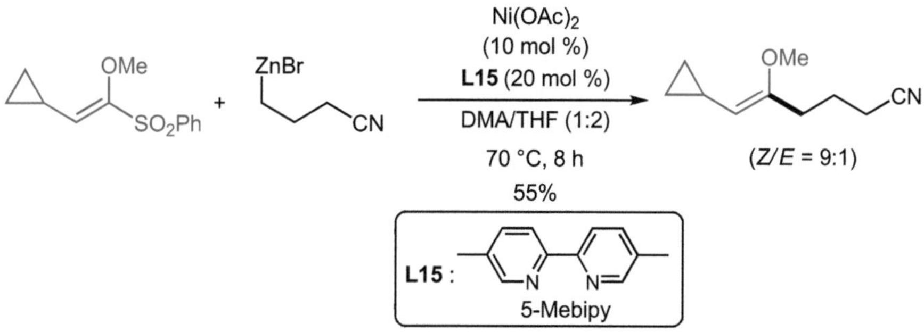

Fig. 2.60: Negishi-type reaction from vinylsulfone and alkylzinc reagents.

2.5.2.2 With alkenyl sulfoximines

In connection with the synthesis of prostacyclin analogues, the reactivity of sulfoximines and aminosulfoxonium salts has been investigated [72]. The use of diethyl ether as the solvent was crucial while a very low conversion was noticed in THF. The presence of $MgBr_2$, as an additive, is also important to observe full conversion. The reaction proceeded with an important level of stereoretention (Fig. 2.61)

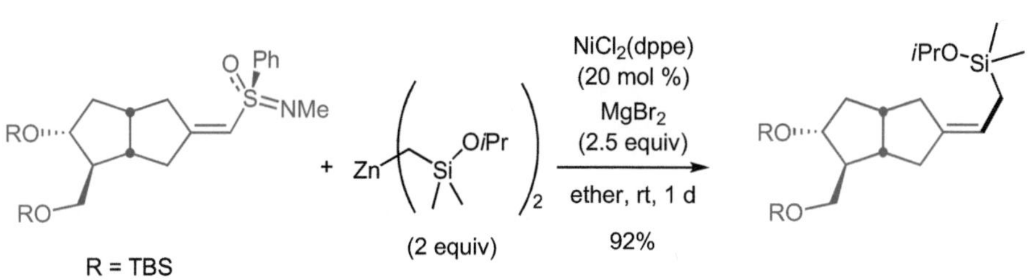

Fig. 2.61: Alkenyl sulfoximine as an efficient cross-coupling partner with alkylzinc reagents.

2.5.3 With alkyl derivatives

The formation of Csp^3-Csp^3 bond starting from an alkyl zinc reagent as the nucleophile and a Csp^3 electrophile in the presence of nickel salts has been extensively studied by the groups of Knochel and Fu [73]. This important coupling reaction was for a long time a challenging process and unsuccessful due to competitive β-hydrogen elimination from the metal-activated species. The use of PyBox ligands represents a real breakthrough and was determinant for the success of the coupling of primary and secondary alkyl bromides and iodides. Unfortunately, cheaper reagents as alkyl tosylates or alkyl chlorides did not react under these conditions. Moreover, hindered reagents as tertiary bromides were also reluctant.

In addition, the control of the stereochemistry has been achieved thanks to enantio-convergent approaches. The conditions for this enantio-control are extremely mild and required only substoechiometric amounts of chiral reagents. The versatility and the scope of applications have been extensively pointed out in different reviews [73–75].

2.5.3.1 With primary alkyl halides and alkylorganozinc species

In a seminal article, Knochel and coworkers reported the efficient cross-coupling of diethylzinc with primary bromides or iodides possessing an unsaturation [76]. The presence of the alkenyl functionalities was primordial to the success of the reaction. After oxidative addition of Ni(0) to the alkyl bromide, the nickel complex **V** could be stabilized intramolecularly by the C = C bond (Fig. 2.62). The transmetallation with diethylzinc could then furnish a new complex **VI** prompt to undergo a reductive elimination.

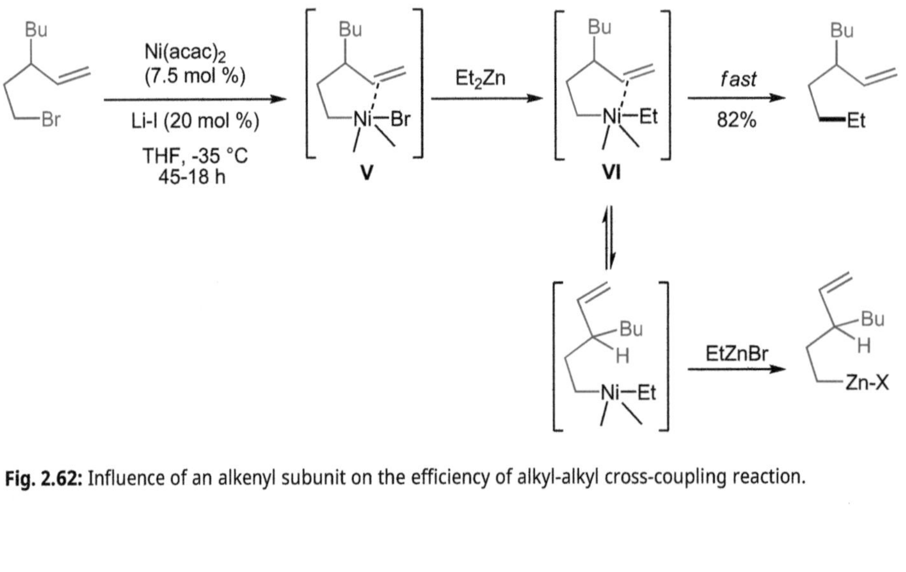

Fig. 2.62: Influence of an alkenyl subunit on the efficiency of alkyl-alkyl cross-coupling reaction.

However, with bulky substrates or in absence of the unsaturated group, such stabilization will be impossible limiting the scope of the reaction.

A few years later, the authors have developed a more general strategy in which *m*-trifluoromethylstyrene or 4-fluorostyrene acts as an external promotor to favor the expected couplings (Fig. 2.63) [45, 77, 78]. The complexation of Ni(II) intermediate with olefins possessing an electron-withdrawing group is known to facilitate the reductive elimination step of the intermediate Ni(II) complex instead of a β-H elimination [79, 80]. Furthermore, the use of a 2:1 mixture of THF and *N*-methylpyrrolidinone (NMP) instead of THF has a strong impact on the yields and reaction times. Finally, addition of tetrabutylammonium iodide in a large excess (up to 3 equiv) was also beneficial in terms of temperature and reaction times, especially with benzylzinc derivatives (Fig. 2.64) [81]. Coordination of iodide to the alkylzinc iodide would deliver a zincate species, which may be more reactive during the transmetallation [45].

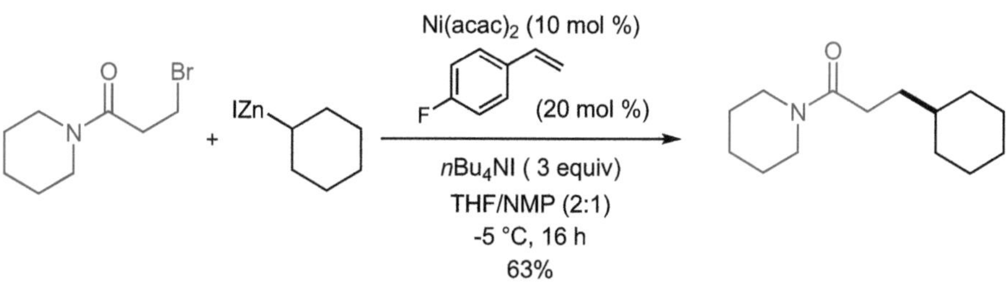

Fig. 2.63: 4-Fluorostyrene as an additive for the effective formation of Csp^3-Csp^3 bonds.

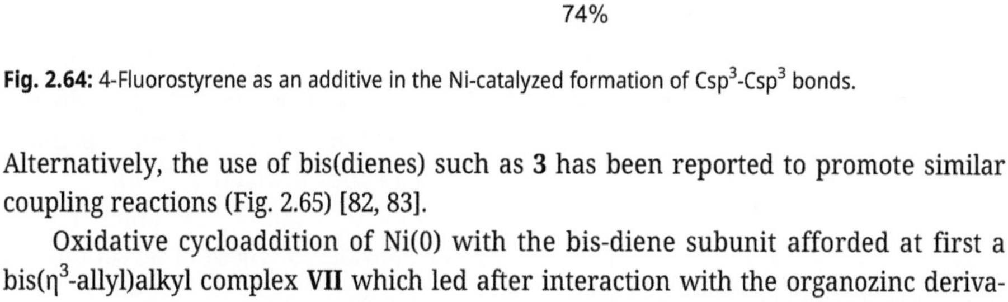

Fig. 2.64: 4-Fluorostyrene as an additive in the Ni-catalyzed formation of Csp^3-Csp^3 bonds.

Alternatively, the use of bis(dienes) such as **3** has been reported to promote similar coupling reactions (Fig. 2.65) [82, 83].

Oxidative cycloaddition of Ni(0) with the bis-diene subunit afforded at first a bis(η^3-allyl)alkyl complex **VII** which led after interaction with the organozinc derivative, to an ate complex **VII** (Fig. 2.66). This **VII** is able to react with the primary alkyl bromide, fluoride or tosylate to furnish a new complex **IX**. The reductive elimination can then furnish the final product along with the regeneration of **VII**.

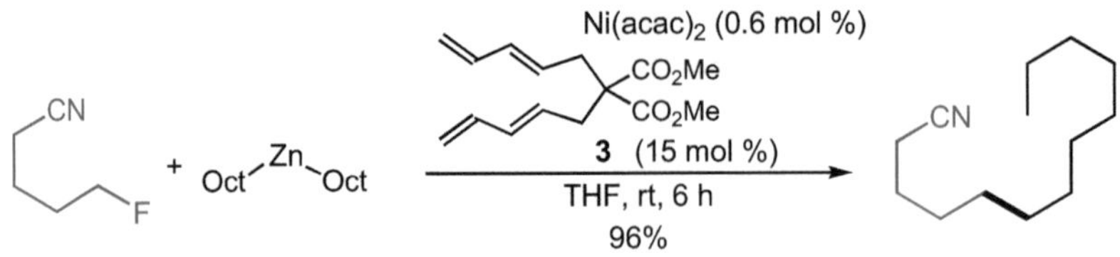

Fig. 2.65: Bis-diene **3** as additive for the alkylation of fluoroderivatives.

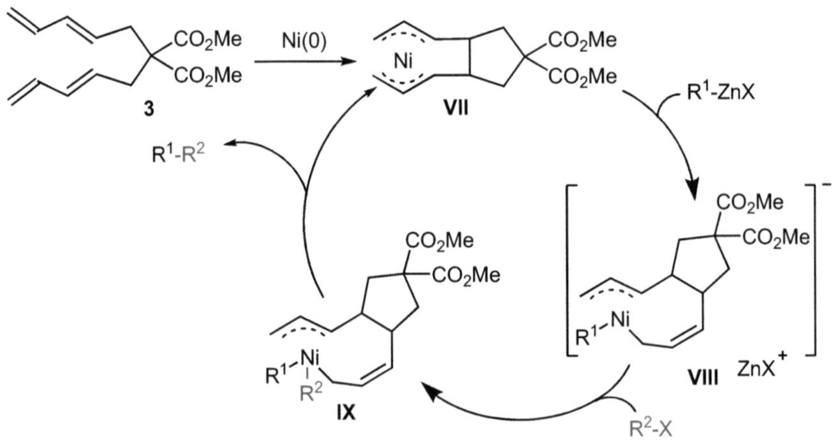

Fig. 2.66: Mechanism proposed for formation of C-C bonds using diene **3** as an additive.

The Negishi cross-coupling of primary alkyl bromides/iodides with primary alkylzinc reagents has been also considered by the group of Fu [84]. In this case, the combination of Ni(cod)$_2$ and a PyBox ligand was the best choice to reach yields in the range of 65–74% (Fig. 2.67).

2.5.3.2 With secondary alkyl halides

Starting from racemic benzylic bromides, a highly valuable catalytic enantioselective Negishi-type reaction was first disclosed by Fu and co-workers in 2005 [85]. Highest enantioselectivities were recorded with indane derivatives when opposed to primary alkylzinc reagents in the presence of nickel(II) and a Pybox ligand (Fig. 2.68).

According to the high yields which were obtained, a simple resolution was excluded and a stereoconvergent process has to be considered. These observations are in favor of a radical process (Fig. 2.69).

Secondary allyl chlorides can also be alkylated with primary alkylzinc bromides according to a similar stereoconvergent process (Fig. 2.70). With symmetrical allyl derivatives, high enantioselectivities were observed but suffered from steric hindrance.

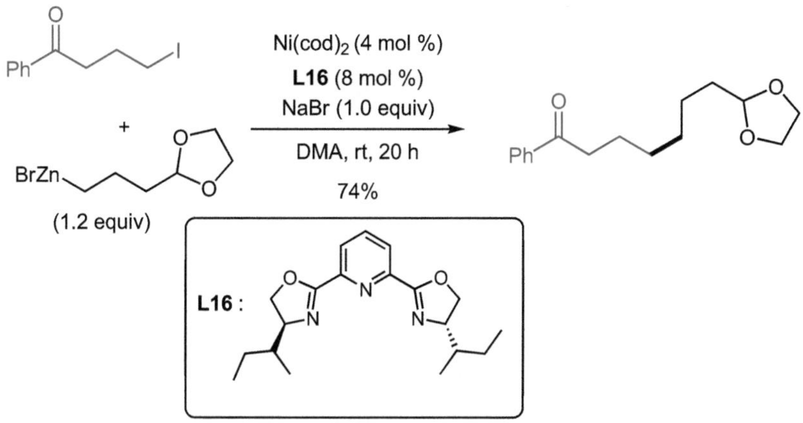

Fig. 2.67: PyBox ligands used during the formation of Csp^3-Csp^3 bonds.

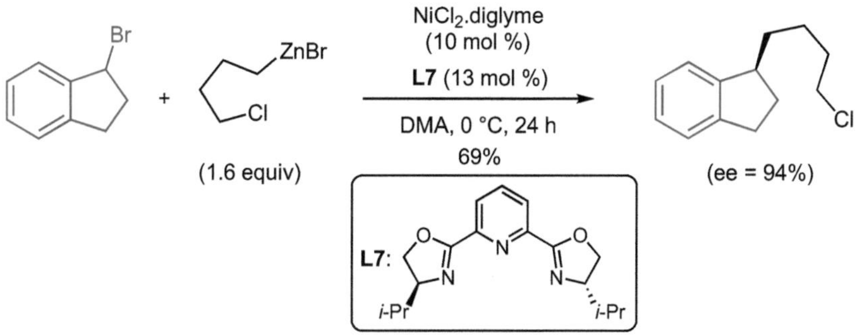

Fig. 2.68: Stereoconvergent cross coupling reaction between bromoindane and zinc reagents.

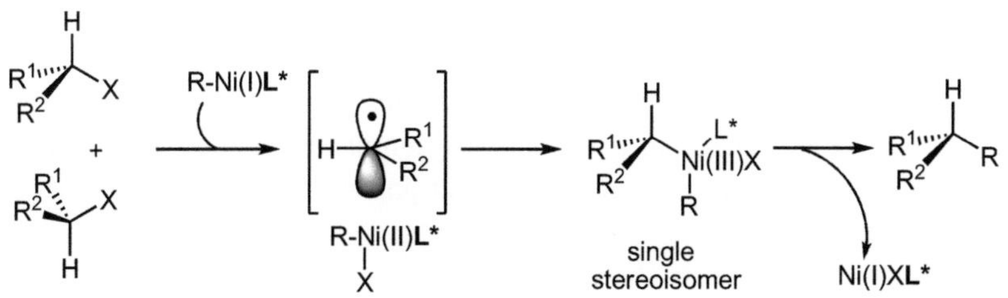

Fig. 2.69: Stereoconvergence during Csp^3-Csp^3 cross-coupling reactions.

Starting from unsymmetrical substrates, the regioselectivity issue can be overcome if the two alkyl groups are sufficiently different in terms of size [86].

Other efficient Csp^3-Csp^3 coupling reactions have been further considered including the synthesis of perfluoroalkyl derivatives (Fig. 2.71) [87]. $NiCl_2$ was the catalyst of

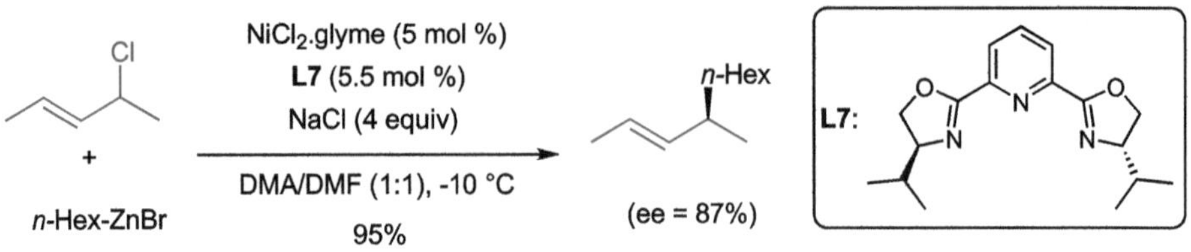

Fig. 2.70: Stereoconvergent cross-coupling between an allyl chloride and a primary alkylzinc reagent.

choice combined with **L7**. The presence of NaBr was also essential to increase the ionic strength of the medium. In its absence, the yield decreased to nearly 30% yield. Some limitations were nevertheless observed, including the lack of reactivity of secondary alkylzinc reagents. Moreover, and despite the presence of a chiral ligand, no indications were mentioned concerning the control of the enantioselectivity, in contrast to the arylation performed with the same perfluoroalkyl substrates (see Section 2.4.1) [40].

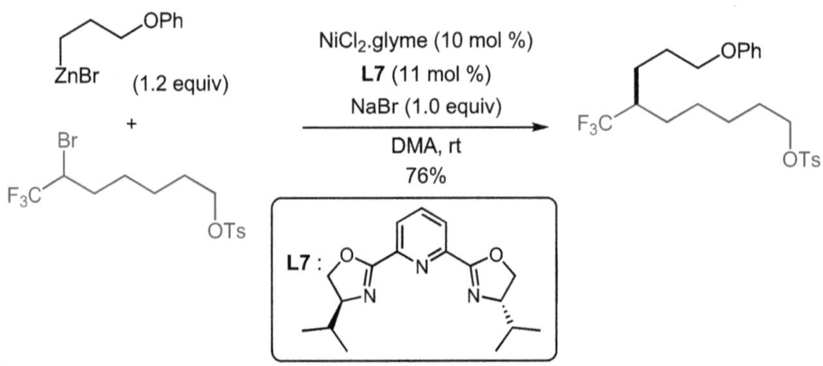

Fig. 2.71: PyBox ligand **L7** used for the asymmetric formation of Csp^3-Csp^3 bonds.

The procedure was extended to cross-coupling reactions between benzylic halides and achiral secondary zinc reagents. After optimization of the conditions, the chiral oxazoline ligand **L17** derived from tertioleucine and possessing an isoquinoline subunit was recognized as the ligand of choice (Fig. 2.72) [88].

An enantioselective synthesis of secondary alkylboronate esters has also been achieved starting from racemic substrates (Fig. 2.73) [89]. It represents an alternative route to the Matteson's procedure using chiral boronates derived from (+)-pinadiol as chiral auxiliaries.

An iterative homologation sequence allowed the formation of enantio-enriched alkylboronate esters with multiple stereocenters. By a rigorous choice of the diamine ligand *(S,S)*-**L18** *versus (R,R)*- **L18** for each step, all diastereomers could be easily obtained with high stereocontrol (Fig. 2.74).

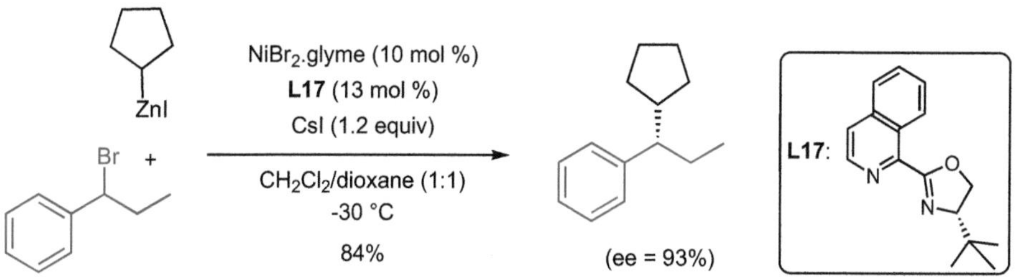

Fig. 2.72: Enantioselective formation of Csp^3-Csp^3 bonds using **L17** as ligand.

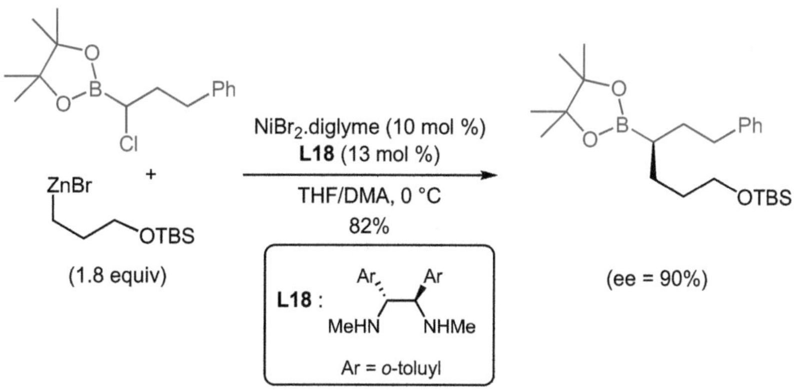

Fig. 2.73: Synthesis of chiral boronates from α-chloroboronates.

This cross-coupling reaction has been extended to the preparation of chiral organosilanes (Fig. 2.75) [90, 91].

2.5.3.3 With propargyl bromides

The synthesis of *gem*-difluoropropargyl compounds has been performed by reaction of the corresponding bromo derivatives with alkylzinc reagents in the presence of Ni(dppf)Cl_2 and *ter*-pyridine [92]. The reaction has been successfully applied to the synthesis of fluoro analogues of prostaglandin F2a (PGF2a) (Fig. 2.76).

The authors investigated also the mechanism of these reactions. Thanks to quenching experiments, observation of selective rearrangements of cyclopropane derivatives and also EPR studies, they confirmed the formation of a transient radical species. Furthermore, by using isolated *N,N,N*-nickel complexes, they highlighted the involvement of these radical species in the process. A plausible catalytic cycle is shown in Fig. 2.77.

The reaction was also carried out starting from racemic *cis*-cyclopropylzinc reagents (2 equiv) and in the presence of chiral PyBox ligands such as *(S,S)-i*Pr-PyBox **L7**.

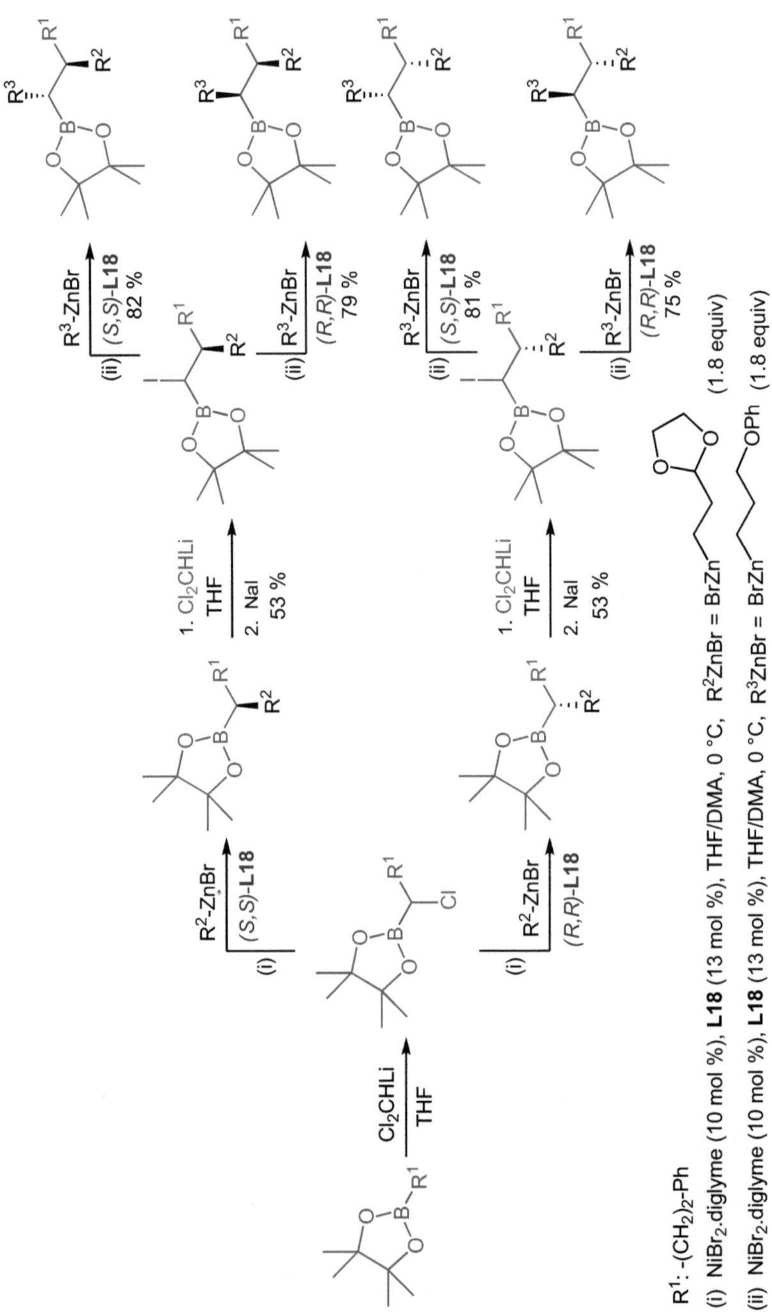

Fig. 2.74: Synthesis of the four different stereoisomers by iterative enantioselective homologation of α-haloboronates.

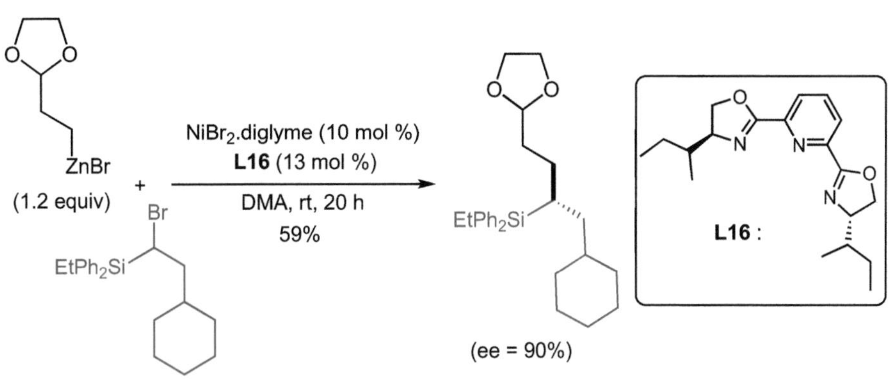

Fig. 2.75: Access to chiral organosilanes by enantioselective Negishi-type reaction.

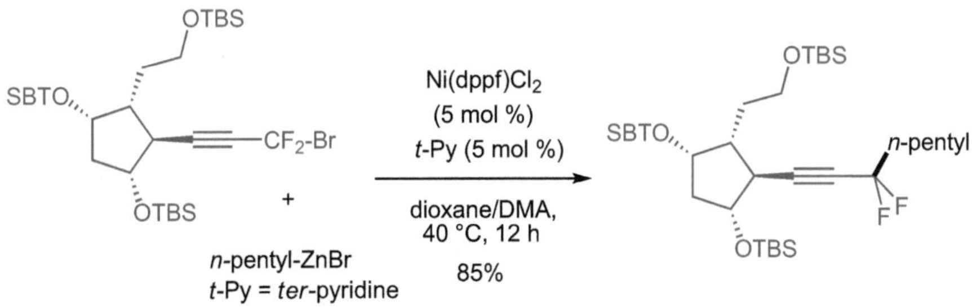

Fig. 2.76: Synthesis of difluoro propargyl derivatives.

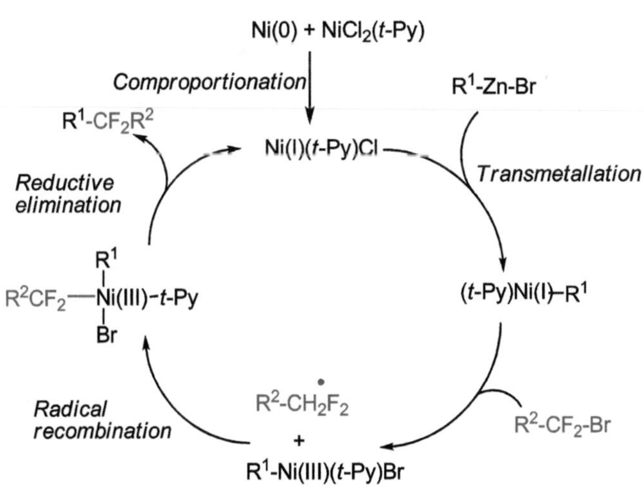

Fig. 2.77: Formation of transient Ni(III) species by combination of radical with Ni(I) catalyst.

Interestingly, a kinetic resolution occurred and led to the expected adducts in high selectivities (Fig. 2.78) [93].

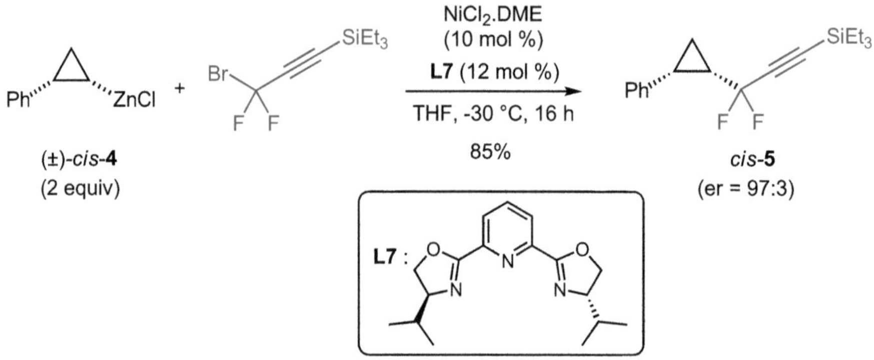

Fig. 2.78: Enantioselective synthesis of difluoropropagyl cyclopropanes by kinetic resolution.

An enantiodivergent dual metal-catalyzed process has been further designed with these reagents. Taking advantage of this kinetic resolution, a first reaction catalyzed with chiral nickel reagent **L7** delivered the expected cyclopropyl adduct **5** with high selectivities; in the same pot, the addition of palladium(II) acetate associated with the achiral SPhos ligand in the presence of a second electrophile, furnished **6,** a different substituted cyclopropane with astonishing enantioselectivities (Fig. 2.79).

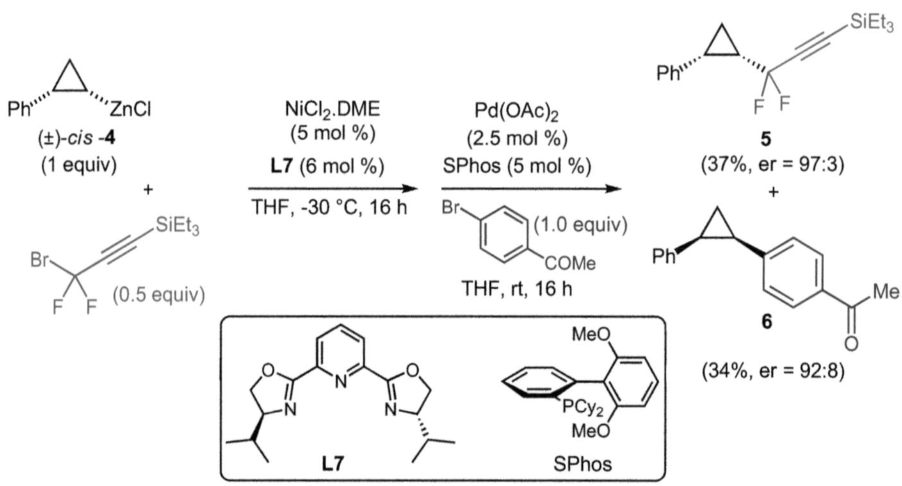

Fig. 2.79: Enantioselective access to two cyclopropane derivatives from a racemic cyclopropylzinc reagent.

To control two vicinal stereogenic centers, the group of Fu demonstrated that β-zincated amides and propargylic halides, both in racemic forms, could react in the presence of a nickel(II)salts and chiral **L19** (derived from aminoindanol), to deliver adduct **9** with very high *trans* diastereoselectivities but also with a very good enantio-control Fig. 2.80). This doubly enantio-convergent approach could be only explained by a catalytic cycle involving radical species with efficient inversion of the initial chiral center of the less reactive coupling reagents (see Fig. 2.69) [94]. This explanation was supported by DFT calculation [95].

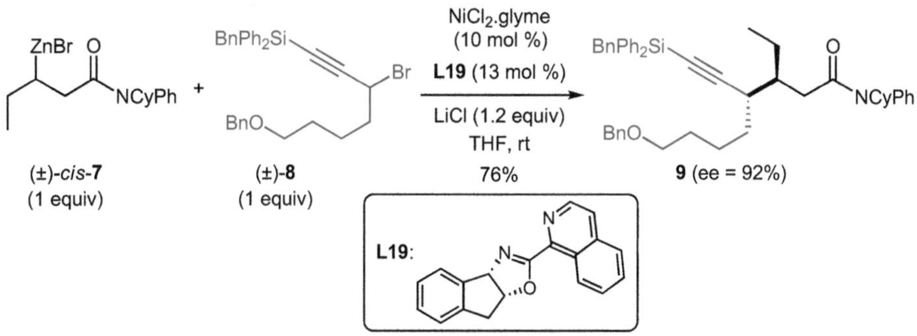

Fig. 2.80: Formation of a single enantiomer from two racemic substrates according a double enantioconvergent process.

2.5.3.4 With α-halosugars

Alkyl organozinc reagents have been reacted with protected α-bromosugar derivatives under nickel catalyzed Negishi cross-coupling reaction. The products were isolated as a mixture of α- and β-anomers depending of the nature of the glycosyl halides. High yields were obtained by using $NiCl_2$ associated with unsubstituted PyBox **L11** as the ligand in *N,N'*-dimethylimidazole (DMI) as the solvent. With pyranosyl chlorides, the reaction occurred nicely with a high diastereoselectivity and retention of configuration at C-1 (Fig. 2.81) [96].

2.5.3.5 With benzylic ethers and esters

Jarvo and coworkers achieved the direct alkylation of benzyl alcohol derivatives. Simple methoxyethers were unreactive. In contrast, 2-methylthioacetic ester reacted with dimethylzinc in the presence of $NiCl_2$ associated with DPE-Phos (Fig. 2.82). The reaction proceeded at rt with complete inversion of configuration and tolerated a wide number of functionalities [97]. It was assumed that the thioether group, by chelating

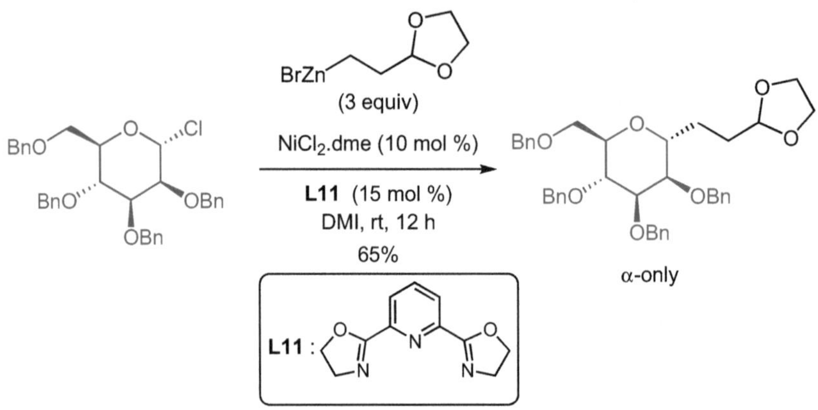

Fig. 2.81: Formation of α-*C*-glycoside derivative by a Negishi cross-coupling reaction.

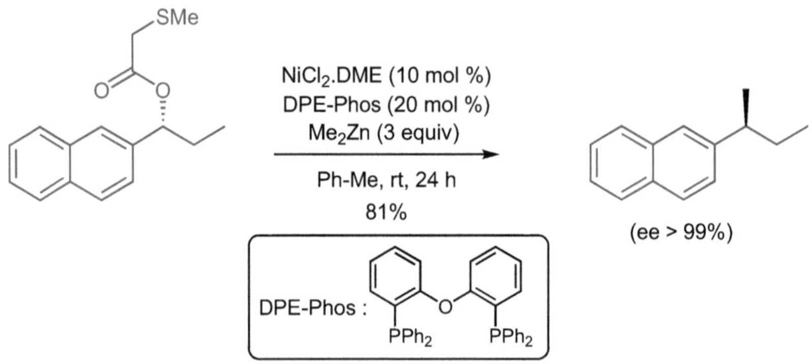

Fig. 2.82: 2-Methylthioacetic ester as substrate for the formation of Csp^3-Csp^3 bond.

the organometallic reagent, could weaken the C–O bond and accelerate the oxidative addition, and finally the transmetallation step.

Moreover, 6-aryl δ-valerolactones submitted to the same conditions gave in high yields and impressive selectivities the corresponding carboxylic acids [98]. Unfortunately, the scope of organozinc reagents is to date limited to dimethylzinc (Fig. 2.83). With other dialkylzincs, the undesired β-hydride elimination was observed as a major competitive process [99].

2.5.3.6 With alkylpyridinium salts

Primary and secondary amines can be easily transformed into alkylpyridimum salts. Subsequently, they can be cross-coupled with alkylzinc reagents in the presence of $Ni(acac)_2$ and *ter*-pyridine ligands (Fig. 2.84). The Csp^3-Csp^3 coupling reaction took place according

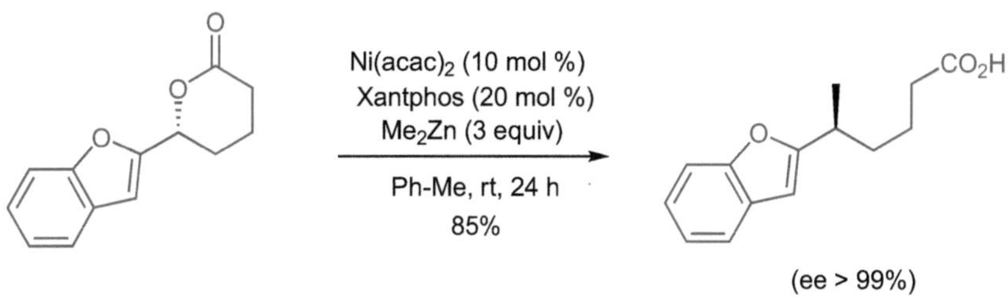

Fig. 2.83: Ring-opening and cross-coupling reaction from 6-aryl δ-valerolactones.

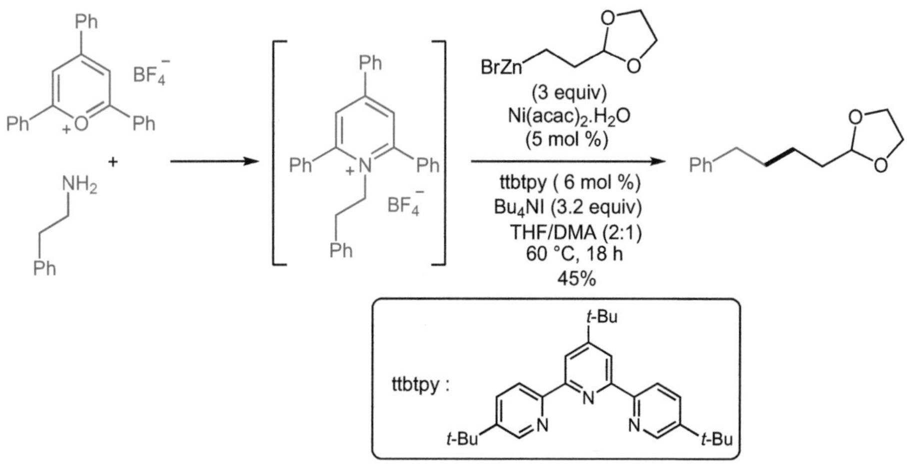

Fig. 2.84: Formation of Csp3-Csp3 bond from alkylpyridinium salts.

to a single electron transfer from nickel(I) intermediate to the *N*-alkylpyridinium. This hypothesis was confirmed through radical trapping experiments performed with TEMPO [100].

2.5.4 With carboxylic acids

2.5.4.1 From alkyl carboxylic acid

Alkyl carboxylic acids can be easily transformed into redox active esters by esterification with a *N*-hydroxyphthalimide under dicyclohexylcarbodiimide (DCC) activation. In the presence of nickel(II) catalyst and *t*-BuBiPy as the most suitable ligand, a cross-coupling with dialkylzinc reagents has been successfully carried out by the Baran's group [58, 101]. The best results in terms of chemoselectivity and yields were observed with TCNHPI for primary and secondary acids (Fig. 2.85). Furthermore, this cross-coupling Csp^3-Csp^3 reaction has been adapted to flow synthesis [102]. The best conditions required $Ni(acac)_2$

and 4,4′-dimethyl-2,2′-bipyridine as the ligand. Compared to batch conditions, the process avoids the need to isolate and transfer sensitive organozinc reagents.

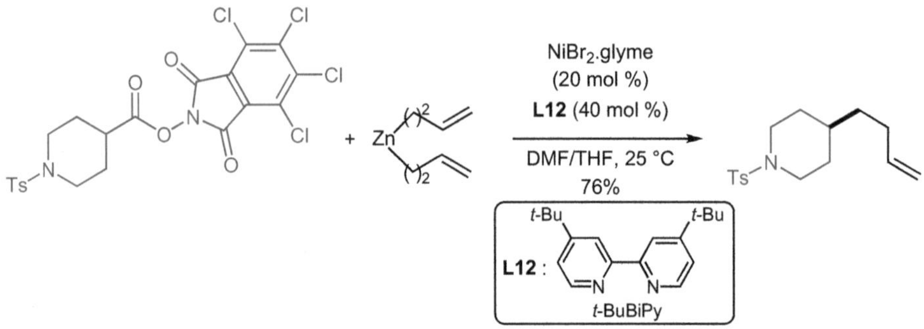

Fig. 2.85: Decarboxylative Csp^3-Csp^3 bond formation from TCNHPI active esters.

2.5.4.2 From α-amino and α-acetoxyacids

α-Aminoacids were easily converted into secondary amines by a two-step procedure. After esterification with *N*-hydroxyphthalimide, the resulting *N*-hydroxyphthalimide (NHP) ester was treated with alkylzinc reagents, nickel(II) bromide and **L20**, a C2-symmetry chiral diamine [103]. Thanks to this ligand an enantio-convergent coupling reaction took place (Fig. 2.86). Additional experiments showed that the stereocontrol of the process depends only on the optical purity of the ligand and not on the enantiomeric excess of the substrate. This method is really appealing compared to previous procedures already described and has found applications for the synthesis of protein kinases inhibitors, peptidomimetics or serotonin inhibitors. A similar reaction has been performed starting from α-acetoxyacids [104]. A plausible pathway involves radicals as already pointed out for the reaction of propargylic halides (see Section 5.4.1).

2.5.4.3 From cyclobutane carboxylic acid

In the aim to prepare [5]-ladderanoic acid, the group of Brown applied the decarboxylative alkyl-alkyl cross-coupling reaction, discovered by Baran and coworkers, to functionalize a polycyclobutane derivative [105]. The cyclobutyl radical zinc intermediate reacted with a dialkylzinc without significant rearrangement or ring-opening and gave, after deprotection of the orthoester moiety, the complex target molecule (Fig. 2.87).

This decarboxylative C-C bond formation has been also successfully applied to the synthesis of unsymmetrical [2]rotoxanes by Leigh and coworkers [106].

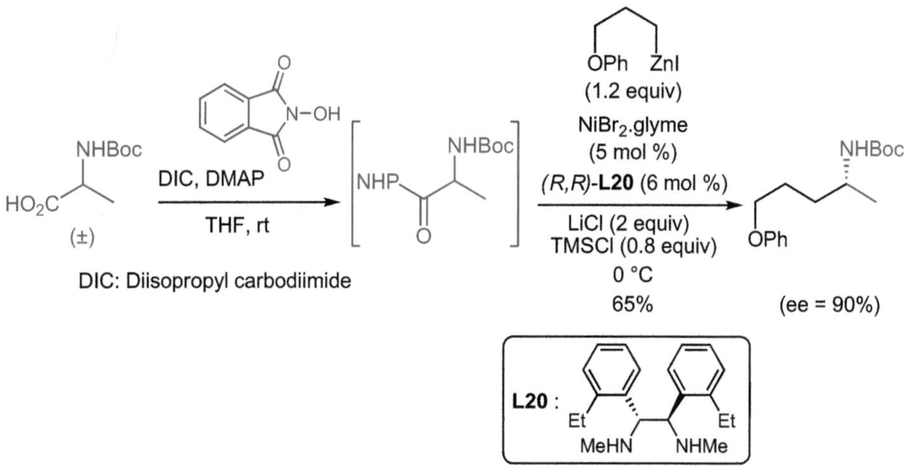

Fig. 2.86: Enantioselective synthesis of chiral amines by decarboxylative cross-coupling reaction of racemic amino acids.

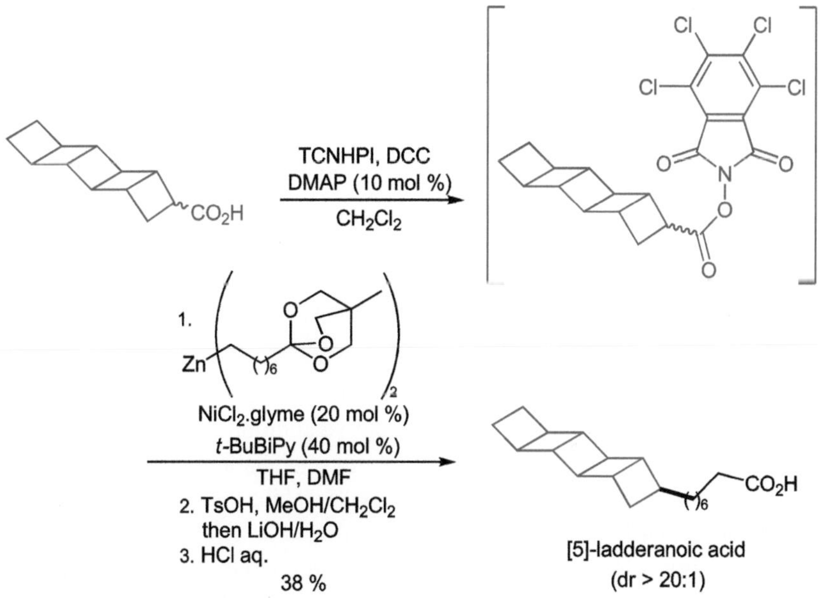

Fig. 2.87: Decarboxylative alkylation and synthesis of [5]-ladderanoic acid.

2.5.5 With *N*-tosylaziridines

The strain inherent to small ring as aziridines has been advantageously used to achieve their ring opening by nucleophilic species. Nearly ten years ago, Doyle et al. described the highly regioselective alkylation of styrenyl *N*-tosylaziridines under nickel(0) catalysis [107]. The reaction required nickel(II) salts stabilized by dimethyl fumarate which were reduced by the organozinc reagent. The reaction allowed the functionalization only of the benzylic position. The curse of the reaction was rationalized by the classical tryptic: oxidative addition/transmetallation and subsequent reductive elimination, while a radical process was not totally excluded.

As an extension of this studies, the same group reported the design of cinsyl, a new protective group to favor the attack on the less substituted position of the azaheterocycles (Fig. 2.88) [108].

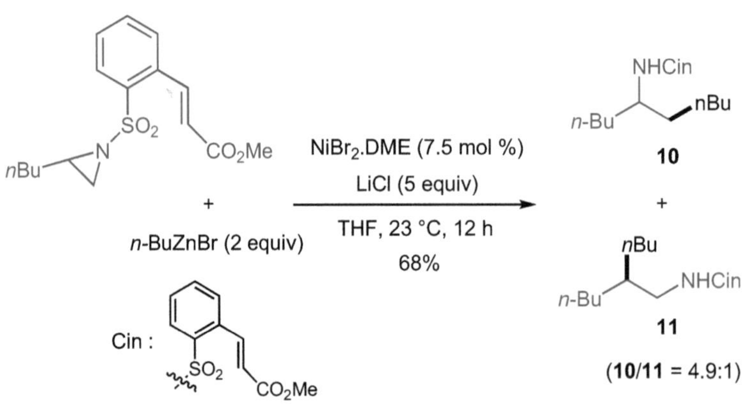

Fig. 2.88: Ring-opening of *N*-sulfonylaziridines.

The authors proposed a catalytic cycle (Fig. 2.89) in which a pre-complexation of the nickel atom on the π-system could direct the oxidative addition to furnish a metalla-azetidine intermediate **X**. Its cleavage during the transmetallation delivers a new intermediate which is quickly converted into complex **XI** stabilized both by the nitrogen atom and the electron-deficient double bond.

Jamison et al. investigated the same process with a combination of nickel(II) chloride and various bipyridines or phenanthridines as a suitable precatalyst system. Reaction of a large excess of alkylzinc on the substituted aziridine **12** afforded the expected compound **13** but also tosylamine (Fig. 2.90). To minimize its formation, probably due to a β-hydrid abstraction, a careful screening of the ligands was considered. Phenanthridine **L21** appeared as the most powerful ligand to maximize the formation of one regioisomer (better than 20:1). The reaction was general with both functionalized zinc reagents and substrates; the yields were in the range of 66–96% [109].

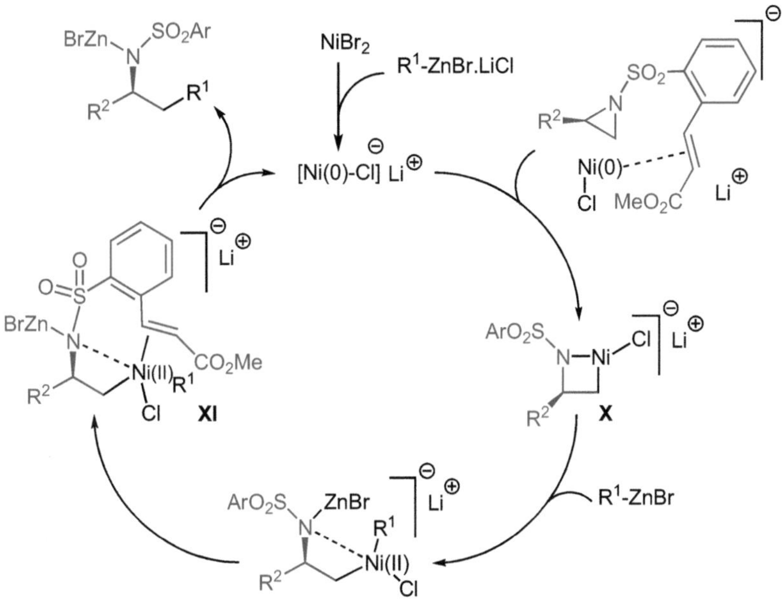

Fig. 2.89: Catalytic cycle involved during the ring-opening of *N*-cinsyl aziridines.

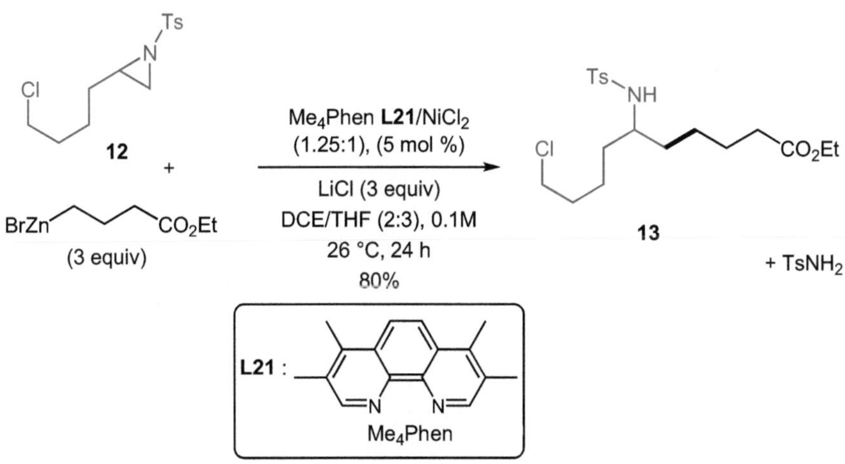

Fig. 2.90: Regioselective ring-opening of *N*-tosylaziridines in the presence of alkylzinc reagents.

It is worth mentioning that the reaction is chemoselective as no cross-coupling reaction was observed with the allyl chloride subunit. Mechanistic studies revealed that a metalla-azetidine was generated *in situ* and can undergo a transmetallation with the alkylzinc reagent. Furthermore, by using deuterium labelled substrates, a catalytic cycle was proposed (Fig. 2.91). After a regioselective oxidative addition, leading to the four-membered ring complex, a transmetallation occurred followed by a reductive elimination with retention of configuration. LiCl was primordial to transform the organozinc reagents into more reactive lithium organozincates and, by this way, to accelerate the transmetallation. To enhance the scope of the reaction, the process was successfully tested on chiral substrates without loss of stereochemistry [110].

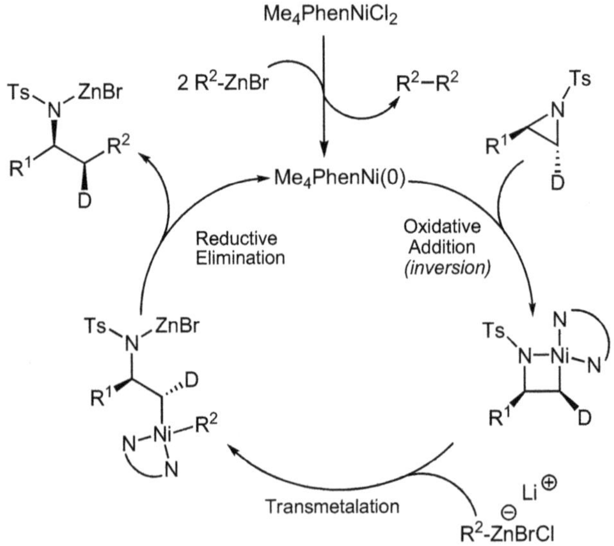

Fig. 2.91: Stereocontrol during the formation of metalla-azetidine intermediate.

Of interest, the tosyl group was easily cleaved under reductive conditions (Mg/$Ti(OiPr)_4$/TMSCl) leading to primary amines.

2.5.6 Miscellaenous reactions

Benzylic allylation of electron-deficient heterocycles has been reported by the group of Newhouse. Starting from pyridines, quinolones, pyrazines and numerous other substrates, they discovered the allylation in benzylic position using a large panel of allylic electrophiles and $Zn(TMP)_2$ as the base [111]. In the presence of a nickel catalyst, such as $Ni(dppf)Cl_2$, the zinc reagent is able to abstract one hydrogen atom, a process which is facilitated by the proximity of the nitrogen atom. Thus, the generated zinc species can undergo, after transmetallation, a cross-coupling reaction with allyl bromide (Fig. 2.92).

The reaction was carried out with numerous substrates with a similar level of efficiency (Fig. 2.93).

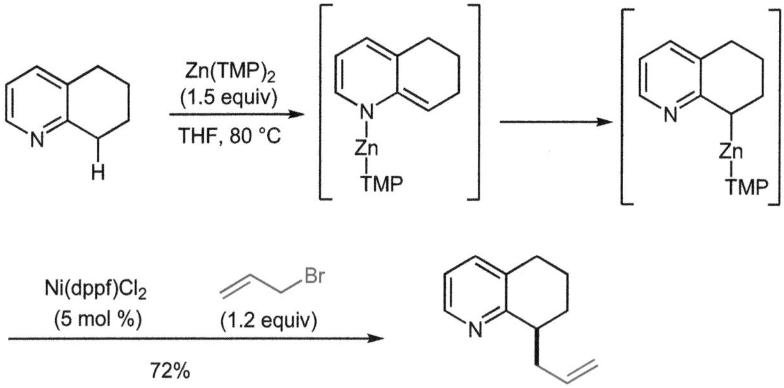

Fig. 2.92: Benzylic allylation of aza-heterocycles.

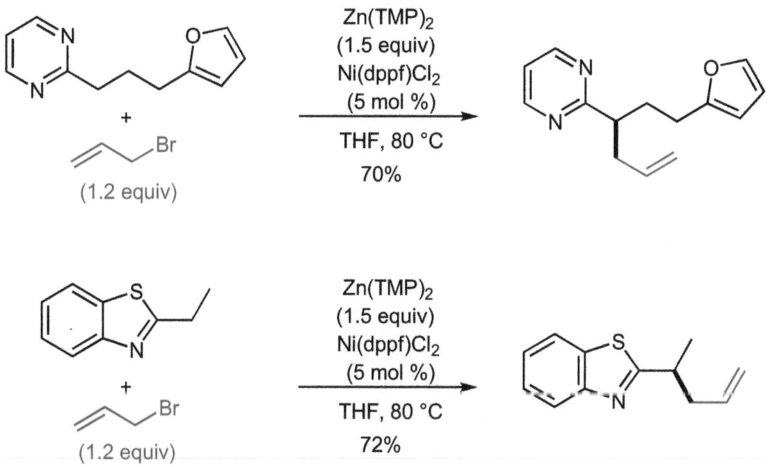

Fig. 2.93: Allylation at the benzylic position of various heterocycles.

2.6 Conclusion

The use of organozinc derivatives associated with nickel catalysts represents a very attractive tool for the formation of C-C bonds, at first, because these organometallic species are easy accessible and compatible with a large number of functionalities. Furthermore, replacement of palladium salts in the Negishi-type cross-coupling reaction by nickel salts represents an improvement in terms of cost for the synthesis of agrochemicals and pharmaceutical compounds. Finally, the use of chiral ligands such as bis-oxazolines allows the formation of Csp^3-Csp^3 bonds with total control of the configuration of the newly created stereogenic centers starting from racemic substrates. No doubt, that the nickel-catalyzed Negishi-type reaction, supported by in-depth mechanistic studies, will know new developments in terms of application.

References

[1] Geist E, Kirschning A, Schmidt T, Nat. Prod. Rep. 2014, 31, 441–448.
[2] Chernyshev VM, Ananikov VP, ACS Catal. 2022, 12, 1180–1200.
[3] Tasker SZ, Standley EA, Jamison TF, Nature 2014, 509, 299–309.
[4] Mondal S, Dumur F, Gigmes D, Sibi MP, Bertrand MP, Nechab M, Chem. Rev. 2022, 122, 5842–5976.
[5] Negishi E-I, King AO, Okukado N, J. Org. Chem. 1977, 42, 1821–1823.
[6] Phapale VB, Cardenas DJ, Chem. Soc. Rev. 2009, 38, 1598–1607.
[7] Negishi E-I, Angew. Chem. Int. Ed. 2011, 50, 6738–6764.
[8] Haas D, Hammann JM, Greiner R, Knochel P, ACS Catal. 2016, 6, 1540–1552.
[9] Chen Y-H, Ellwart M, Malakhov V, Knochel P, Synthesis 2017, 49, 3215–3223.
[10] Liu X, Wang J, Li J, Synlett 2022, 33, 1688–1694.
[11] Campeau L-C, Hazari N, Organometallics 2019, 38, 3–35.
[12] Everson DA, Jones BA, Weix DJ, J. Am. Chem. Soc. 2012, 134, 6146–6159.
[13] Poremba KE, Dibrell SE, Reisman SE, ACS Catal. 2020, 10, 8237–8246.
[14] Zhu C, Yue H, Jia J, Rueping M, Angew. Chem. Int. Ed. 2021, 60, 17810–17831.
[15] Yi L, Ji T, Chen K-Q, Chen X-Y, Rueping M, CCS Chem. 2022, 4, 9–30.
[16] Hyodo I, Tobisu M, Chatani N, Chem. Asian J. 2012, 7, 1357–1365.
[17] Qi X, Diao T, ACS Catal. 2020, 10, 8542–8556.
[18] Manolikakes G, Munoz Hernandez C, Schade MA, Metzger A, Knochel P, J. Org. Chem. 2008, 73, 8422–8436.
[19] Xi Z, Zhou Y, Chen W, J. Org. Chem. 2008, 73, 8497–8501.
[20] Peng Z, Sun H, Du A, Li X, Z. Anorg. Allg. Chem. 2015, 641, 838–841.
[21] Nakamura Y, Yoshikai N, Ilies L, Nakamura E, Org. Lett. 2012, 14, 3316–3319.
[22] Zhu F, Wang Z-X, J. Org. Chem. 2014, 79, 4285–4292.
[23] Sun AD, Leung K, Restivo AD, LaBerge NA, Takasaki H, Love JA, Chem. Eur. J. 2014, 20, 3162–3168.
[24] Li B-J, Li Y-Z, Lu X-Y, Liu J, Guan B-T, Shi Z-J, Angew. Chem. Int. Ed. 2008, 47, 10124–10127.
[25] Manolikes SM, Ellwart M, Stathakis CI, Knochel P, Chem. Eur. J. 2014, 20, 12289–12297.
[26] Hofmayer MS, Lutter FH, Grokenberger L, Hammann JM, Knochel P, Org. Lett. 2019, 21, 36–39.
[27] Wang C, Ozaki T, Takita R, Uchiyama M, Chem. Eur. J. 2012, 18, 3482–3485.
[28] Tobisu M, Chatani N, Acc. Chem. Res. 2015, 48, 1717–1726.
[29] Dai W-C, Yang B, Xu S-H, Wang Z-X, J. Org. Chem. 2021, 86, 2235–2243.
[30] Xie L-G, Wang Z-X, Angew. Chem. Int. Ed. 2011, 50, 4901–4904.
[31] Blakey SB, MacMillan DWC, J. Am. Chem. Soc. 2003, 125, 6046–6047.
[32] Wu D, Tao J-L, Wang Z-X, Org. Chem. Front. 2015, 2, 265–273.
[33] Huang Y-M, J. Chem. Res. 2022, http://doi.org/10.1177/17475198211063806.
[34] Melzig L, Metzger A, Knochel P, J. Org. Chem. 2010, 75, 2131–2133.
[35] Aukland MH, Talbot FJT, Fernandez-Salas JA, Ball M, Pulis AP, Procter DJ, Angew. Chem. Int. Ed. 2018, 57, 9785–9789.
[36] Yamamoto K, Otsuka S, Nogi K, Yorimitsu H, ACS Catal. 2017, 7, 7623–7628.
[37] Nassar Y, Rodier F, Ferey V, Cossy J, ACS Catal. 2021, 11, 5736–5761.
[38] Wu J, Yang Z, J. Org. Chem. 2001, 66, 7875–7878.
[39] Kondoh A, Aoki T, Terada M, Chem. Eur. J. 2017, 23, 2769–2773.

[40] Liang Y, Fu GC, J. Am. Chem. Soc. 2015, 137, 9523–9526.
[41] Barde E, Guérinot A, Cossy J, Synthesis 2019, 51, 178–184.
[42] Tarui A, Shinohara S, Sato K, Omote M, Ando A, Org. Lett. 2016, 18, 1128–1131.
[43] Choi J, Fu G, J. Am. Chem. Soc. 2012, 134, 9102–9105.
[44] Do H-Q, Chandrashekar ERR, Fu GC, J. Am. Chem. Soc. 2013, 135, 16288–16291.
[45] Jensen AE, Knochel P, J. Org. Chem. 2002, 67, 79–85.
[46] Tao J-L, Yang B, Wang Z-X, J. Org. Chem. 2015, 80, 12627–12634.
[47] Oelke AJ, Sun J, Fu GC, J. Am. Chem. Soc. 2012, 134, 2966–2969.
[48] Schley ND, Fu GC, J. Am. Chem. Soc. 2014, 136, 16588–16593.
[49] Monteith JJ, Rousseaux SAL, Org. Lett. 2021, 23, 9485–9489.
[50] Minami H, Nogi K, Yorimitsu H, Synlett 2021, 32, 1542–1546.
[51] Konev MO, Jarvo ER, Angew. Chem. Int. Ed. 2016, 55, 11340–11342.
[52] Zuo Z, Ahneman DT, Chu L, Terrett JA, Doyle AG, MacMillan DWC, Science 2014, 345, 437–440.
[53] Skubi KL, Blum TR, Yoon TP, Chem. Rev. 2016, 116, 10035–10074.
[54] Cornella J, Edwards JT, Qin T, Kawamura S, Wang J, Pan C-M, Gianatassio R, Schmidt M, Eastgate MD, Baran PS, J. Am. Chem. Soc. 2016, 138, 2174–2177.
[55] Parida SK, Mandal T, Das S, Hota SK, De Sarkar S, Murarka S, ACS Catal. 2021, 11, 1640–1643.
[56] Chen TG, Zhang H, Mykhailiuk PK, Merchant RR, Smith CA, Qin T, Baran PS, Angew. Chem. Int. Ed. 2019, 58, 2454–2458.
[57] Primer DN, Molander GA, J. Am. Chem. Soc. 2017, 139, 9847–9850.
[58] Qin T, Cornella J, Li C, Malins LR, Edwards JT, Kawamura S, Maxwell BD, Eastgate MD, Baran PS, Science 2016, 352, 801–805.
[59] Merchant RR, Edwards JT, Qin T, Kruszyk MM, Bi C, Che G, Bao D-H, Qiao W, Sun L, Collins MR, Fadeyi OO, Gallego GM, Mousseau JJ, Nuhant P, Baran PS, Science 2018, 360, 75–80.
[60] Nambo M, Maekawa Y, Crudden CM, ACS Catal. 2022, 12, 3013–3032.
[61] Sämann C, Schade MA, Yamada S, Knochel P, Angew. Chem. Int. Ed. 2013, 52, 9495–9499.
[62] Edwards JT, Merchant RR, McClymont KS, Knouse KW, Qin T, Malins LR, Vokits B, Shaw SA, Bao D-H, Wei F-L, Zhou T, Eastgate MD, Baran PS, Nature 2017, 545, 213–218.
[63] Chen Y-H, Tüllmann CP, Ellwart M, Knochel P, Angew. Chem. Int. Ed. 2017, 56, 9236–9239.
[64] Smith JM, Qin T, Merchant RR, Edwards JT, Malins LR, Liu Z, Che G, Shen Z, Shaw SA, Eastgate MD, Baran PS, Angew. Chem. Int. Ed. 2017, 56, 11906–11910.
[65] Shimada Y, Haraguchi R, Matsubara S, Synlett 2015, 26, 2395–2398.
[66] Melzig L, Gavryushin A, Knochel P, Org. Lett. 2007, 9, 5529–5532.
[67] Liu X, Jia J, Rueping M, ACS Catal. 2017, 7, 4491–4496.
[68] Melzig L, Metzger A, Knochel P, Chem. Eur. J. 2011, 17, 2948–2956.
[69] Lee K, Counceller CM, Stambuli JP, Org. Lett. 2009, 11, 1457–1459.
[70] Counceller CM, Eichman CC, Proust N, Stambuli JP, Adv. Synth. Catal. 2011, 353, 79–83.
[71] Gong L, Zhang Q, Xie D, Zhang W, Xu S-Y, Zhang X, Niu D, Chem. Commun. 2021, 57, 12273–12276.
[72] Erdelmeier I, Bülow G, Woo C-W, Decker J, Raabe G, Gais H-J, Chem Eur. J. 2019, 25, 8371–8386.
[73] Choi J, Fu GC, Science 2017, 356, 152.
[74] Rudolph A, Lautens M, Angew. Chem. Int. Ed. 2009, 48, 2656–2670.
[75] Kranthikumar R, Organometallics 2022, 41, 667–379.
[76] Devasagayaraj A, Stüdemann T, Knochel P, Angew. Chem. Int. Ed. 1995, 34, 2723–2725.
[77] Giovannini R, Stüdemann T, Dussin G, Knochel P, Angew. Chem. Int. Ed. 1998, 37, 2387–2390.
[78] Giovannini R, Stüdemann T, Devasagayaraj A, Dussin G, Knochel P, J. Org. Chem. 1999, 64, 3544–3553.
[79] Yamamoto T, Yamamoto A, Ikeda S, J. Am. Chem. Soc. 1971, 93, 3350–3359.
[80] Sustmann R, Lau J, Zipp M, Tetrahedron Lett. 1986, 27, 5207–5210.
[81] Piber M, Jensen AE, Rottländer M, Knochel P, Org. Lett. 1999, 1, 1323–1326.

[82] Terao J, Todo H, Watanabe H, Ikumi A, Kambe N, Angew. Chem. Int. Ed. 2004, 43, 6180–6182.
[83] Terao J, Kambe N, Acc. Chem. Res. 2008, 41, 1545–1554.
[84] Zhou J, Fu GC, J. Am. Chem. Soc. 2003, 125, 14726–14727.
[85] Arp FO, Fu GC, J. Am. Chem. Soc. 2005, 127, 10482–10483.
[86] Son S, Fu GC, J. Am. Chem. Soc. 2008, 130, 2756–2757.
[87] Liang Y, Fu GC, Angew. Chem. Int. Ed. 2015, 54, 9047–9051.
[88] Binder JT, Cordier CJ, Fu GC, J. Am. Chem. Soc. 2012, 134, 17003–17006.
[89] Schmidt J, Choi J, Liu AT, Slusarczyk M, Fu GC, Science 2016, 354, 1265–1269.
[90] Schwarzwalder GM, Matier CD, Fu GC, Angew. Chem. Int. Ed. 2019, 58, 3571–3574.
[91] Yi H, Mao W, Oestreich M, Angew. Chem. Int. Ed. 2019, 58, 3575–3578.
[92] An L, Xu C, Zhang X, Nature Commun. 2017, 8, 1460.
[93] An L, Tong F-F, Zhang S, Zhang X, J. Am. Chem. Soc. 2020, 142, 11884–11892.
[94] Huo J, Gorsline BJ, Fu GC, Science 2020, 367, 559–564.
[95] Zhang C, Lu Y, Zhao R, Chen X-Y, Wang Z-X, Org. Chem. Front. 2020, 7, 3411–3419.
[96] Gong H, Sinisi R, Gagné MR, J. Am. Chem. Soc. 2007, 129, 1908–1909.
[97] Wisniewska HM, Swift EC, Jarvo ER, J. Am. Chem. Soc. 2013, 135, 9083–9090.
[98] Tollefson EJ, Dawson DD, Osborne CA, Jarvo ER, J. Am. Chem. Soc. 2014, 136, 14951–14958.
[99] Tollefson EJ, Hanna LE, Jarvo ER, Acc. Chem. Res. 2015, 48, 2344–2353.
[100] Plunkett S, Basch CH, Santana SO, Watson MP, J. Am. Chem. Soc. 2019, 141, 2257–2262.
[101] Laudadio G, Palkowitz MD, El-Hayek Ewing T, Baran PS, ACS Med. Chem. Lett. 2022, 13, 1413–1420.
[102] Herath A, Molteni V, Pan S, Loren J, Org. Lett. 2018, 20, 7429–7432.
[103] Yang Z-P, Freas DJ, Fu GC, J. Am. Chem. Soc. 2021, 143, 2930–2937.
[104] Gao Y, Zhang B, Levy L, Zhang H-J, He C, Baran PS, J. Am. Chem. Soc. 2022, 144, 10992–11002.
[105] Hancock EN, Kuker EL, Tantillo DJ, Brown MK, Angew. Chem. Int. Ed. 2020, 59, 436–441.
[106] Echavarren J, Gall MAY, Haertsch A, Leigh DA, Marcos V, Tetlow DJ, Chem. Sci. 2019, 10, 7269.
[107] Huang C-Y, Doyle AG, J. Am. Chem. Soc. 2012, 134, 9541–9544.
[108] Nielsen DK, Huang C-Y, Doyle AG, J. Am. Chem. Soc. 2013, 135, 13605–13609.
[109] Jensen KL, Standley EA, Jamison TF, J. Am. Chem. Soc. 2014, 136, 11145–11152.
[110] Jensen KL, Nielsen DU, Jamison TF, Chem. Eur. J. 2015, 71, 7379–7383.
[111] Zhang P, Wang J, Robertson ZR, Newhouse TR, Angew. Chem. Int. Ed. 2022, 61, e202200602.

Julien Legros and Bruno Figadère

3 Organozinc reagents and iron

3.1 Introduction

Among transition metals used in catalysis for bond formation, iron exhibits a decisive advantage over most of the other metals: it is one of the most abundant metal on earth truly inexpensive and, in addition non-toxic. Synthetic methods involving iron salts in organic chemistry have tremendously expanded over the last few decades, including cross-coupling reactions, substitutions, additions, C-H bond activations, eliminations, oxidations, reductions, to only cite a few ones [1]. Herein, we focus on C-C bond formation, involving organozinc derivatives and iron catalysis. If iron-catalyzed Grignard cross-coupling reactions are nowadays predominant transformations, and even used at an industrial level, the development of organozinc reagents as partners for such C-C bond formation, has been increasing in the recent years. Iron-catalyzed carbometallation across strained olefins was first reported by Nakamura et al. in the early 2000's, utilizing diethylzinc and a chiral ligand in presence of a catalytic amount of $FeCl_3$ in THF/THP (Fig. 3.1) [2]. Moreover, the authors early demonstrated the enantioselectivity of the carbometallation of cyclopropenone acetal by zinc reagents, since the expected alkylated product was obtained with a good enantiomeric excess (ee = 85%) (Fig. 3.1).

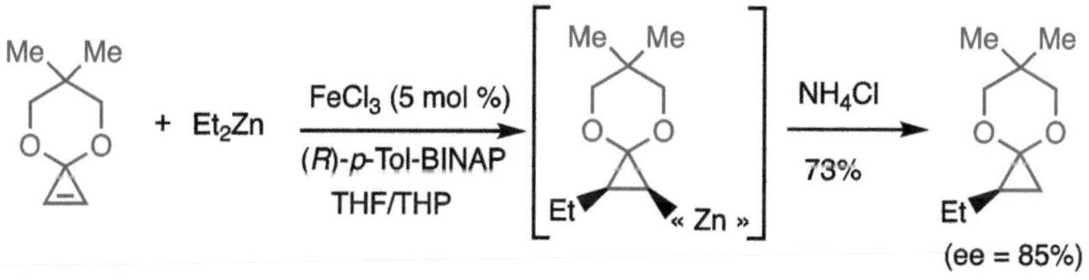

Fig. 3.1: Fe-catalyzed carbometallation with Et_2Zn by Nakamura et al.

After a few reports on the iron-catalyzed cross-coupling reactions of organozinc reagents with different electrophiles, iron-catalyzed C-H bond activation has gained much more importance and has become a versatile tool in organic synthesis. These works have paved the way for numerous developments of iron catalysts [mostly $FeCl_3$ and $Fe(acac)_3$ as precatalysts], which have been partially reported in general reviews [3–7].

Julien Legros, COBRA, Normandie Univ, UNIROUEN, INSA Rouen and CNRS, 1 rue Lucien Tesnière, 76821, Mont-Saint-Aignan, France
Bruno Figadère, BioCIS, Univ Paris-Saclay, CNRS, Bat. Henri Moissan, 17 av. des Sciences, 91400, Orsay, France

https://doi.org/10.1515/9783110728859-003

3.2 Cross-coupling reactions of alkylzinc reagents

Alkyl organozinc reagents are reported to react with a few categories of carbonyl derivatives, namely acyl chlorides.

3.2.1 With acyl chlorides

The synthesis of ketones through "addition" of an organometallic reagent onto carboxylic derivatives is often doomed to failure due to the facile over-addition onto the so formed carbonyl products, leading to tertiary alcohols. In this context, in the 1980s Marchese et al. demonstrated that Grignard reagents, in the presence of $Fe(acac)_3$ catalyst, were able to perform the formal nucleophilic substitution onto carbonyl derivatives to afford a variety of ketones [8–12]. Alternatively, diarylzincs can also be valuable coupling partners, and acyl chlorides are very efficient electrophiles in the iron-catalyzed cross-coupling with alkyl organozinc reagents. Thus, primary dialkylzinc reagents react, at −10 °C, with acyl chlorides to provide the expected ketones in good yields (Fig. 3.2) [13].

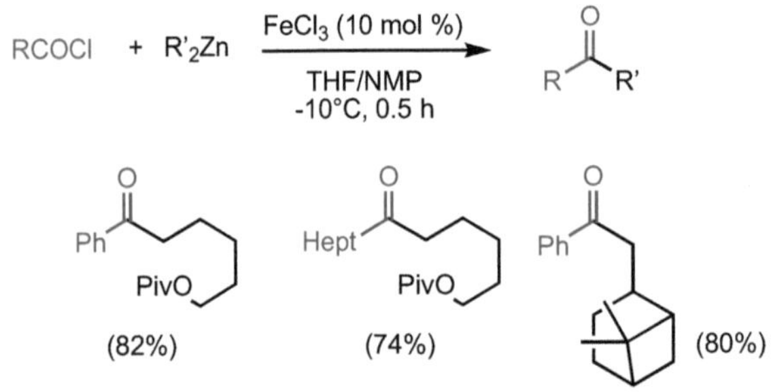

Fig. 3.2: Fe-catalyzed cross-coupling reaction between alkyl organozinc reagents and acyl chlorides.

3.3 Cross-coupling reactions of arylzinc reagents

Aryl organozinc reagents are powerful nucleophilic partners that react with a wide variety of electrophiles through Fe-catalyzed cross-coupling reactions.

3.3.1 With aryl derivatives

With aryl halides

The first successful cross-coupling was reported by Bedford et al. between di-*p*-tolylzinc reagent and 2-halogenopyridines (Fig. 3.3, eq. 1) or pyrimidines (Fig. 3.3, eq. 2) catalyzed by an iron catalyst bearing two bis-phosphine ligands (Fe cat.), albeit in moderate yields (Fig. 3.3) [14]. However, it is worth noting that in the case of the 2,5-dibromopyridine the reaction is chemoselective as the sole 5-bromo-2-*p*-tolylpyridine was isolated.

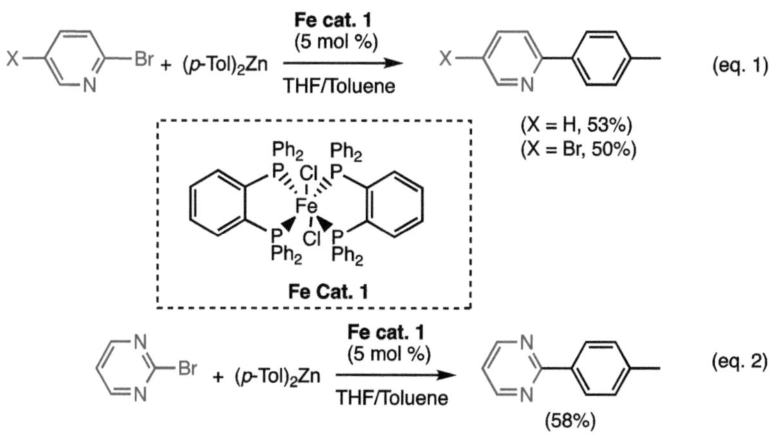

Fig. 3.3: Cross-coupling reaction of di-*p*-tolylzinc reagent with 2-halopyridines or pyrimidines.

While homocoupling of zinc reagents is most often a non-desired side-reaction in these cross-coupling reactions, it is worth noting that it is also possible to perform an efficient hetero cross-coupling of arylmethyl zinc reagents with dialkylzinc reagents in the presence of $Fe(acac)_3$ and 1,2-dibromoethane as the oxidizing agent. Primary or secondary aliphatic di-organozinc reagents were both substrates for this reaction. It is noteworthy that good yields were obtained under mild reaction conditions (no ligand, room temperature, 3–6 h) (Fig. 3.4) [15].

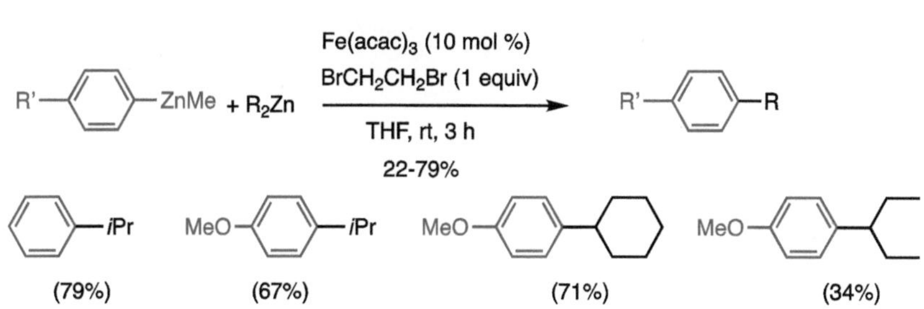

Fig. 3.4: Fe-catalyzed hetero cross-coupling reaction of ArZnMe and R_2Zn.

3.3.2 With alkyl derivatives

With alkyl halides

Several reports appeared in the literature dealing with cross-coupling reactions of aryl organozinc reagents with alkyl halides (chlorides, bromides and iodides) in the presence of iron trichloride. One of the earliest results has been reported by Nakamura and co-workers who used $FeCl_3$ in presence of 1,2-bis-(diphenylphosphino)benzene (dppbz) as a ligand (Fig. 3.5) [16]. The choice of dppbz, instead of tetramethylethylene diamine (TMEDA), which was used previously [17], allowed the coupling of low reactive polyfluorinated arylzinc reagents [16].

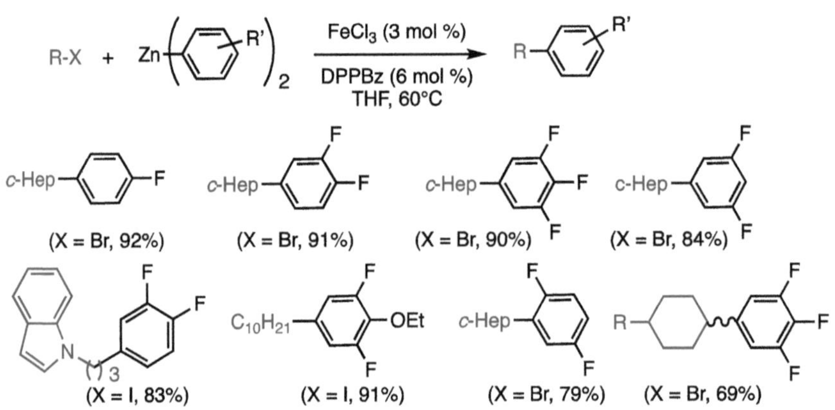

Fig. 3.5: Iron catalyzed cross-coupling reaction between aryl organozinc reagents and alkyl iodides or bromides.

To avoid the loss of an aryl substituent, a non-transferable group, such as $TMSCH_2$, could be introduced on the zinc atom (Fig. 3.6) [17].

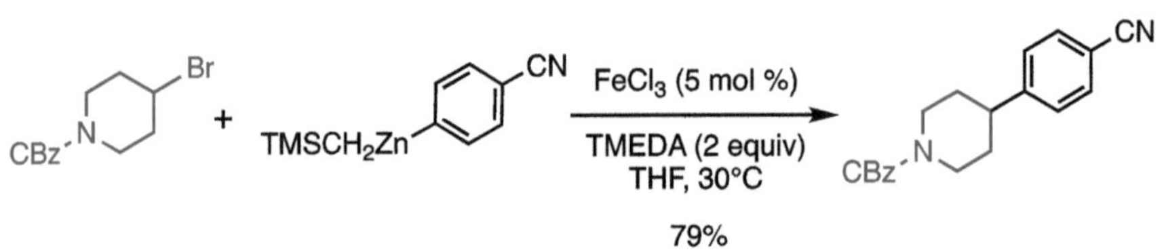

Fig. 3.6: Iron catalyzed cross-coupling reaction between a mixed arylzinc reagent and a secondary alkyl bromide.

Recently, Nakamura and co-workers applied this method to the preparation of arylglycosides by coupling glycosyl bromides with di-arylzinc reagents in presence of iron(II) chloride and a bulky bis-phosphine ligand, in THF (Fig. 3.7) [18]. This reaction was used to transfer several aryl and heteroaryl substituents (12 examples) *via* a plausible radical mechanism [18]. Under these reaction conditions, it is worth noting that acetate groups were well tolerated.

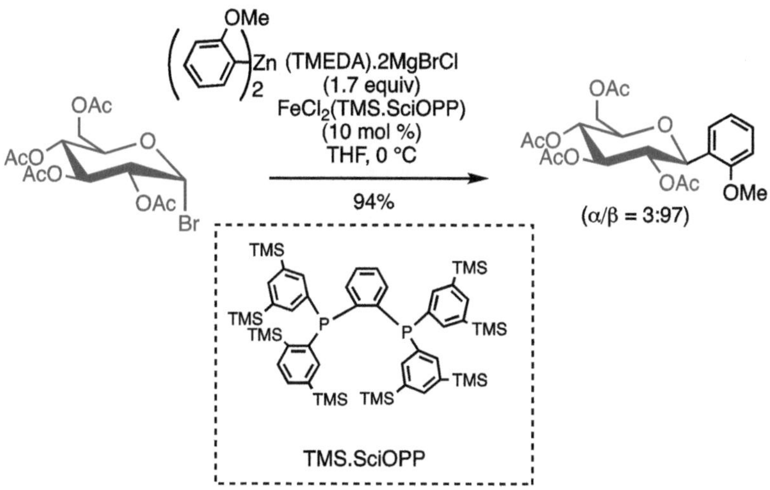

Fig. 3.7: Iron catalyzed cross-coupling reaction between arylzinc reagent and glycosyl bromides.

A few years later, secondary alkyl iodides, bearing β-fluorine atoms, were reported to react with di-phenylzinc, in the presence of iron dichloride (10 mol%) and TMEDA in THF/toluene at 0 °C, to afford the expected cross-coupled products in good yields (68%) (Fig. 3.8) [19].

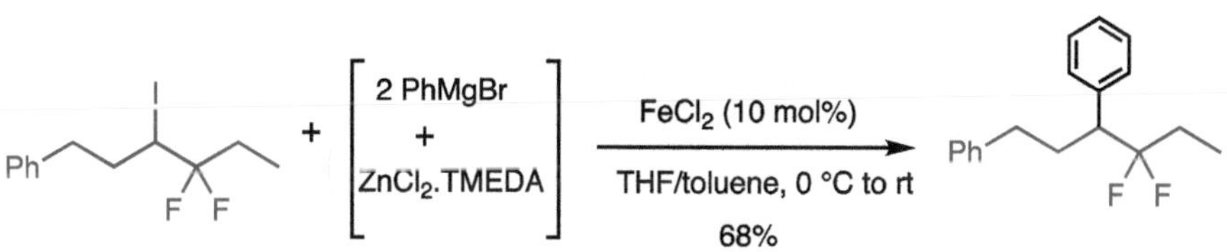

Fig. 3.8: Iron catalyzed cross-coupling reaction between diphenylylzinc reagent and a secondary alkyl iodide.

With alkyl tosylates

A cross-coupling reaction between arylzinc reagents and alkyl tosylates was reported by Nakamura et al. using $FeCl_3$ and TMEDA as a catalytic system (Fig. 3.9) [20]. Interestingly, primary and secondary alkyl tosylates gave the expected products in good yields (Fig. 3.9, eq. 1), and heteroarylzinc reagents, such as thiophenyltrimethylmethylsilylzinc reagent (Fig. 3.9, eq. 2), are also reactive under these reaction conditions, while cyanide and methyl carboxylate groups did not react.

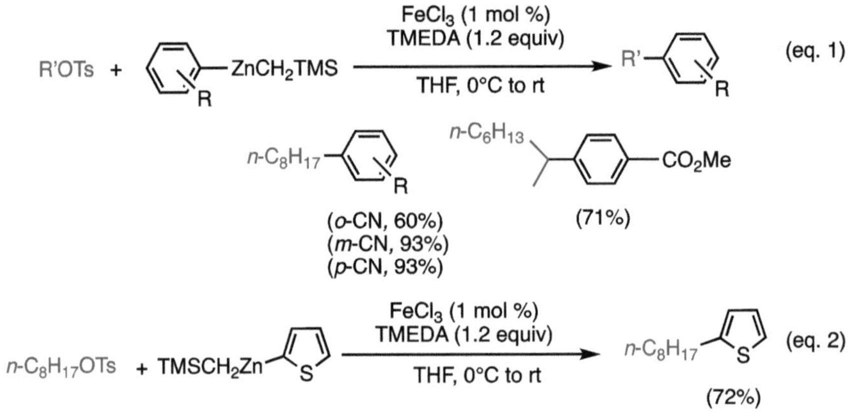

Fig. 3.9: Iron catalyzed cross-coupling reaction between arylzinc reagents and alkyl tosylates.

3.3.3 With allyl and propargyl derivatives

With allyl and propargyl halides

Among electrophiles, allyl and propargyl halides and their derivatives (phosphonates, sufonates) can react with organometallic nucleophiles (such as Grignard reagents) through two different pathways, either a direct S_N2 substitution, or by a S_N2' pathway [21]. With arylzinc reagents, at the best of our knowledge, only one example was disclosed in the literature, where diphenylzinc reagent reacted with allyl bromide in the presence of iron(II), complexed with the monophosphine **L1**, to lead to allylbenzene in 44% isolated yield (Fig. 3.10) [22].

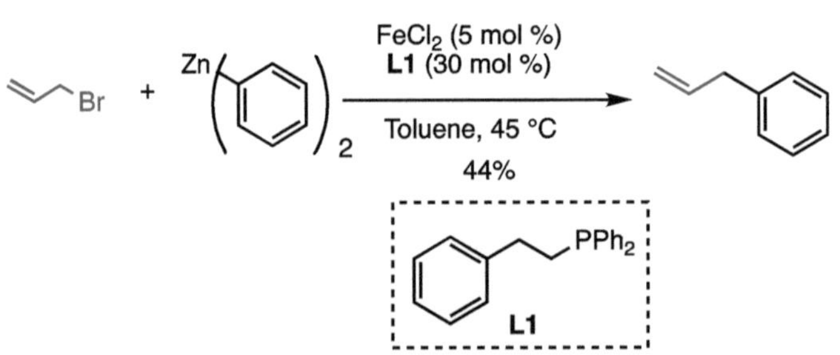

Fig. 3.10: Iron catalyzed cross-coupling reaction between diphenylzinc reagent and allyl bromide.

With allyl ethers

Heterobicyclic alkenes such as **1** undergo a carbometallation in the presence of $FeCl_3$ and diphenylzinc, without inducing the ring-opening when electron-deficient phosphine-based ligands are used (Fig. 3.11) [23]. Without a ligand, a S_N2' reaction takes place and the ring-opening occurs to produce the corresponding derivative **3** in good yield.

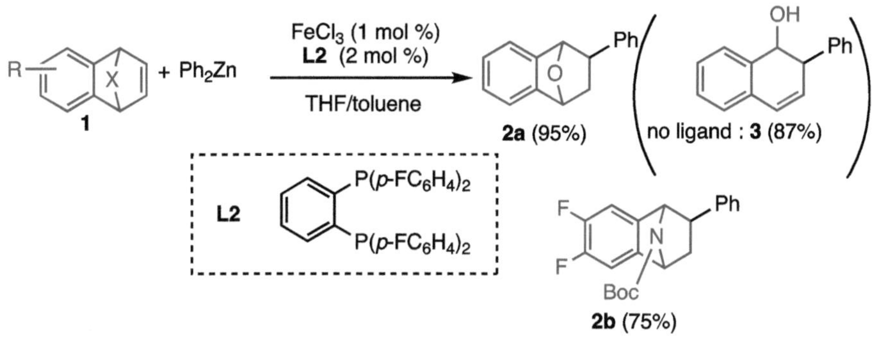

Fig. 3.11: Carbometallation *vs* S_N2' coupling between diphenylzinc reagent and oxa- or azabicyclic alkenes.

3.3.4 With benzyl halides

The cross-coupling reaction of benzylic electrophiles remained a difficult task for a long time because of the competing homocoupling of the halide partner. However, the cross-coupled products could be obtained by utilizing a di-arylzinc reagent and an iron catalyst. Thus, Bedford et al. reported that di-arylzinc reagents reacted with benzyl halides (Cl, Br) in toluene and in the presence of an iron complex (such as bis-dppbz iron(II) complex), to afford the hetero cross-coupling products $ArCH_2Ar'$ (58–92%) (Fig. 3.12) [24]. It is worth noting that the reaction is chemoselective, since methyl carboxylate, cyanide, and bromide are tolerated on the benzyl halide.

Bedford et al. showed that the bis-diphenylphosphinobenzene (dppbz) ligand was a highly efficient catalyst among the tested diphosphine ligands [25]. However, some diphosphinothiophenes can produce iron catalysts that show significantly higher activity and selectivity than the dppbz analogues, with lower amount of homocoupling products. The same authors developed also (diphenylphosphino)ethane ligands, cheaper than the dppbz ones, and delivering chemoselectively the hetero cross-coupling products $ArCH_2Ar'$ in high yields (Fig. 3.13) [26].

It is worth noting that the Negishi cross-coupling reaction between diphenylzinc and benzyl bromides can also be catalyzed by iron(II) complexed with the monophosphine **L1** (Fig. 3.14) [22].

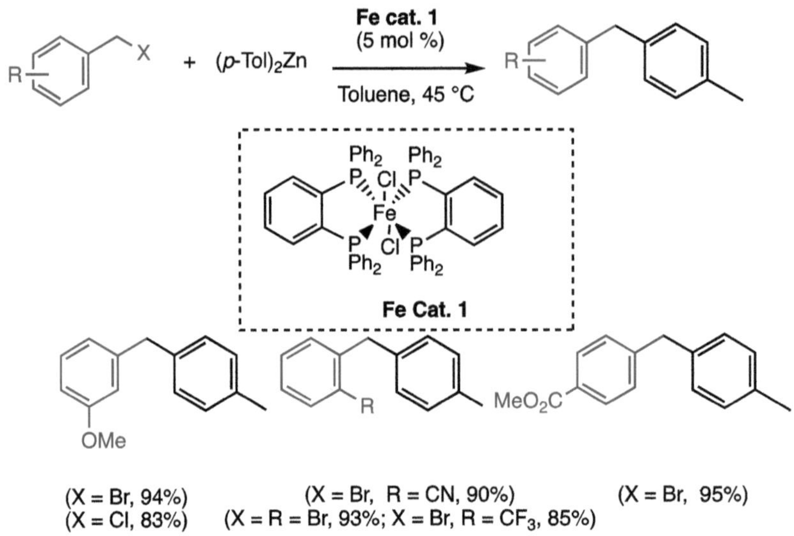

Fig. 3.12: Iron-catalyzed cross-coupling reaction between arylzinc reagents and benzyl bromides.

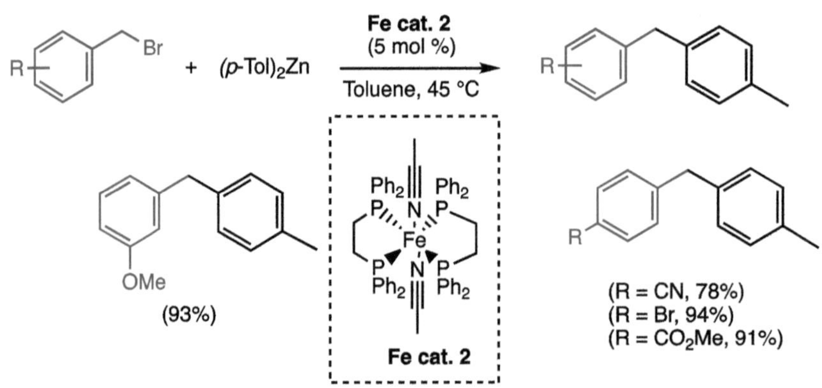

Fig. 3.13: Iron-catalyzed cross-coupling reaction between di-*p*-tolylzinc reagent and benzyl bromides.

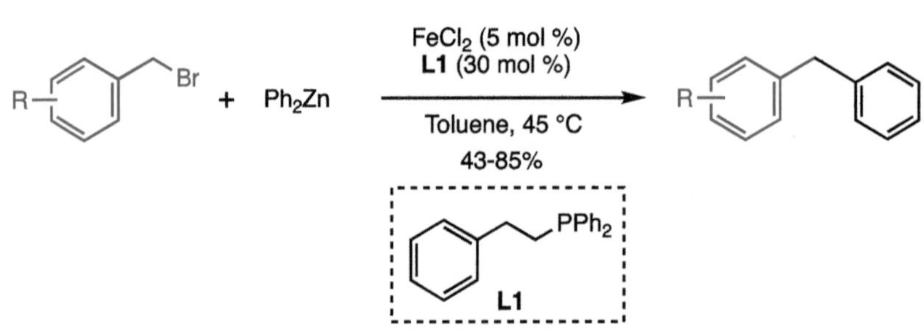

Fig. 3.14: Iron-catalyzed cross-coupling reaction between diphenylzinc reagent and benzyl bromides.

3.4 Cross-coupling reactions of alkenylzinc reagents

Alkenylzinc reagents have been much less used than the aryl zinc reagents counterparts in cross-coupling reactions. Thus, few reports have just appeared in the literature, as shown below.

3.4.1 With alkyl halides

Alkenylzinc reagents have been studied almost exclusively with alkyl electrophiles. Nakamura et al. reported that vinylzinc reagents reacted with primary and secondary alkyl bromides or chlorides in THF, in the presence of $FeCl_3$ (5 mol %) and TMEDA (3.5 equiv), at 30 °C, to afford the expected coupled products in excellent yields (Fig. 3.15) [27].

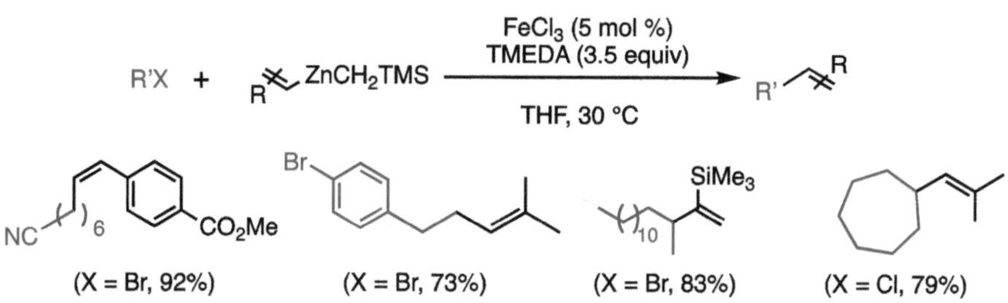

Fig. 3.15: Iron-catalyzed cross-coupling between vinylzinc reagents and alkyl bromides and chlorides.

Nakamura et al. applied this method to prepare vinyl-glycosides by coupling glycosyl bromides with di-vinylzinc reagents in THF, in the presence of iron(II) chloride and a bulky bis-phosphine ligand (Fig. 3.16) [18].

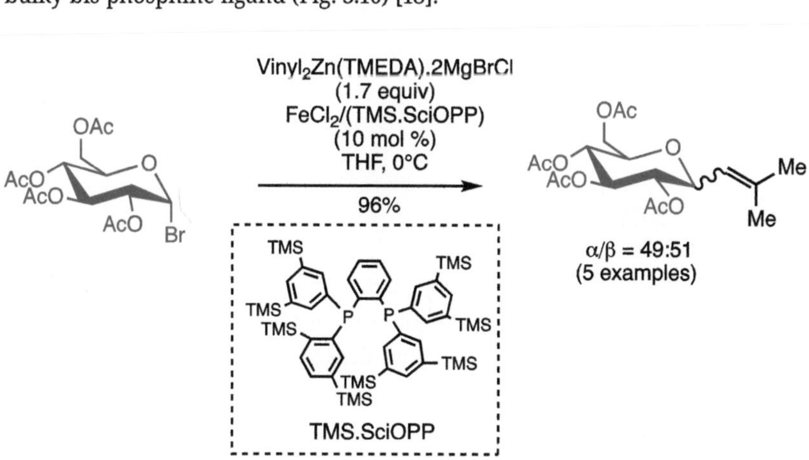

Fig. 3.16: Iron-catalyzed cross-coupling reaction between vinylzinc reagents and glycosyl bromides.

3.5 Cross-coupling reactions of alkynylzinc reagents

Alkynylzinc reagents are rarely used in iron-catalyzed cross-coupling reactions, but remain an interesting alternative the Sonogashira cross-coupling reactions, because of the reaction conditions.

3.5.1 With aryl halides

Tsai et al. reported that enynes can be obtained by the cross-coupling reactions of vinyl iodides with terminal alkynes (either prepared *in situ* or directly added) in the presence of $FeCl_3.6H_2O$ and a cationic 2,2′-bipyridyl ligand in water, in the presence of KOH as a base, and zinc(0) (Fig. 3.17) [28]. Here, an alkynylzinc intermediate was involved, which would have been formed by transmetallation of alkynyl potassium species and iron salts.

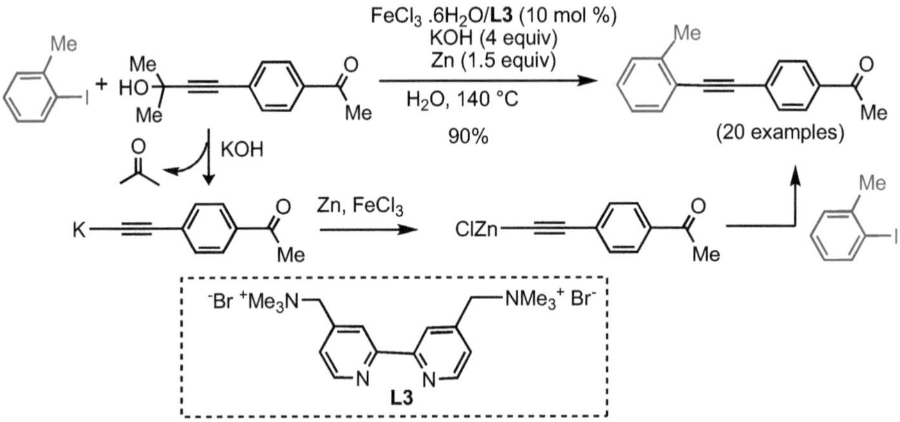

Fig. 3.17: $FeCl_3$-catalyzed cross-coupling reaction between alkynyl reagents and aryl iodides.

3.6 C-H bond activation

C-H bond activation has become a versatile approach to create C-C bonds. Several metals have been used for such transformation, but iron remains an attractive metal because of its low cost, innocuous effect and often because the chemoselectivity observed while performing these reactions.

3.6.1 Iron-catalyzed arylation of C(sp^2)-H bonds

With heteroaryls

Iron-catalyzed C(sp^2)-H arylation using diphenylzinc reagent, prepared from phenyl Grignard reagent (2 equiv ArMgBr + $ZnCl_2$ + TMEDA), was serendipitously discovered by Nakamura and co-workers in 2008 [29]. They found that the reaction can be performed in THF at 0 °C, and that TMEDA and 1,10-phenanthroline together with 1,2-dichloro*iso*-butane (DCIB), as a co-oxidant, were required in order to obtain the cross-coupling products in good yields. It is noteworthy that benzoquinoline, 2-phenylpyridine, 2-phenylpyrimidine, 4-phenylpyrimidine and 1-phenyl-1*H*-pyrazole were good substrates for this new transformation (Fig. 3.18) [29].

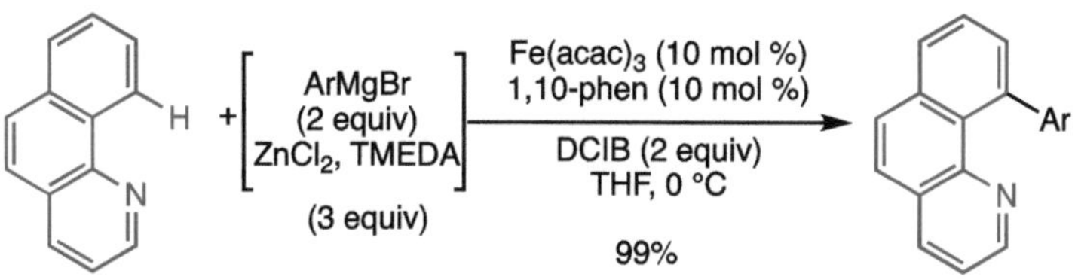

Fig. 3.18: C-C bond formation between arylzinc reagent and benzoquinoline by Fe-catalyzed C-H bond activation.

With aromatic imines

The most impressive results have been reported by Nakamura et al. who described the C-C bond formation by *N*-oriented C-H bond activation of aromatic imines leading to the direct *ortho*-arylated ketones (Fig. 3.19) [30]. Later on, the same authors reported that DCIB could be replaced by molecular oxygen as the oxidant, albeit with slightly lower yields [31]. They noticed that due to steric hindrance, *ortho*-substituted arylzinc reagents did not deliver the expected coupled products.

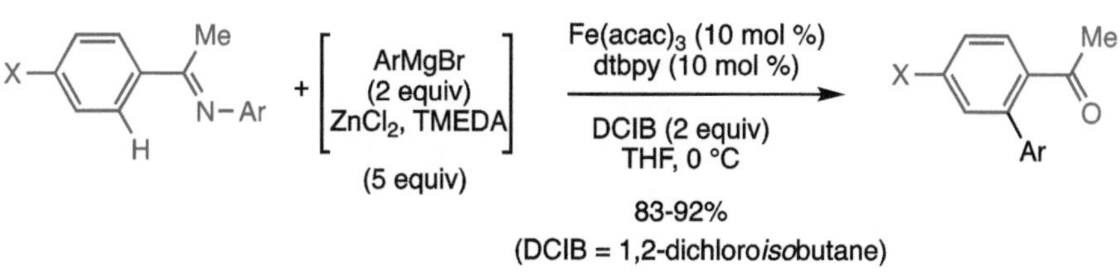

Fig. 3.19: C-C bond formation of aryl imines with arylzinc reagents by Fe-catalyzed *N*-oriented C-H bond activation.

With benzamides

Later on, the same authors reported the iron-catalyzed mono *ortho*-arylation of benzamides with diphenylzinc reagent (prepared from the corresponding Grignard reagent), in THF at 0 °C, in the presence of TMEDA, di-*tert*-butylpyridine (dtbpy) and 1,2-dichloro *iso*butane (DCIB) as the co-oxidant (Fig. 3.20) [32].

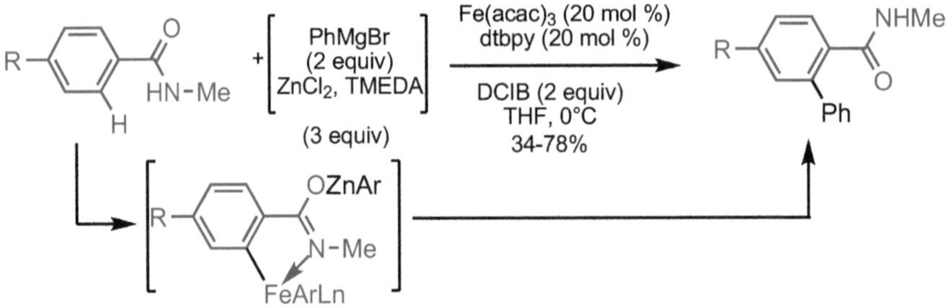

Fig. 3.20: C-C bond formation between benzamides and diphenylzinc reagent by Fe-catalyzed *N*-oriented C-H bond activation.

In 2014, Ackerman et al. showed that 1,2,3-triazole moieties were able to assist the C-H bond activation of benzamides, to achieve the arylation. The presence of the 1,2,3-triazole moiety leads to the stabilization of the low-valent iron intermediate in the presence of the dppe ligand (Fig. 3.21) [33].

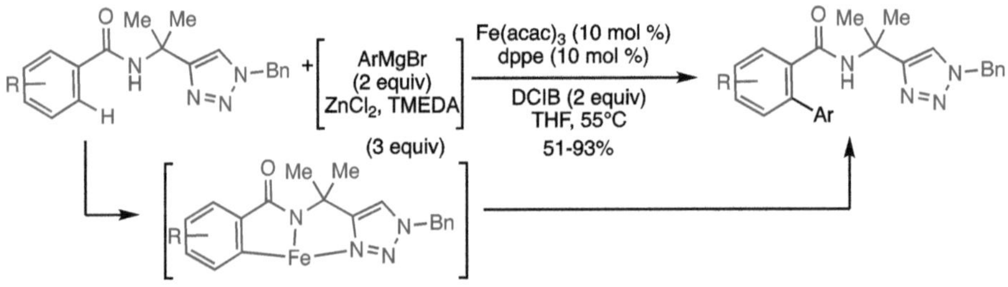

Fig. 3.21: C-C bond formation between1,2,3-triazole-benzamides and diarylzinc reagents by Fe-catalyzed *N*-oriented C-H bond activation.

In 2019, Nakamura et al. reported the hetero-arylation of arenes and olefins bearing a *N*-(quinoline-8-yl)amide group by using $Fe(acac)_3$, a bisphosphine ligand (dppen), an organozinc reagent, $Zn(CH_2SiMe_3)_2$ as the base, dichloropropane (DCP) as a mild oxidizing agent and an heteroarene (thiophene, benzothiophene) as the electrophile (Fig. 3.22) [34]. The cross-coupled products were obtained in good yields. Cyclic and acyclic alkenamides react stereospecifically giving only the *Z* products. The reaction has been suggested to proceed through consecutive C-H activation steps involving the C-H activation of amide group-assisted arene at first, followed by the C-H deprotonation of the heteroarene to form an intermediate connected to iron coordination (Fig. 3.23).

This C-C bond formation can be rationalized through the mechanism as described below (Fig. 3.23). Intermediate **I,** formed by deprotonation of 8-quinolylbenzamide,

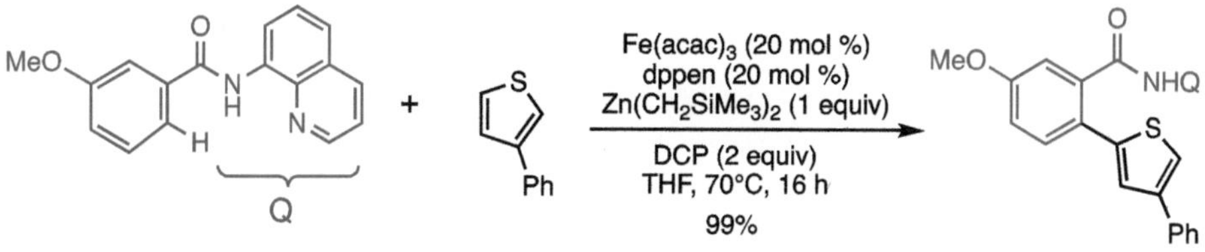

Fig. 3.22: C-C bond formation *via* two-fold C-H Fe-catalyzed activation of 8-quinolylbenzamide.

leads to intermediate **II** by C-H bond activation, which reacts with heteroarene by a second C-H bond activation to afford species **III**, which leads to intermediate **IV** by intramolecular C-H bond activation, followed by a reductive elimination to afford species **V**. This latter reacts with 8-quinolylbenzamide and, in the presence of DCP, affords species **I** and the expected cross-coupled product.

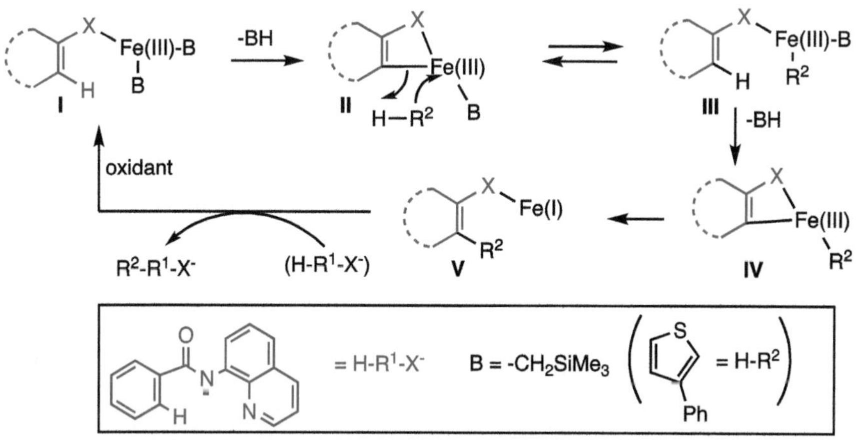

Fig. 3.23: Mechanism of C-C bond formation *via* two-fold C-H Fe-catalyzed activation of 8-quinolylbenzamide.

With benzylamines

Ackermann and co-workers showed also that electron-rich benzylamines bearing a tri-substituted 1,2,3-triazole were good substrates for this cross-coupling reaction (Fig. 3.24) [35].

With acrylamides

Interestingly, the same authors reported that the previous reaction conditions could be applied to acrylamide derivatives to obtain exclusively the *Z*-selective arylation product (Fig. 3.25) [33].

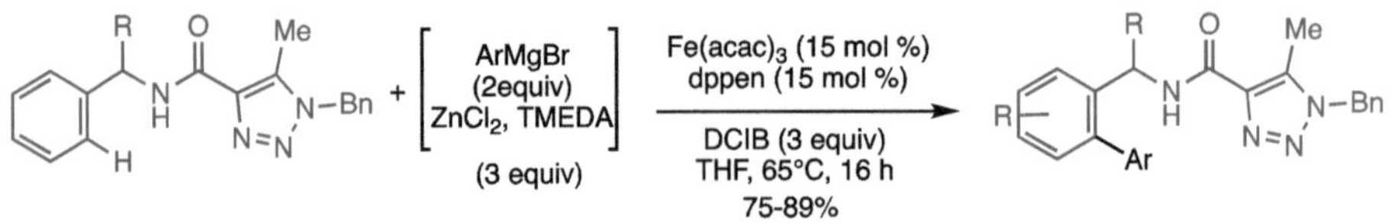

Fig. 3.24: C-C bond formation between 1,2,3-triazole-benzylamines and diarylzinc reagents by Fe-catalyzed *N*-oriented C-H bond activation.

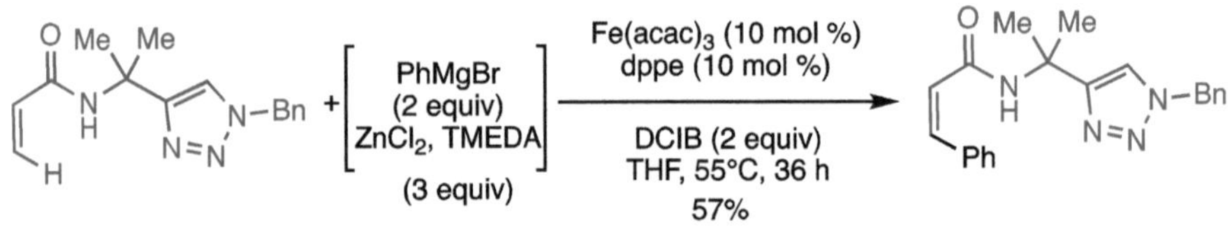

Fig. 3.25: C-C bond formation between 1,2,3-triazole-acrylamide and diphenylzinc reagent by Fe-catalyzed *N*-oriented C-H bond activation.

3.6.2 Iron-catalyzed alkylation of C(sp^2)-H bonds

With benzamides

In 2015, Nakamura et al. reported the iron-catalyzed alkylation of arenes possessing a bidentate directing group such as an 8-quinolylamine, with a monoalkylzinc reagent in the presence of a bis-phosphine ligand (Fig. 3.26) [36].

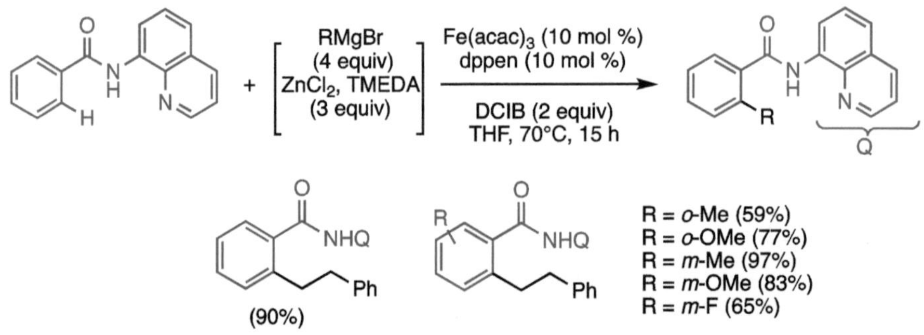

Fig. 3.26: C-C bond formation between 8-quinolylbenzamide and mono-alkylylzinc reagents by Fe-catalyzed *N*-oriented C-H bond activation.

The same year, Ackermann et al. reported the C-H methylation of arenes and alkenes using an 1,2,3-triazole directing group of benzamides. They found that $FeCl_3$ and dppe ligand were well suited for the reaction with dimethylzinc, and diethylzinc, used as the alkylating reagents (Fig. 3.27) [37].

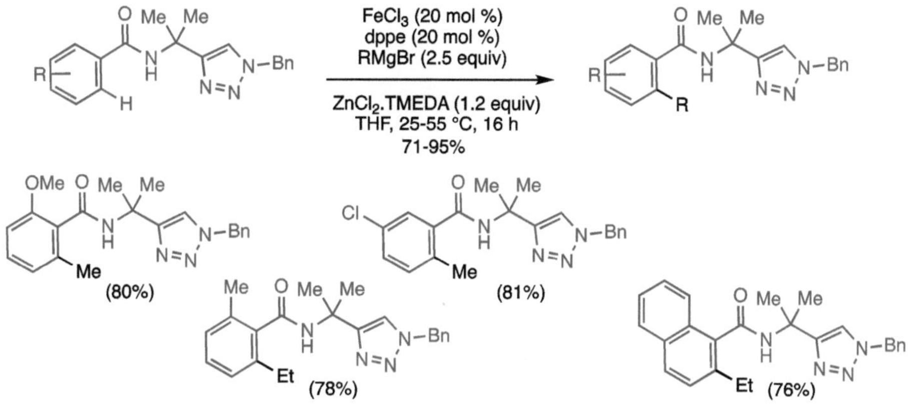

Fig. 3.27: C-C bond formation between 1,2,3-triazole-benzylamines and dimethyl- or diethylzinc reagent by Fe-catalyzed C-H bond activation.

Nakamura et al. reported the iron-catalyzed *ortho*-allylation of aromatic carboxamides bearing an 8-quinolylamide moiety as a directing group. Iron(III) and a diphosphine ligand were necessary as well as 1.2 equivalents of (t-BuCH$_2$)$_2$Zn as a base, in the presence of allyl ethers, the sole mono *ortho*-allylated product was produced, and the C-H di-alkylated product was not observed (Fig. 3.28) [38].

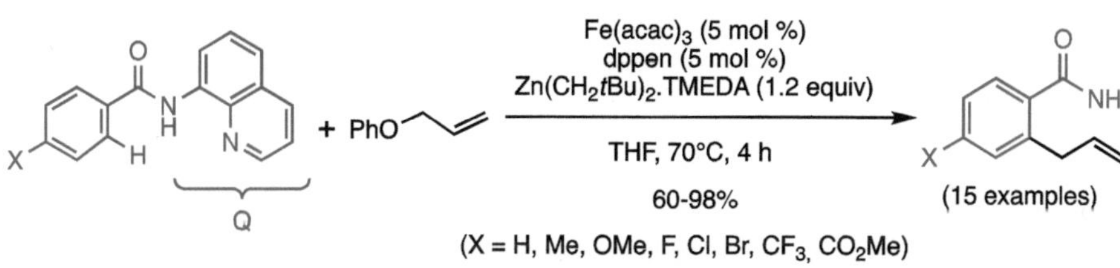

Fig. 3.28: C-C bond formation between 8-quinolylamide and dialkylzinc reagent by Fe-catalyzed C-H bond activation.

The same authors also reported the iron-catalyzed alkylation of arenes, alkenes and heteroarenes bearing an 8-quinolylamide moiety as the directing group with primary, secondary alkyl tosylates and halides. Fe(acac)$_3$/diphosphine and p-anisylzinc bromide, as the base, were required in conjunction with sodium iodide (1.5 equiv) to suppress the arylation side-product. Several evidences in favor of the radical character of this reaction (e.g. loss of stereochemistry of the chiral center with a secondary tosylate, or the ring-opening of cyclopropylmethyl bromide) (Fig. 3.29) [39].

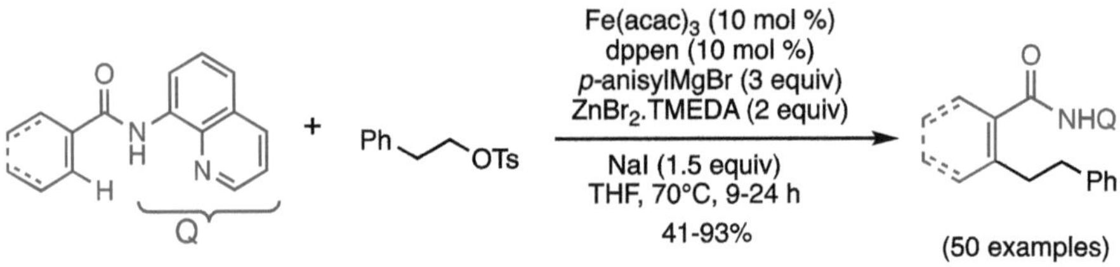

Fig. 3.29: C-C bond formation between 8-quinolylamide and mono-alkylzinc reagent by Fe-catalyzed bond C-H activation.

With benzylamines

Interestingly, electron-rich benzylamines bearing a tri-substituted 1,2,3-triazole are also good substrates for their methylation by dimethylzinc. However, if the aryl moiety was not *ortho*-substituted, the bi-alkylated was obtained (82%), (Fig. 3.30) [35].

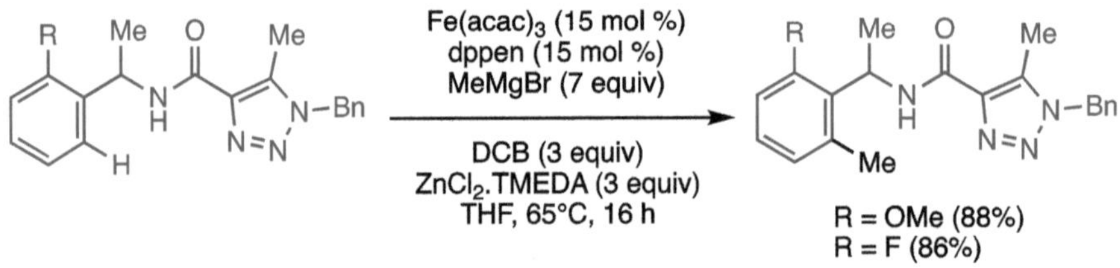

Fig. 3.30: C-C bond formation between 1,2,3-triazole-benzylamines and dimethylzinc reagent by Fe-catalyzed C-H bond activation.

With acrylamides

When applied to acrylamide derivatives, the desired mono-alkylated product was obtained in high yield and high *Z* selectivity. It is worth noting that in these cases, the monoalkylzinc halide (prepared from the corresponding Grignard reagent and zinc dichloride) was found to be a better alkyl donor than the corresponding dialkylzinc reagent (Fig. 3.31) [36].

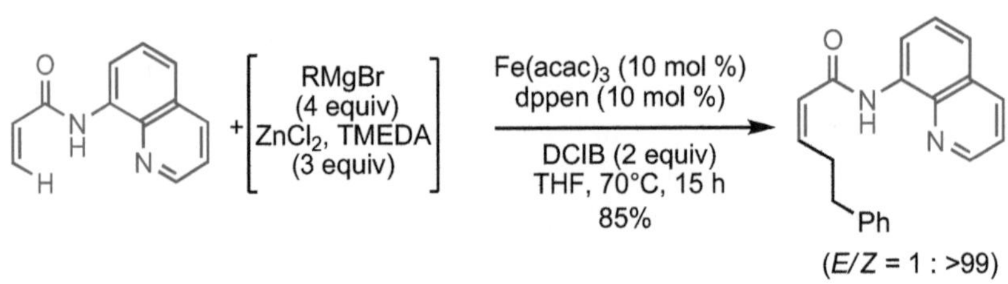

Fig. 3.31: C-C bond formation between 8-quinolylacrylamide and mono-alkylzinc reagent by Fe-catalyzed C-H bond activation.

3.6.3 Iron-catalyzed arylation of C(sp^3)-H bonds

When Nakamura et al. studied the cross-coupling reaction between 4-iodotoluene with diphenylzinc in THF, they also isolated 2-phenyl THF (in low yield), arising from the abstraction of a hydrogen to THF by a formed tolyl radical, followed by a cross-coupling reaction (Fig. 3.32, eq. 1). This hypothesis lies on previous report by Kochi who studied the reaction of nickel complexes with aromatic halides and evidenced an intermediate paramagnetic ion pair intermediate, [M(I)ArX$^{\bullet-}$], that further gives an aryl radical, observed by ESR spectroscopy. Thus, the authors decided to apply this method to iodobenzyl pyrrolidines, which, by an intramolecular 1,5-hydrogen transfer, gave the arylated product as expected (Fig. 3.32, eq. 2) [40].

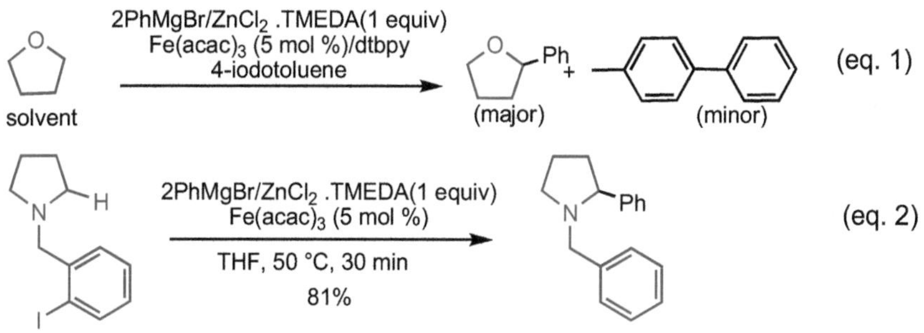

Fig. 3.32: C-H phenylation of iodobenzyl pyrrolidine by 1,5-hydrogen transfer under iron catalysis.

Later on, the authors applied the iron-catalyzed arylation of C(*sp*^{3})-H to carboxamides possessing an 8-aminoquinolyl directing substituent. The reaction worked well with diarylzinc reagents in the presence of a bisphosphine ligand and 1,2-dichloro*iso*butane (DCIB) as a co-oxidant, in THF at 50 °C (Fig. 3.33) [41].

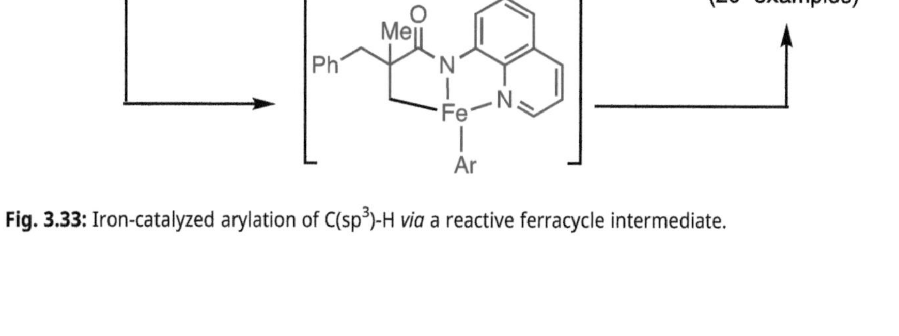

Fig. 3.33: Iron-catalyzed arylation of C(sp^3)-H *via* a reactive ferracycle intermediate.

Ackermann and co-workers found that carboxamides bearing a triazole group were also good substrates for the arylation by a C-H bond activation, albeit, under the same reaction conditions than those used for the 8-amino quinolyl carboxamides, the biaryl homo-coupling products were also isolated (Fig. 3.34) [33].

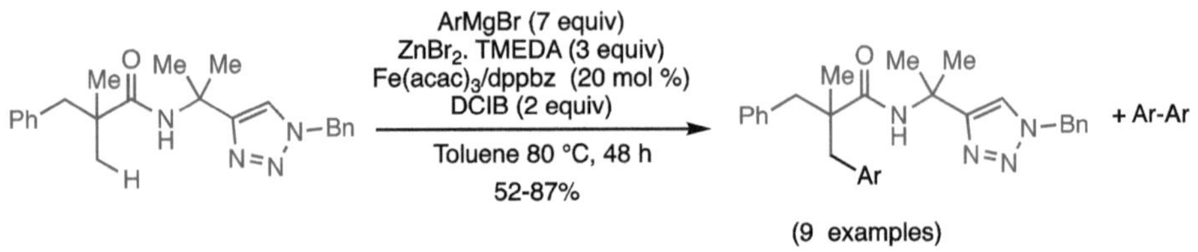

Fig. 3.34: Iron-catalyzed arylation of C(sp^3)-H of carboxamides bearing a 1,2,3-triazole directing group.

In 2018, Nakamura et al. developed a regioselective γ–C-H arylation of 2-iodoalkylarenes using diphenylzinc and a catalytic amount of Fe(acac)$_3$ using a *N*-heterocyclic carbene as a ligand, and the reaction was achieved in fluorobenzene (Fig. 3.35) [42].

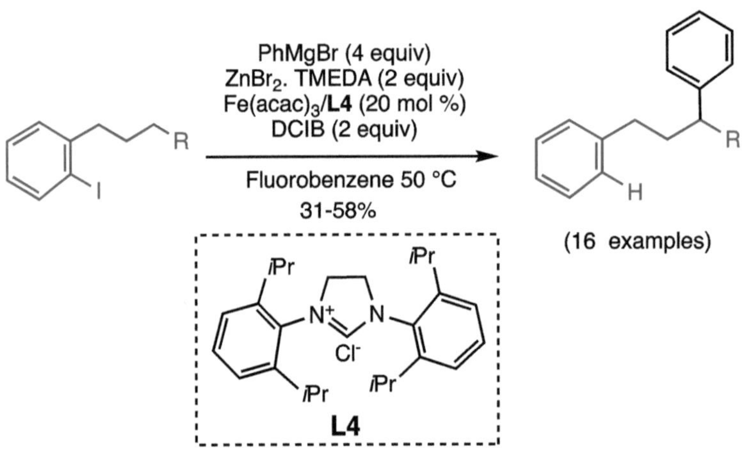

Fig. 3.35: Remote arylation of 2-iodoalkylbenzene by diphenylzinc under iron catalysis.

3.6.4 Iron-catalyzed alkylation of C(sp^3)-H bonds

Ackermann et al. reported the methylation of an alkyl carboxamide possessing an 1,2,3-triazole moiety, as a directing group, by β-C-H bond activation. This methylation was performed by using dimethylzinc, DCIB as a co-oxidant, and a bisphosphine as the ligand (Fig. 3.36) [37].

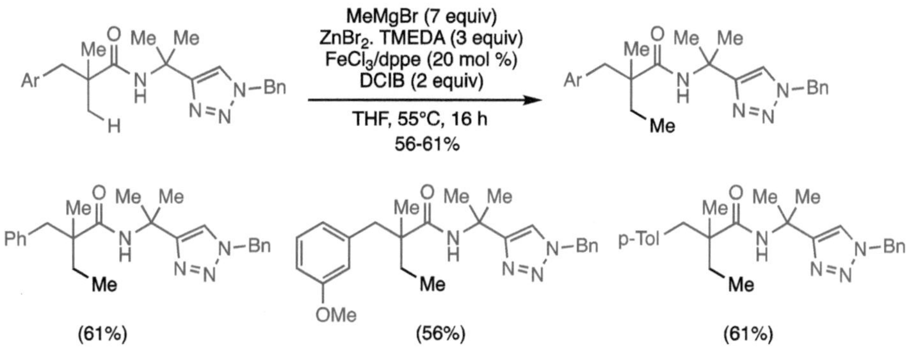

Fig. 3.36: Triazole-assisted iron-catalyzed β-methylation of carboxamides by dimethylzinc.

3.6.5 Iron-catalyzed addition of C(sp^2)-H bonds to alkynes, alkenes and allenes

Carboxamides have been extensively used in the studies of iron-catalyzed addition of C(sp^2)-H bonds to alkynes, allenes and alkenes. Nakamura et al. reported the cross-coupling reaction of carboxamides, bearing an 8-aminoquinolyl directing group, with alkynes, *via* a C-H bond iron-catalyzed activation. Depending on the nature of the zinc reagent, and the presence or not of a co-oxidant, different products were obtained. Mono-organozinc halide along with dichloropropane (DCP), as the co-oxidant, provided the addition product, while the diorganozinc reagent in the presence of DCP led to the cyclized product. In the absence of DCP, mono-organozinc reagents led to the desired indenones (Fig. 3.37) [43]. It is worth noting that a broad range of diaryl-, dialkyl alkynes, enynes and 1,3-diynes are well-suited for this reaction.

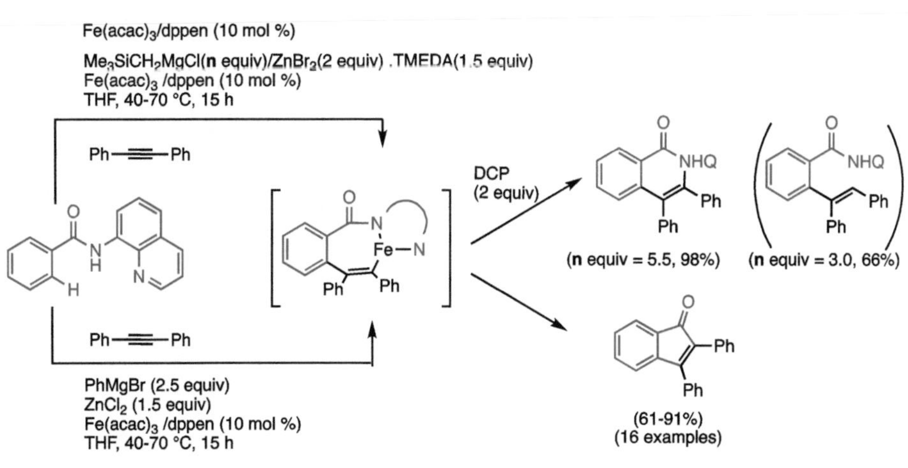

Fig. 3.37: Organozinc reagents regulated the selective formation of oxidative annulation products.

The same authors developed a directed alkylation of carboxamides through the carbometallation of olefins, giving high yields of the mono-alkylated products (Fig. 3.38) [44].

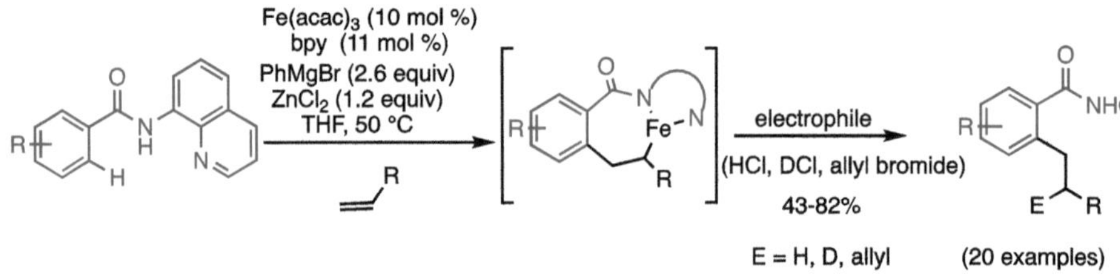

Fig. 3.38: *Ortho*-alkylation of carboxamides by mono organozinc reagents catalyzed by iron(III).

Ackermann and co-workers reported that carboxamides bearing an 1,2,3-triazole directing group reacted with allenyl acetates, through a C-H bond iron-catalyzed activation, in the presence of diarylzinc and a bisphosphine (Fig. 3.39) [45].

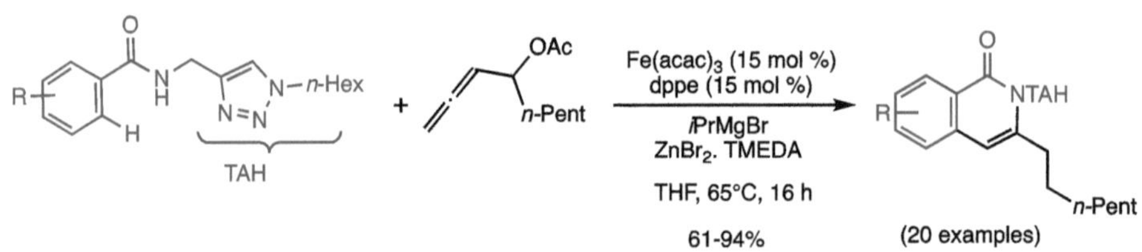

Fig. 3.39: Iron-catalyzed allene annulation with mono alkylzinc reagent.

The same authors developed also a directed alkylation of carboxamides bearing an 1,2,3-triazole directing group, through a carbometallation of propargyl acetates, giving high yields of the cyclized products (Fig. 3.40) [46].

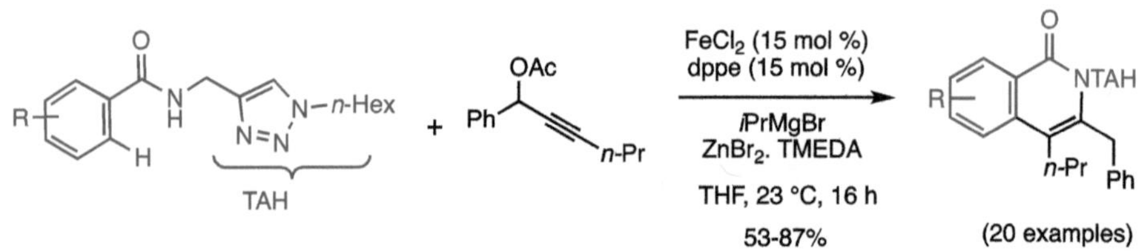

Fig. 3.40: Iron-catalyzed C-H propargyl acetates annulation with mono alkylzinc reagent.

3.7 Mechanistic considerations

Whereas, the cross-coupling reactions with Grignard reagents (Kumada cross-coupling) have been the focus of many investigations [47–50], highlighting the role of "ate organo-iron" complexes [50], the Negishi cross-coupling (with organozinc reagents), has been much less studied. However, in 2012, Bedford and co-workers reported the isolation of an iron(I) complex that was proven to be catalytically competent in cross-coupling reaction of aryl zinc reagents with various partners (e.g. primary, secondary alkyl bromides, 2-bromo-pyridine) (Fig. 3.41) [51].

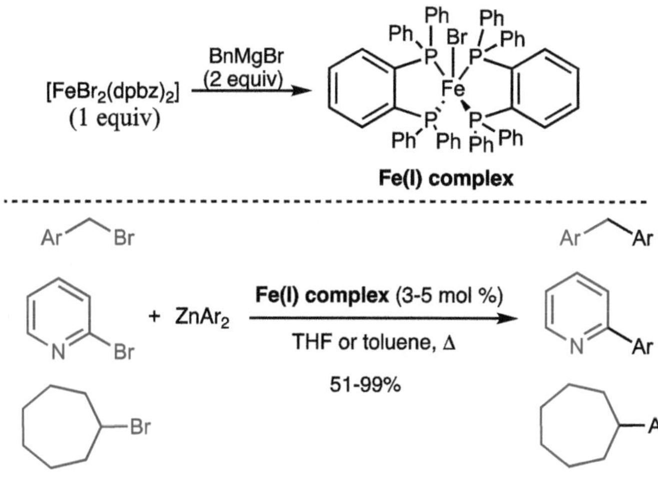

Fig. 3.41: Negishi cross-coupling by the induced by an iron(I) species.

3.8 Conclusion

Cross-coupling reactions using iron catalysts are becoming common tools in organic chemistry, because of the low toxicity of the catalysts, and because of their low costs. The C-C bond activation and cross-coupling reactions are getting more and more of interest for the community. A strong development has started with the discovery that zinc reagents can be used in place of Grignard reagents for these transformations. As pointed out in the different sections of this chapter, these cross-coupling reactions are highly chemoselective and tolerate numerous functional groups (e.g., cyanide, methyl carboxylate, halide). The most recent promising result is the C-H bond activation which clearly highlights the high potential of zinc reagent/iron catalyst to perform the construction of C-C bond in complex molecules, opens the path for forming C-C bonds to access polyfunctionalized compounds of interest.

References

[1] Rana S, Biswas JP, Paul S, Paik A, Maiti D, Chem. Soc. Rev. 2021, 50, 243–472.
[2] Nakamura M, Hirai A, Nakamura E, J. Am. Chem. Soc. 2000, 122, 978–979.
[3] Wei B, Knochel P, Synthesis 2022, 54, 246–254.
[4] Sherry BD, Fürstner A, Acc. Chem. Res. 2008, 41, 1500–1511.
[5] Fürstner A, Bull. Soc. Chem. Jpn. 2021, 94, 666–777.
[6] Bakas NJ, Neidig ML, ACS Catal. 2021, 11, 8493–8503.
[7] Guérinot A, Cossy J, Top. Curr. Chem. 2016, 374, 1–74.
[8] Cardellicchio C, Fiandanese V, Marchese G, Ronzini L, Tetrahedron Lett. 1985, 26, 3595–3598.
[9] Cardellicchio C, Fiandanese V, Marchese G, Ronzini L, Tetrahedron Lett. 1987, 28, 2053–2056.
[10] Fiandanese V, Marchese G, Naso F, Tetrahedron Lett. 1988, 29, 3587–3590.
[11] Babudri F, D'Ettole A, Fiandanese V, Marchese G, Naso F, J. Organomet. Chem. 1991, 405, 53–58.
[12] Dell'Anna MM, Mastrorilli P, Nobile CF, Marchese G, Taurino MR, J. Mol. Catal. Chem. 2000, 161, 239–243.
[13] Reddy CK, Knochel P, Angew. Chem. Int. Ed. 1996, 35, 1700–1701.
[14] Bedford RD, Hall MA, Hodges GR, Huwe M, Wilkinson MC, Chem. Commun. 2009, 6430–6432.
[15] Cahiez G, Foulgoc L, Moyeux A, Angew. Chem. Int. Ed. 2009, 48, 2969–2972.
[16] Hatakeyama T, Kondo Y, Fujiwara Y-I, Takaya H, Ito S, Nakamura E, Nakamura M, Chem. Commun. 2009, 1216–1218.
[17] Nakamura M, Ito S, Matsuo K, Nakamura E, Synlett 2005, 1794–1798.
[18] Adak L, Kawamura S, Toma G, Takenaka T, Isozaki K, Takaya H, Orita A, Hc LI, Shing KM, Nakamura M, J. Am. Chem. Soc. 2017, 139, 10693–11069.
[19] Lin X, Zheng F, Qing FL, Organometallics 2012, 31, 578–582.
[20] Ito S, Fujiwara Y, Nakamura E, Nakamura M, Org. Lett. 2009, 11, 4306–4309.
[21] Figadère B, Franck X. S_N2 versus S_N2' reactions. In: Silverman GS, Rakita PE Eds. Handbook of Grignard Reagents. CRC Press, 1996.
[22] Brown CA, Nile TA, Mahon MF, Webster RL, Dalton Trans. 2015, 44, 12189–12195.
[23] Ito S, Itoh T, Nakamura M, Angew. Chem. Int. Ed. 2011, 50, 454–457.
[24] Bedford RB, Huwe M, Wilkinson MC, Chem. Commun. 2009, 600–602.
[25] Clifton J, Habraken ERM, Pringle PG, Manners I, Cat. Sci. Technol. 2015, 5, 4350–4353.
[26] Bedford RB, Carter E, Cogswell PM, Gower NJ, Haddow MF, Harvey JN, Murphy DM, Neeve EC, Nunn J, Angew. Chem. Int. Ed. 2013, 52, 1285–1288.
[27] Hatakeyama T, Nakagawa N, Nakamura M, Org. Lett. 2009, 11, 4496–4499.
[28] Hung TT, Huang CM, Tsai FY, Chem. Catal. Chem. 2012, 4, 540–545.
[29] Norinder J, Matsumoto A, Yoshikai N, Nakamura E, J. Am. Chem. Soc. 2008, 130, 5858–5859.
[30] Yoshikai N, Matsumoto A, Norinder J, Nakamura E, Angew. Chem. Int. Ed. 2009, 48, 2925–2959.
[31] Yoshikai N, Matsumoto A, Norinder J, Nakamura E, Synlett 2010, 313–316.
[32] Ilies L, Asako S, Nakamura E, Asian J. Org. Chem. 2012, 1, 142–145.
[33] Gu Q, Al Mamari HH, Graczyk K, Diers E, Ackermann L, Angew. Chem. Int. Ed. 2014, 53, 3868–3871.
[34] Doba T, Matsubara T, Ilies L, Shang R, Nakamura E, Nat. Catal. 2019, 2, 400–406.
[35] Shen Z, Cera G, Haven T, Ackermann L, Org. Lett. 2017, 19, 3795–3798.
[36] Ilies L, Ichikawa S, Asako S, Matsubara T, Nakamura E, Adv. Synth. Catal. 2015, 357, 2175–2179.
[37] Graczyk K, Haven T, Ackermann L, Chem. Eur. J. 2015, 21, 8812–8814.
[38] Asako S, Ilies L, Nakamura E, J. Am. Chem. Soc. 2013, 135, 17755–17757.
[39] Ilies L, Matsubara T, Ichikawa S, Asako S, Nakamura E, J. Am. Chem. Soc. 2014, 136, 13126–13129.
[40] Yoshikai N, Mieczkowski A, Matsumoto A, Nakamura E, J. Am. Chem. Soc. 2010, 132, 5568–5569.
[41] Shang R, Ilies L, Matsumoto A, Nakamura E, J. Am. Chem. Soc. 2013, 135, 6030–6032.
[42] Zhou B, Sato H, Ilies L, Nakamura E, ACS Catal. 2018, 8, 8–11.

[43] Ilies L, Arslanoglu Y, Matsubara T, Nakamura E, Asian J. Org. Chem. 2018, 7, 1327–1329.
[44] Ilies L, Zhou Y, Yang H, Matsubara T, Shang R, Nakamura E, ACS Catal. 2018, 8, 11478–11482.
[45] Mo J, Müller T, Oliveira JCA, Ackermann L, Angew. Chem. Int. Ed. 2018, 57, 7719–7723.
[46] Mo J, Müller T, Oliveira JCA, Demeshko S, Meyer F, Ackermann L, Angew. Chem. Int. Ed. 2019, 58, 12874–12878.
[47] Rousseau L, Herrero C, Clémancey M, Imberdis A, Blondin G, Lefèvre G, Chem. Eur. J. 2020, 26, 2417–2428.
[48] Rousseau L, Touati N, Binet L, Thuéry P, Lefèvre G, Inorg. Chem. 2021, 60, 7991–7997.
[49] Wowk V, Rousseau L, Lefèvre G, Organometallics 2021, 40, 3253–3266.
[50] Martin R, Fürstner A, Angew. Chem. Int. Ed. 2004, 43, 3955–3957.
[51] Adams CJ, Bedford RB, Carter E, Gower NJ, Haddow MF, Harvey JN, Huwe M, Cartes MA, Mansell SM, Mendoza C, Murphy DM, Neeve EC, Nunn J, J. Am. Chem. Soc. 2012, 134, 10333–10336.

Pauline Schiltz, Mengyu Gao and Corinne Gosmini

4 Organozinc reagents and cobalt

4.1 Introduction

Transition metal-catalyzed cross-coupling reactions to form C-C bonds are among the most powerful methods and have received considerable attention from organic chemists over the last years [1]. They have become indispensable tools for the construction of organic molecules. These approaches involve an organometallic species as the nucleophilic reagent, and an electrophilic compound such as an organic or a pseudo-halide [1]. Classically, various organometallic reagents such as organomagnesium, organoboron, organostannane or organozinc reagents can be employed [1]. Although Grignard reagents are among the most reactive and most frequently employed reagents to form carbon-carbon bonds with various catalysts, organozinc reagents offer a high functional group tolerance compared to Grignard reagents and a good reactivity in the presence of appropriate catalysts [2]. Although the most common catalysts for cross-coupling reactions are palladium and nickel [3], the use of more eco-compatible and economical catalysts, such as cobalt [4–8], is important in modern chemistry. Since the early 2000's, cobalt-catalyzed carbon-carbon bond forming reactions have received particular attention since Kharasch's first seminal works, reported, in 1943, on cobalt-catalyzed C-C bond formation with organomagnesium reagents [9]. The inexpensive catalyst systems based on cobalt salts are also appropriated to couple organozinc species allowing the presence of reactive functions such as ester, nitrile, ketone and others [5, 6]. Moreover, it is worth mentioning that cobalt can also be used to form organozinc compounds. This chapter describes the different methods to synthesize various cobalt-catalyzed organozinc reagents and their cobalt-catalyzed cross-coupling reactions to form C-C bonds.

4.2 Formation of arylzinc reagents

The common preparation of arylzinc species from aryl bromides and chlorides is achieved via the preliminary formation of aryl lithium or magnesium compounds followed by transmetallation with zinc halides. In the presence of reactive functional groups, a very low temperature is required. To avoid the preparation of an intermediate organometallic reagent, an electrochemical method using a sacrificial zinc anode was reported to directly synthesize arylzinc species from an aryl chloride or bromide

Corinne Gosmini, Laboratoire de Chimie Moléculaire, Ecole Polytechnique, CNRS, UMR 9168, Route de Saclay, 91128 Palaiseau Cedex, France
Pauline Schiltz, Mengyu Gao, Laboratoire de Chimie Moléculaire, Ecole Polytechnique, CNRS, UMR 9168, Route de Saclay, 91128 Palaiseau Cedex, France

https://doi.org/10.1515/9783110728859-004

in the presence of cobalt chloride, associated with pyridine as the ligand, in DMF or acetonitrile as the solvent [10]. Right now, only arylzinc species can be synthesized using a cobalt catalyst. However, different arylzinc reagents bearing electron-donating or -withdrawing groups are formed with very simple experimental conditions.

To prevent the use of pyridine, which can be sometimes damaging in the cross-coupling reaction, a pyridine-free electrochemical process was also reported by the Gosmini's group in the presence of cobalt bromide in pure acetonitrile from aryl bromides [11] (Fig. 4.1).

Although these electrochemical processes were favorably compared with known chemical processes, they were considered difficult to handle, they need some specialized devices and were poorly used by organic chemists. Therefore, purely chemical reactions were extended from these electrochemical processes replacing electricity by zinc dust to reduce cobalt bromide in order to generate zinc bromide in acetonitrile [12]. These new and versatile preparations of arylzinc bromides from various aryl bromides generally give good to excellent yields.

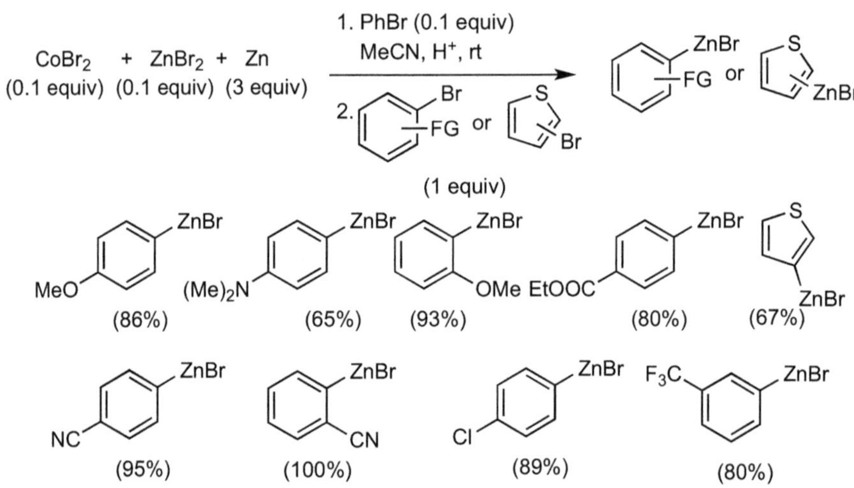

Fig. 4.1: Co-catalyzed formation of functionalized arylzinc compounds from aryl and thienyl bromides.

The authors proposed a mechanism where the catalytic cycle was initiated by the reduction of $CoBr_2$ by zinc dust, activated previously by traces of acid. The resulting Co(I)Br underwent an oxidative addition with aryl halides to afford the trivalent cobalt complex ArCo(III)Br_2. This latter species was reduced into ArCo(II)Br by the excess of zinc dust and transmetalled with $ZnBr_2$ formed in the previous steps to lead to the arylzinc compound and regenerate the Co(II) species (Fig. 4.2).

In 2005, the Gosmini's group extended the chemical method to aryl and thienyl chloride adding pyridine in the medium and increasing the amount of catalyst to

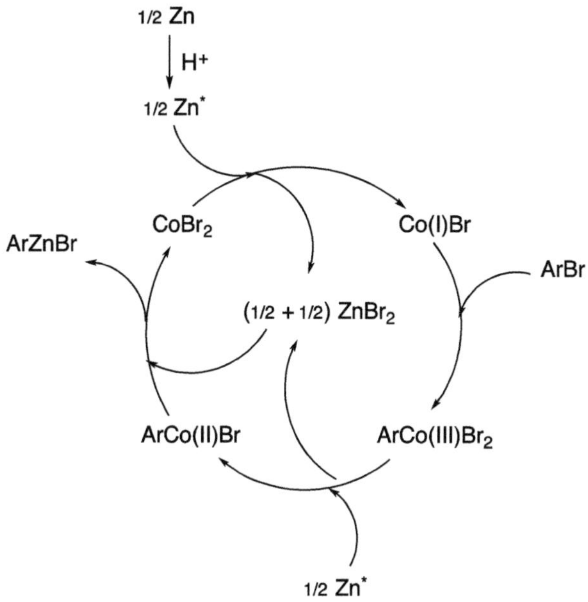

Fig. 4.2: Mechanism for the cobalt-catalyzed formation of ArZnBr.

33 mol % [13]. However the replacement of $CoBr_2$ by $CoBr_2$(bpy) can decrease the reaction time and the amount of catalyst if the reaction is conducted at 50 °C. These conditions can be extended to aryl triflates [14] (Fig. 4.3).

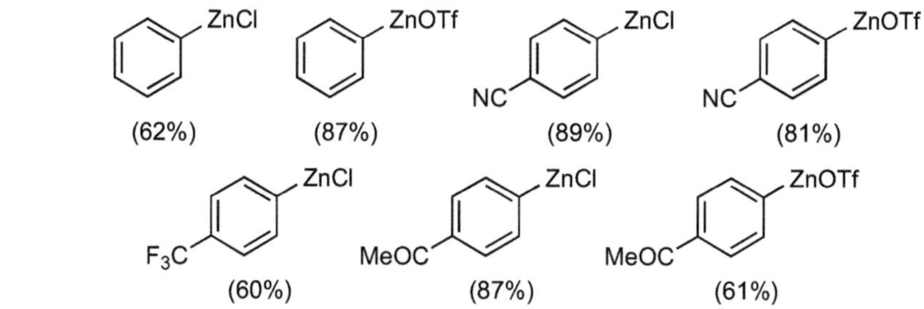

Fig. 4.3: Co-catalyzed formation of functionalized arylzinc compounds from aromatic chlorides and triflates.

The cobalt-catalyzed activation of C-SMe of various methylthio-substituted *N*-heterocycles also led to the synthesis of the corresponding organozinc species in good yields [15] (Fig. 4.4).

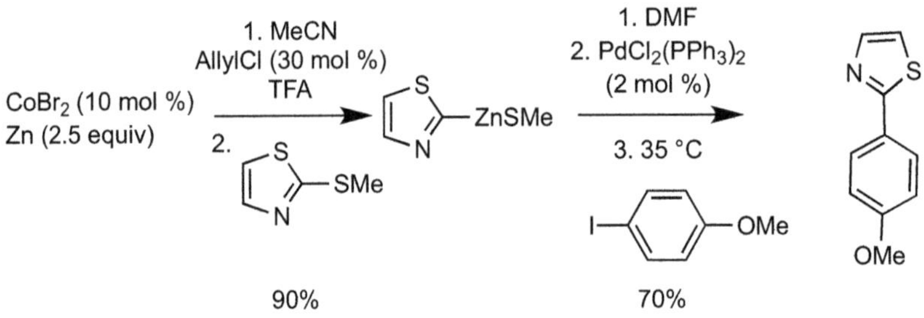

Fig. 4.4: Pd-catalyzed cross-coupling of organozinc species from 2-methylthiothiazole with iodoanisole.

Later on, the Yoshikai's group reported that a cobalt/XanPhos-catalyzed, LiCl-mediated system allowed the preparation of arylzinc species directly from aryl iodides, bromides and chlorides in THF [16] (Fig. 4.5).

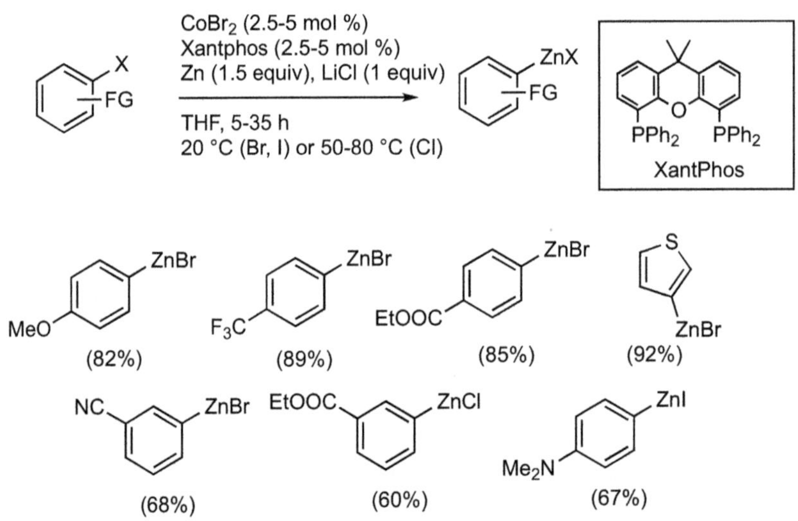

Fig. 4.5: Cobalt/XanPhos-catalyzed, LiCl-mediated preparation of arylzinc reagents from aryl iodides, bromides and chlorides.

The application of the previous conditions did not allow the formation of pyridylzinc compounds from the corresponding pyridyl halides. However, using a polydentate *N*-heterocyclic ligand such as *N*-(1-(pyridin-2-yl)ethyliden)quinolin-8-amine with a cobalt catalyst allowed the reaction to be efficiently carried out in acetonitrile or THF in the absence of pyridine as a co-solvent [17] (Fig. 4.6).

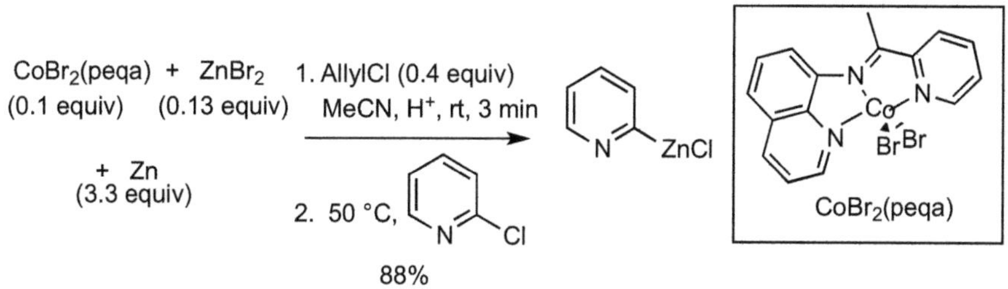

Fig. 4.6: Cobalt-catalyzed pyridylzinc chloride using *N*-(1-(pyridin-2-yl)ethyliden)quinolin-8-amine (peqa) as the ligand.

It is worth mentioning that other organometallic reagents such as arylstannanes [18] and arylboronates [19] can be obtained from aryl bromides *via* the corresponding intermediate arylzinc species synthesized using $CoBr_2$ in acetonitrile by a simple transmetallation with tributylstannyl chloride and haloboronic ester respectively in good to excellent yields.

4.3 Coupling reactions of arylzinc reagents

4.3.1 With aryl halides, heteroaryl halides and vinyl halides

The first cobalt-catalyzed cross-coupling reaction of arylzinc halides with aromatic compounds was reported by Gosmini and Begouin in 2009 with 2-chloropyrimidines and 2-chloropyrazines [20] (Fig. 4.7). The advantage of this procedure was the use of the same cobalt halide as the catalyst both for the preparation of the organozinc species and the cross-coupling with the diazine chloride. Depending on the nature of the substrate, this process allowed the synthesis of a wide variety of 2-aryldiazines in a Barbier-type reaction.

This procedure can also be applied to triazine halides [21] but not to other (hetero) aromatic halides. In the same manner, the cobalt-catalyzed cross-coupling reaction of arylzinc species with various methylthio-substituted *N*-heterocycles was successful [15] (Fig. 4.8).

Recently, based on this cobalt-catalyzed cross-coupling reaction of arylzinc with aryl halides, cost-effective and simple synthetic methods for the synthesis of thermally activated delayed fluorescence (TADF) emitters, incorporating 1,3,5-triazine moieties or containing 4,6-bis-phenyl phenothiazine as donor units and 2-thiophen-1,3,5-triazine as the acceptor unit, were reported [22] (Fig. 4.9).

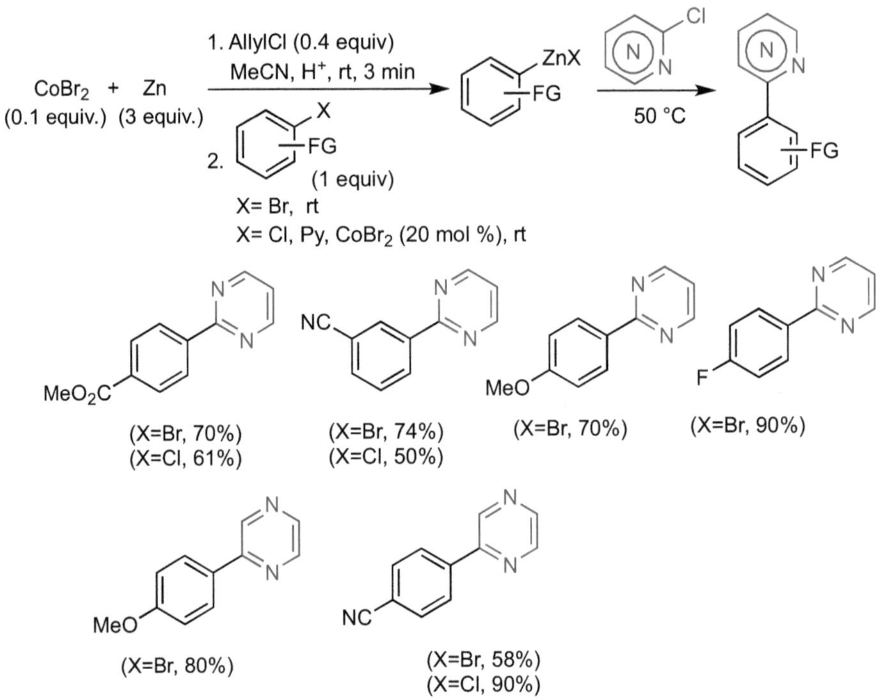

Fig. 4.7: Cobalt-catalyzed cross-coupling between arylzinc halides, 2-chloropyrimidine and 2-chloropyrazine.

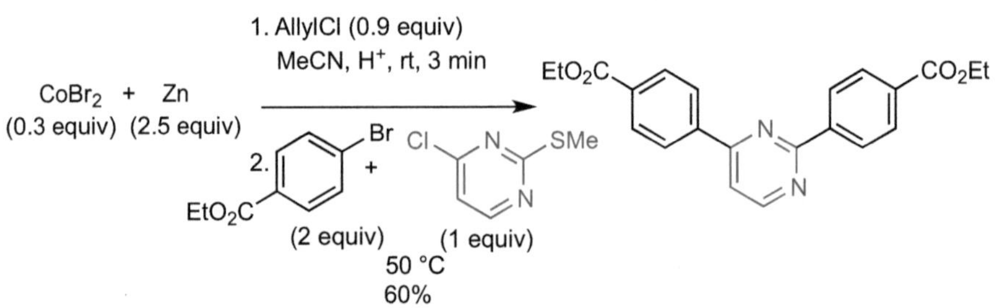

Fig. 4.8: Cobalt-catalyzed formation of a 2,4-diarylpyrimidine from 2-methylthio-4-chloropyrimidine.

Moreover, it is worth mentioning that the formation of symmetrical biaryl compounds from Co-catalyzed synthesis of arylzinc reagents was possible by oxidative homocoupling in the presence of air or *p*-benzoquinone depending on the nature of the functional group [23].

In 2016, Knochel et al. reported a smooth cross-coupling of various arylzinc chlorides with heteroaryl chlorides or bromides within a few hours at room temperature in the presence of $CoCl_2{\cdot}2LiCl$ which was conveniently soluble in THF [24] (Fig. 4.10).

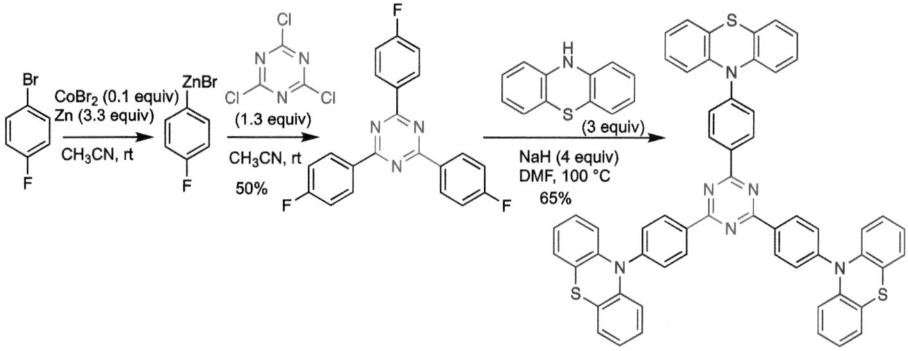

Fig. 4.9: Synthesis of the TADF molecule TRZ 3(Ph-PT).

In order to suppress side reactions, the addition of sodium formate as a ligand led to a more selective cross-coupling reaction and the yield was improved.

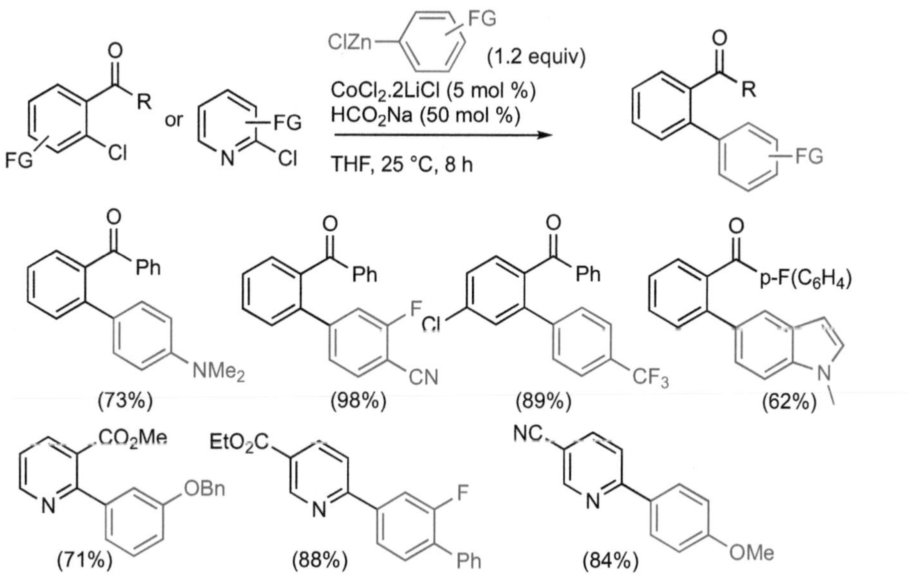

Fig. 4.10: Cobalt-catalyzed cross-coupling of various arylzinc chlorides with heteroaryl halides.

Thereby, polyfunctional naphthyridines were prepared in THF using arylzinc chlorides, a catalytic amount of $CoCl_2$ and sodium formate (HCO_2Na) as ligand [25] (Fig. 4.11). Among them, one was highly fluorescent with tunable emission from blue to yellow and long excited-state lifetimes.

In 2017, Knochel et al. reported a robust and broadly applicable $CoCl_2$-catalyzed cross-coupling reaction between functionalized aryl or heteroarylzinc pivalates with

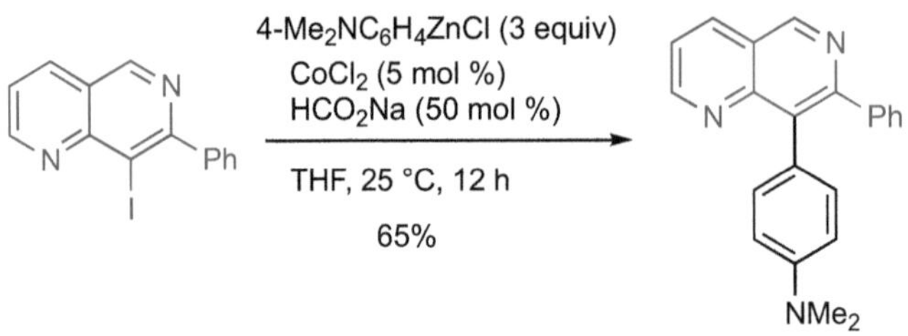

Fig. 4.11: Co-catalyzed 8-iodo-1,6-naphthyridine withp-$Me_2NC_6H_4ZnCl$.

various electron-poor aryl and heteroaryl halides (X = Cl, Br, I) [26, 27] (Fig. 4.12). In this case, the cobalt-catalyzed cross-coupling reaction was practical and scalable using only 1–5 mol % of $CoCl_2$. Alkenyl iodides or bromides were also good partners to undergo cobalt-catalyzed cross-coupling reaction with arylzinc pivalates.

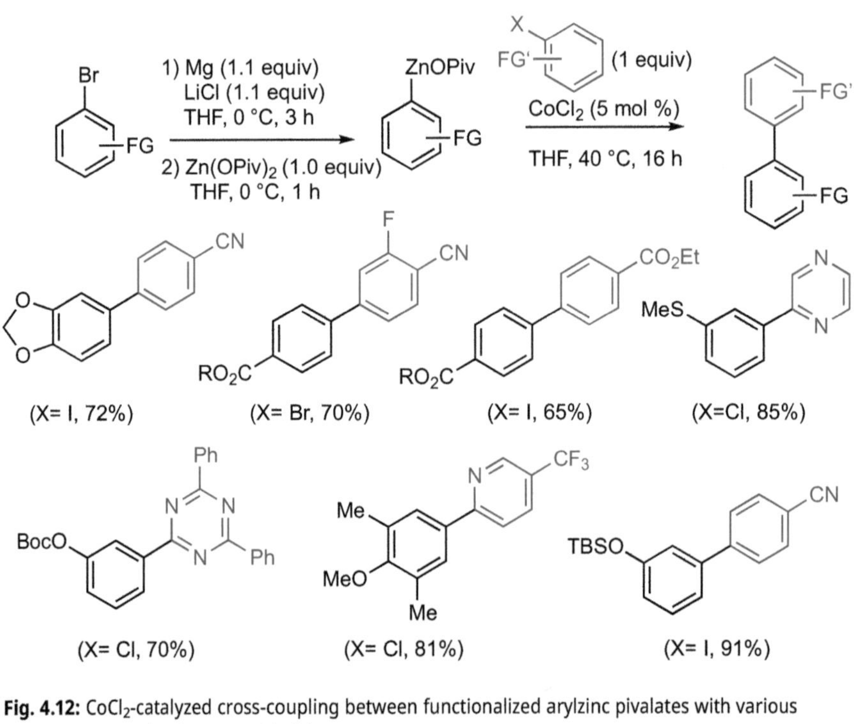

Fig. 4.12: $CoCl_2$-catalyzed cross-coupling between functionalized arylzinc pivalates with various electron-poor aryl or heteroaryl halides.

In general, the use of organozinc chlorides instead of zinc pivalates gave significantly lower yields of cross-coupling products without ligand.

4.3.2 With alkyl halides

The first cobalt-catalyzed cross-coupling of arylzinc species involving alkyl compounds was reported by Gosmini et al. in 2003. Aryl bromides were directly coupled with allylic acetates in the presence of zinc and $CoBr_2$ without the pre-formation of the corresponding arylzinc species in acetonitrile [28]. In this case, the allylic acetate, which was present in the reaction mixture, reacted directly by substitution with the intermediate arylzinc compound. Good yields were only obtained with aryl bromides bearing electron-donating or -withdrawing groups if the allylic acetate was not substituted on the double bond. Concerning the mechanism, fast complexation of the reduced Co(II) is taking place with the allylic acetate to form a (η^2–allylOAc)-Co(I) complex which reacted with the intermediate arylzinc species to form the aryl-allyl compound upon its SN_2 reaction (Fig. 4.13). In 2007, Knochel et al. reported the cobalt-catalyzed cross-coupling between diarylzinc and allyl chlorides or alkyl phosphates [29]. This reaction was carried out in *N*-methylpyrrolidone (NMP) in the presence of $Co(acac)_2$ at 0 °C. The diarylzinc was formed directly from the corresponding aryl iodide. In 2008, the $CoBr_2$-mediated cross-coupling reaction of arylzinc species in acetonitrile with reactive alkyl compounds was extended to benzyl chlorides by Gosmini et al. under mild conditions to prepare a variety of diarylmethanes in good to excellent yields [30]. This reaction was also possible according to a Barbier-type reaction when the aryl halide was more reactive than the benzyl chloride (Fig. 4.14).

Later on, Inoue et al. reported the cobalt-catalyzed cross-coupling reaction of arylzinc reagents with ethyl bromodifluoroacetate. This reaction afforded a variety of aryldifluoroacetates in good yields in the presence of two equivalents of ArZnX-TMEDA complexes and with *trans*-1,2-bis(dimethylamino)-cyclohexane as a $CoCl_2$ ligand [31] (Fig. 4.15).

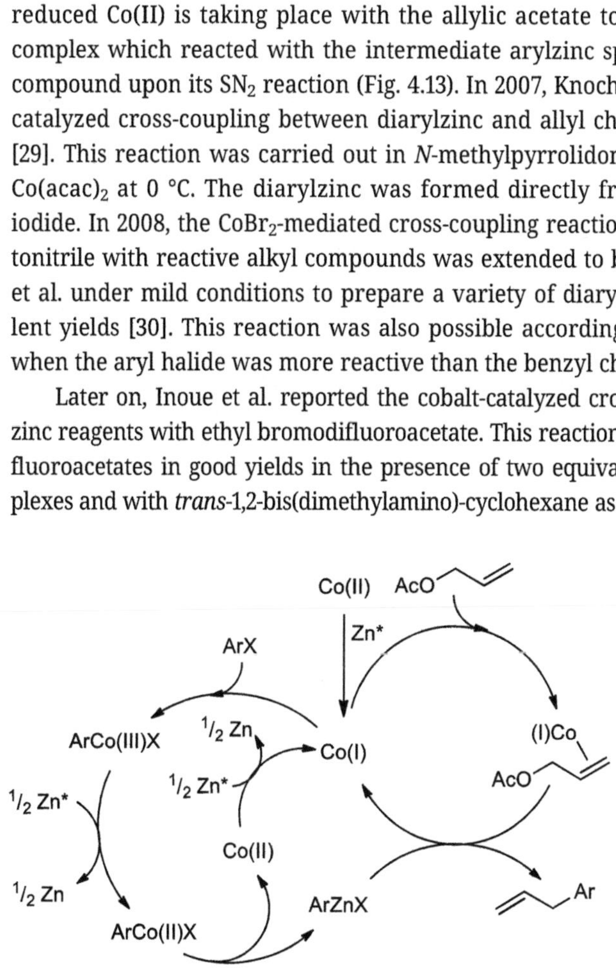

Fig. 4.13: Proposed mechanism for the cobalt-catalyzed cross-coupling reaction of arylzinc species with allylic acetate.

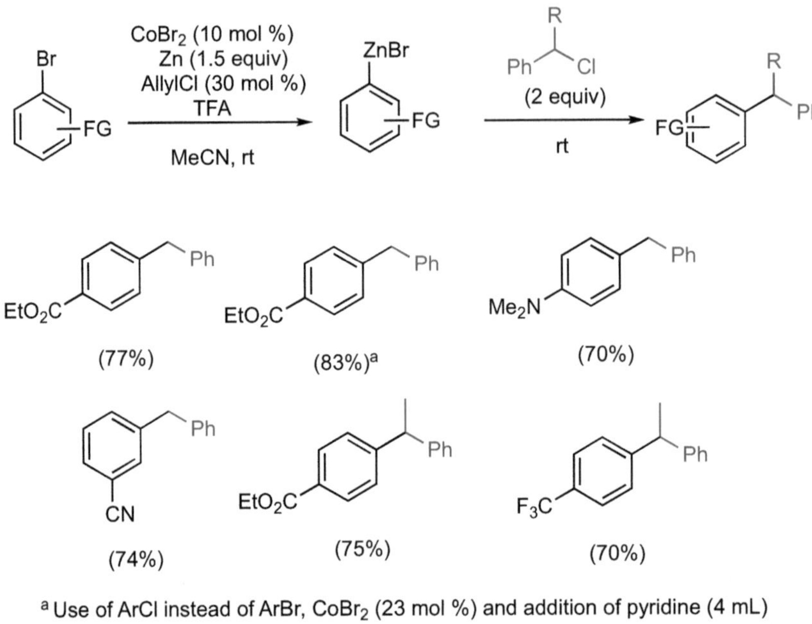

Fig. 4.14: Co-catalyzed formation of functionalized diarylmethanes.

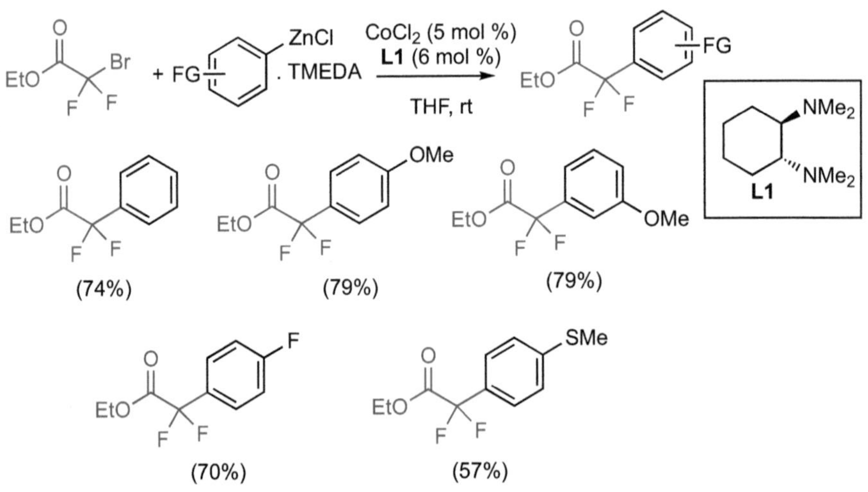

Fig. 4.15: Cobalt-catalyzed cross-coupling reaction of arylzinc reagents with ethyl bromodifluoroacetate.

All the reactions, described above, involved a reactive alkyl compound. However in 2015, Knochel et al. reported cobalt-catalyzed Negishi-type cross-couplings of polyfunctional diaryl- or diheteroarylzinc reagents with primary and secondary alkyl bromides or iodides using THF-soluble $CoCl_2 \cdot 2LiCl$ and TMEDA as the ligand [32] (Fig. 4.16). No rearrangement of secondary alkyl iodides was observed.

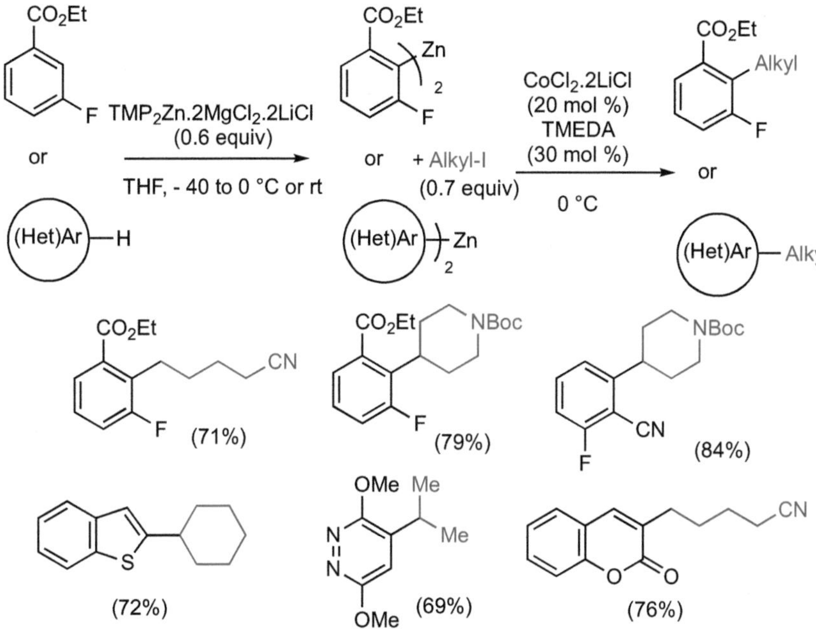

Fig. 4.16: Cobalt-catalyzed Negishi cross-coupling of polyfunctional diaryl- or diheteroarylzinc reagents with primary and secondary alkyl bromides and iodides.

The same authors also reported that α-bromolactones bearing a substituent in the β-position performed a highly *trans*-diastereoselective arylation with arylzinc chlorides in the presence of 10–20 mol % of $CoCl_2$ and 10–20 mol % of PPh_3 under mild conditions [33].

Bian et al. reported the first cobalt-catalyzed enantioselective Negishi-type cross-coupling reaction of racemic α-bromoesters with arylzinc species [34] (Fig. 4.17). Different α-arylalkanoic esters bearing various functional groups such as ether, halide, thioether, silyl, amine, ester, acetal, amide, olefin and heteroaromatic were synthesized with excellent enantioselectivities and yields, using a cobalt-bisoxazoline catalyst.

Different drugs can be synthesized using this catalyst. The (*R*)-xanthorrhizol can be obtained using this method as the key step. An efficient synthesis of the (*S*)-preclamol was also developed using the same conditions for the cobalt-catalyzed asymmetric catalytic cross-coupling of a α-bromoester with arylzinc species as one of the key steps [35] (Fig. 4.18).

Finally, a novel enantioselective synthesis of (*R*)-cinacalcet obtained with 99% enantiomeric excess has been achieved on a gram-scale by a cobalt-catalyzed asymmetric cross-coupling of racemic 4-methoxybenzyl 2-bromopropanoate with 1-bromonaphtalene [36].

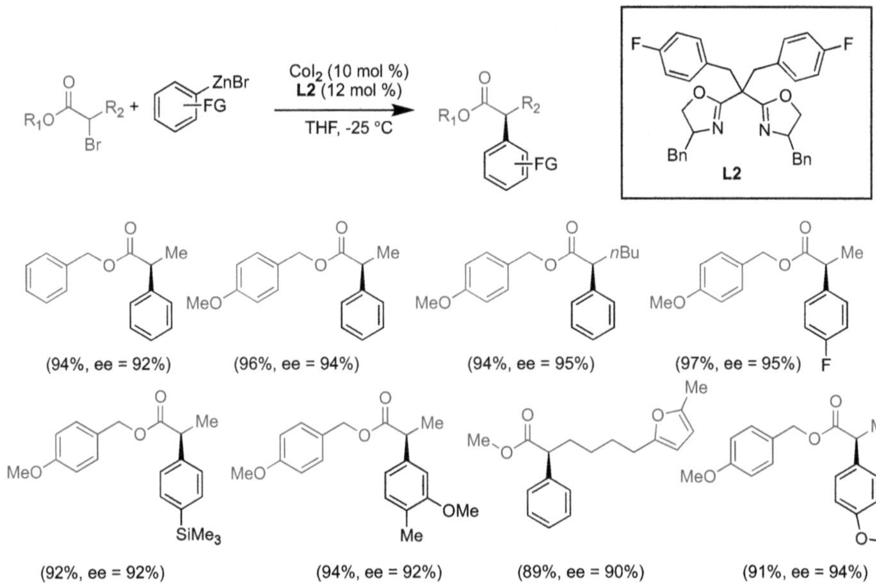

Fig. 4.17: First cobalt-catalyzed enantioselective cross-coupling reaction of racemic α-bromoesters with arylzinc species.

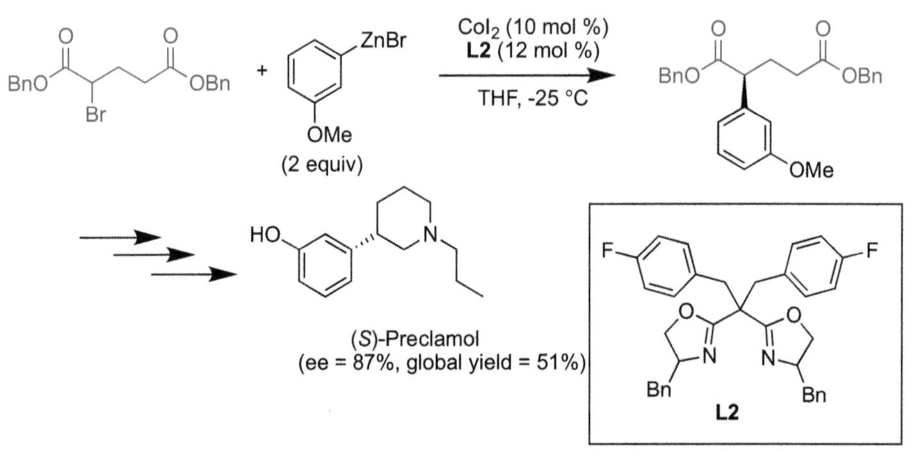

Fig. 4.18: Synthesis of the (*S*)-preclamol.

4.3.3 With alkynyl halides

The phenylacetylene is an important motif used for the synthesis of materials and polymers [37–39]. The most important method for the synthesis of these compounds is the cross-coupling reaction of an alkyne moiety as the nucleophile and an aryl compound as the electrophile in the presence of Pd as the catalyst. The reverse reaction using an alkyne as an electrophile has been less developed regardless the catalyst used. Moreover, few cobalt-catalyzed cross-coupling reactions were reported in this direction. In 2014, Gosmini et al. reported the cobalt-catalyzed cross-coupling reaction of arylzinc halides with bromoalkynes. In this reaction, the same cobalt catalyst, $CoBr_2$(Phen), was used for both the synthesis of the arylzinc species and the cross-coupling reaction with haloalkynes (bromide or chloride) bearing an alkyl or an aryl group after filtration of the organozinc species [40] (Fig. 4.19).

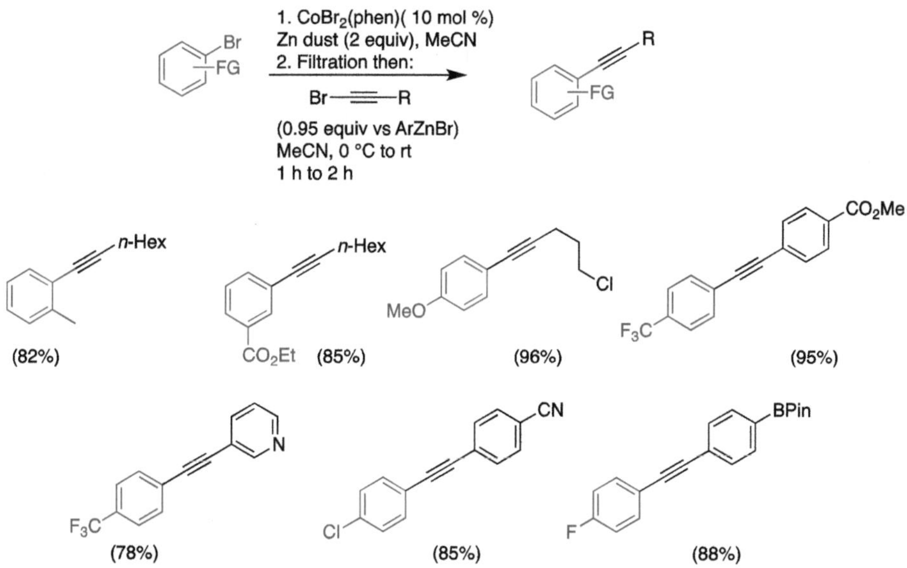

Fig. 4.19: Cobalt-catalyzed cross-coupling reaction of arylzinc reagents with alkynyl bromides.

A large number of functional groups were tolerated and the steric hindrance of the alkyne had no influence in comparison with the Sonogashira or Heck-Cassar protocols. As previously reported for the formation of biaryls from arylzinc pivaltes, Knochel et al. extended the coupling of these latter compounds with bromoalkynes; however, the reaction has to be carried out at −40 °C instead of 40 °C but only 5 mol % of $CoCl_2$ were necessary [26].

4.3.4 With acyl halides

Many reactions have been developed to synthesize arylketones which are found in many natural products and pharmaceutical molecules. Aromatic ketones can be formed from an organometallic species and an acyl chloride in particular from an arylzinc species in the presence of copper [41] or palladium [42]. However in 2003, Gosmini et al. reported the first preparation of aromatic ketones in good yields from acyl chlorides and arylzinc bromides using a cobalt catalyst at room temperature [43]. Arylzinc bromides prepared chemically *via* a cobalt catalysis underwent coupling with acyl chlorides without any additional catalyst unlike the corresponding electrochemical method. By utilizing this method, the preparation of arylzinc species was longer and cobalt had the time to disproportionate before the addition of acyl chloride. With the less reactive acid anhydrides, the reaction was also possible in a Barbier-type reaction with a simple $CoBr_2$ as catalyst and zinc [44]. In this case, the arylzinc intermediate reacted rapidly with the acid anhydride present in the medium. Using ethyl chloroformate, a wide range of symmetrical diaryl ketones were synthesized by the same group from the corresponding aryl bromide through the decomposition of the chloroformate to furnish a carbonyl intermediate [45] (Fig. 4.20).

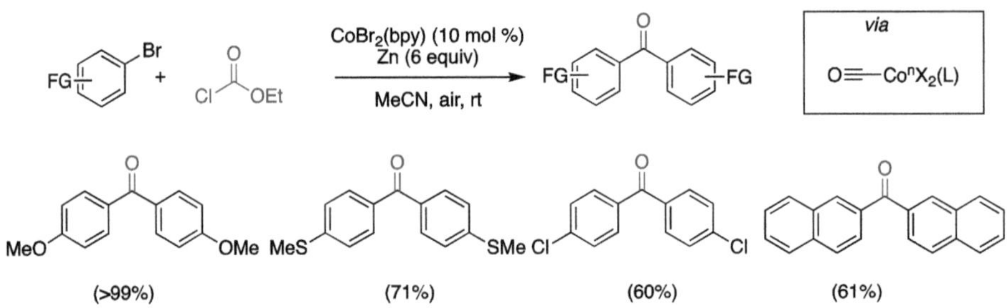

Fig. 4.20: Synthesis of symmetrical diaryl ketones from the corresponding aryl bromide and ethyl chloroformate.

This Barbier-type reaction involved an arylzinc intermediate. More recently, the Gosmini's group also described a cobalt-catalyzed Negishi-type cross-coupling of amides to access ketones from arylzinc [46] (Fig. 4.21) and alkylzinc [47] species. The same simple catalytic system ($CoBr_2$), without the use of ligand to form the arylzinc compound, was used for the cross-coupling with *N*-benzoyl glutarimides at room temperature according to a one-pot procedure.

Recently, Knochel et al. also reported a cobalt-catalyzed acylation of various primary, secondary and tertiary alkyl, benzyl and heteroaryl *S*-pyridyl thioesters with heteroarylzinc pivalates using $CoCl_2$ associated to 4,4′-di-*tert*-butyl-2,2′-dipyridyl (dtbbpy) as the ligand and the reaction was performed in THF [48] (Fig. 4.22).

Different thioesters were prepared from various functionalized carboxylic acids under neutral conditions. Moreover, this method allowed the synthesis of α-chiral ketones with high stereo-retention (ee = 94% to >99%) starting from a chiral (*S*)-pyridyl thioester. This method constitutes the first step of the synthesis of the antilipidemic drug fenofibrate.

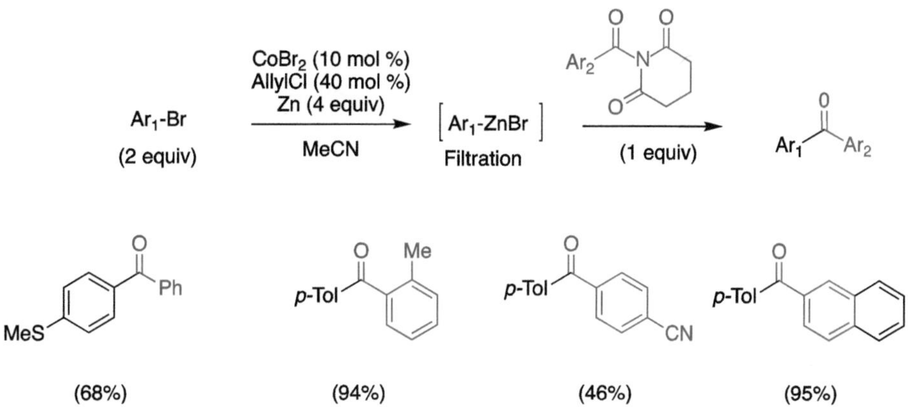

Fig. 4.21: Cobalt-catalyzed Negishi-type cross-coupling of amides and arylzinc species.

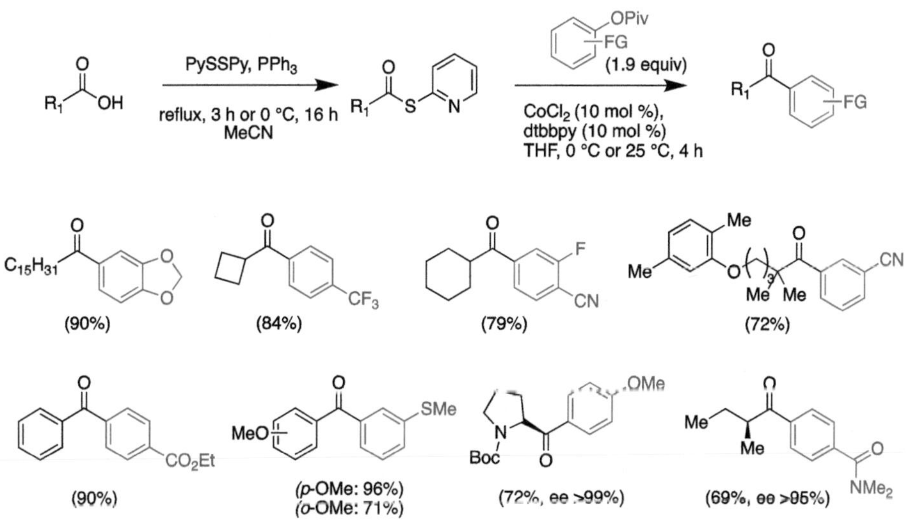

Fig. 4.22: Cobalt-catalyzed acylation of various primary, secondary and tertiary alkyl, benzyl and heteroaryl *S*-pyridyl thioesters by heteroarylzinc pivalates.

4.3.5 With nitriles

Benzonitriles play an important role in natural products, pharmacy, dyes and electronic materials [49–52]. The electrophilic cyanation involving a nucleophilic organometallic is a good alternative to transition metal-catalyzed nucleophilic reaction using toxic cyanide reagents to access aryl nitriles, but this method is not commonly employed. Although Grignard reagents have been recently explored in cyanation reactions, the cross-coupling reaction of arylzinc compounds with *N*-cyano-*N*-phenyl-*p*-methylbenzenesulfonamide

(NCTS) was a complementary route to aryl nitriles, in which an arylzinc compound was generated *in situ* from the corresponding aryl bromides catalyzed by a cobalt catalyst [53] (Fig. 4.23). In the second step, the addition of a catalytic amount of zinc dust was necessary to achieve good efficiency.

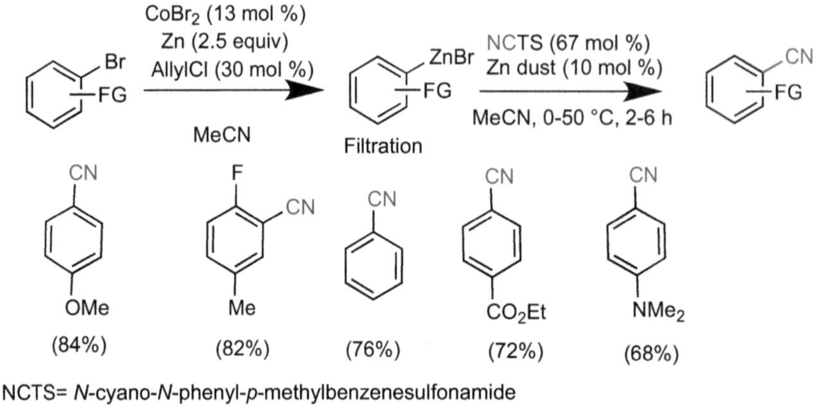

Fig. 4.23: Formation of various benzonitriles by Co-catalyzed cross-coupling of arylzinc compounds with *N*-cyano-*N*-phenyl-*p*-methylbenzenesulfonamide (NCTS).

This reaction was a remarkable alternative to more classical methods using expensive catalysts and led to the cross-coupling products with moderate to excellent yields in mild and bench-friendly conditions, and with excellent functional group Tolerance.

4.3.6 With N-hydroxyphtalimide esters

Organohalides have been one of the most employed electrophilic partners, but recently aliphatic carboxylic acids or their derivatives have also been used due to their wide availability and stability compared to alkyl halides. Although some cross-coupling of organozinc species with unactivated alkyl halides were reported as indicated above, a cobalt-catalyzed decarboxylative Negishi-type cross-coupling was also described with *N*-(acyloxy)phthalimides (NHPI esters) under mild conditions. This reaction proceeds in good yields at room temperature, in DMF, without any additive and external ligand [54] (Fig. 4.24).

Different redox active aliphatic esters such as piperidine *N*-hydroxyphthalimide esters reacted not only with diarylzinc compounds but also with dialkenyl zinc reagents. The use of the tetrachloro-substituted *N*-hydroxyphthalimide ester (TCNPH) was also effective but resulted in the cross-coupled products in lower yields. Dialkynyl zinc reagents were also efficient to form Csp^3-Csp bonds but sometimes alkynyl zinc pivalates gave better yields. In this reaction, the involvement of a radical intermediate

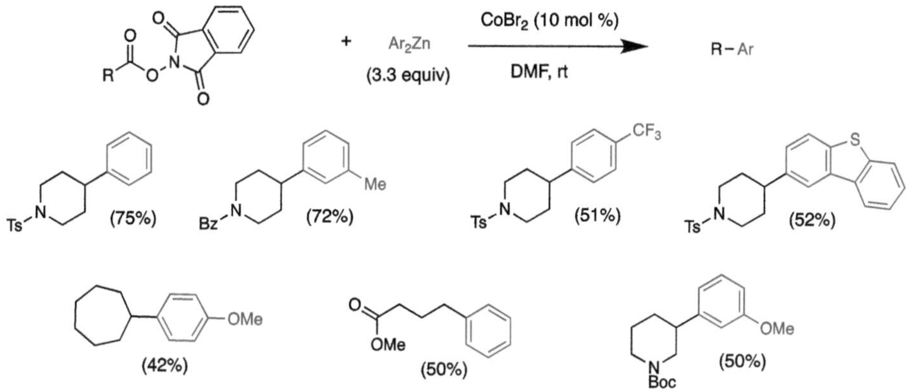

Fig. 4.24: Cobalt-catalyzed decarboxylative Negishi cross-coupling.

was demonstrated. A single-electron transfer to the *N*-hydroxyphthalimide (NHPI) ester induced decarboxylative fragmentation to deliver an alkyl radical.

Although the first example of cobalt-catalyzed cross-coupling reaction involving aromatic organometallic reagents was described using Grignard reagents by Kharasch and Fuchs [9], various arylzinc species revealed also efficient in cobalt-catalyzed cross-coupling reactions since the beginning of the twenty-first century. The advantage of using zinc reagents compared to Grignard reagents is the functional group tolerance. Moreover, in some cases, the catalyst used for the generation of the arylzinc species can be also used for the cross-coupling reaction without a supplementary one.

4.4 Coupling reactions of alkenylzinc and alkynylzinc reagents

Arylzinc reagents have been of interest in cobalt-catalyzed C-C bond formation due to their compatibility with a variety of functional groups [24, 26, 30, 32]. Various arylzinc species such as arylzinc halides (ArZnX, X = Hal), diarylzinc reagents (Ar_2Zn) or arylzinc pivalates were used to construct Csp^2-Csp^2 or Csp^2-Csp^3 bonds. These last compounds reported by Knochel et al. also present an air and moisture stability. The Knochel group has then extended the procedure to alkynylzinc pivalates (RZnOPiv). These alkynylzinc pivalates can be generated by deprotonation with EtMgBr followed by a transmetallation with $Zn(OPiv)_2$, and be handled easily due to the solidification of the organozinc pivalates after solvent evaporation [55].

Knochel et al. reported a range of bench-stable alkynylzinc pivalates which were coupled with various electron-poor aryl and heteroaryl halides utilizing a catalytic system which consists of $CoCl_2{\cdot}2LiCl$ and TMEDA (Fig. 4.25). Remarkably when the alkynylzinc pivalate, derived from the natural product lynestrenol was used, the coupling

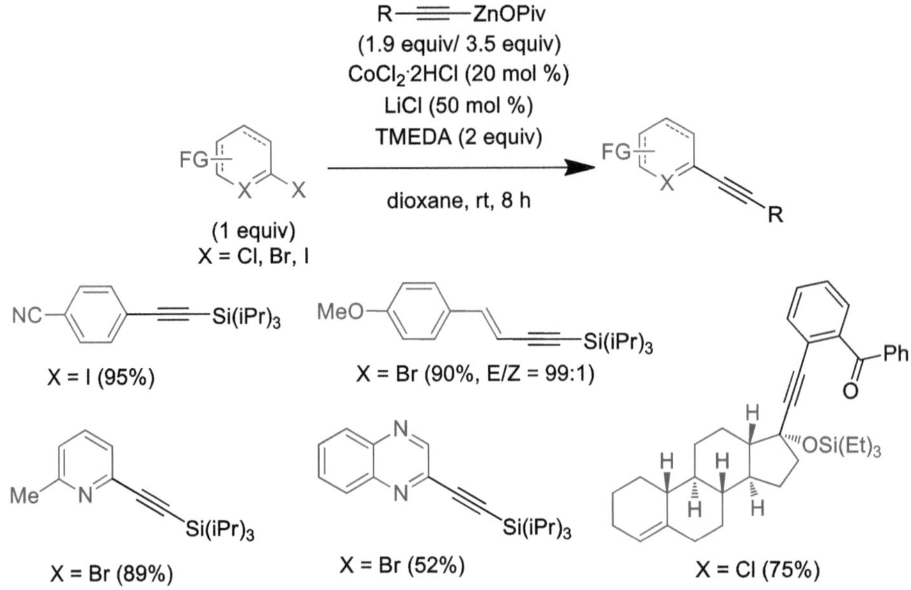

Fig. 4.25: Cobalt-catalyzed cross-couplings between alkynylzinc pivalates with (hetero)aryl and alkenyl halides.

product was isolated in good yield. Moreover, cross-coupling reactions with alkenyl halides occurred with retention of configuration [55].

Besides the cross-coupling reaction of alkynylzinc pivalates with (hetero)aryl and alkenyl halides, the same group demonstrated that alkynylzinc pivalates could also perform well with various 1,2-, 1,3-, and 1,4-substituted cyclic iodides or bromides in high and predictable diastereoselectivity, with $CoCl_2$ (20 mol %) and *trans-N,N,N',N'*-tetramethylcyclohexane-1,2-diamine as a catalytic system [56] (Fig. 4.26). This method was also utilized to prepare alkynyl-substituted glycosides with an excellent α-selectivity.

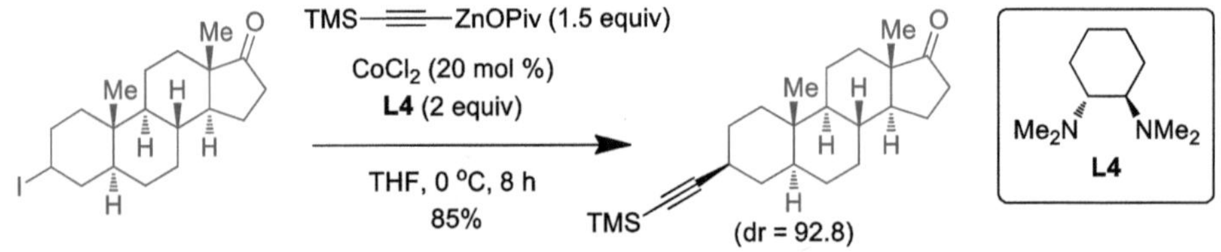

Fig. 4.26: Diastereoselective cross-coupling of steroid derivatives with alkynylzinc pivalate.

Conjugated dienes are important structural motifs in numerous natural products and widely applied in organic synthesis [57]. Recently, Knochel et al. reported a novel alkenylations between alkenylzinc pivalates and various functionalized alkenyl acetates in the presence of $CoBr_2$ (5 mol %) and 2,2′-bipyridyl (5 mol %) at ambient temperature [58] (Fig. 4.27). Either *iso*propenylzinc pivalate or α-arylvinylzinc pivalates with alkenyl acetates provided the substituted dienes in moderate yields.

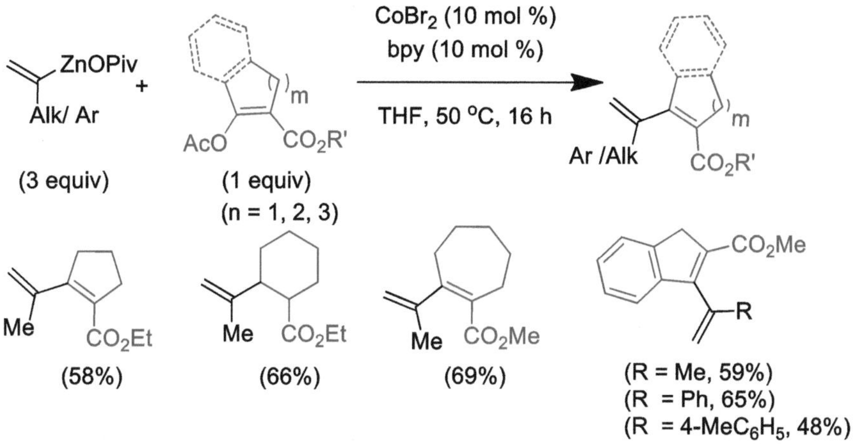

Fig. 4.27: Cobalt-catalyzed cross-coupling of alkenyl zinc pivalates with alkenyl acetates.

4.5 Coupling reactions of allylzinc reagents

Synthesis of cyclic compounds with high enantioselectivities is key in the synthesis of natural and bioactive molecules [59, 60]. For these enantioselective reactions, noble metals are often required [61–63], and investigations using nickel [64], cobalt [65, 66], copper [67] and iron [68] catalysts were achieved. The cobalt-catalyzed asymmetric allylation of heterobicyclic alkenes was described by the Fan's group in 2018 [69] (Fig. 4.28). The method combined a transition metal (cobalt) and a Lewis acid ($ZnCl_2$), and this co-catalytic system was particularly suitable for the asymmetric ring-opening of heterobicyclic alkenes by *in situ* generation of allylic organozinc halides to obtain products of oxabenzonorbornadienes. Various allylic halides were used such as allyl chloride, allyl bromide and allyl iodide. However, even if the allyl iodide is highly active, a lower yield and enantioselectivity were obtained compared to the allyl chloride and bromide.

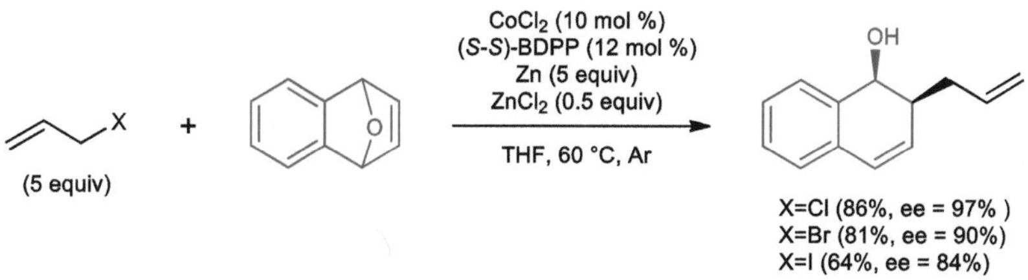

Fig. 4.28: Cobalt-catalyzed asymmetric coupling reactions of heterobicyclic alkenes with *in situ*-generated allylzinc halides.

4.6 Coupling reactions of alkylzinc reagents

4.6.1 With aryl and heteroaryl halides

In 2008, the Knochel group has reported several methods to prepare benzylic zinc halides and demonstrated their reactivity in Negishi-type cross-couplings [70, 71], and in 2009 the Gosmini's group reported a novel strategy to synthesize aryl and heteroaryl methane derivatives with 1,3,5-triazines [21]. 1,3,5-Triazines are known to be useful compounds particularly as pharmaceuticals. For example, they are known as therapeutic agents to treat cancer [72, 73], diabetes [74], dyslipidemia [75], microbial and bacterial infections [76, 77], or inflammatory diseases [78]. 1,3,5-Triazines can also be used in cross-coupling reactions [79]. The Gosmini's process has proven to be efficient to couple 2-chloro-4,6-dimethoxy-1,3,5-triazine with various functionalized benzylic zinc halides *via* a Barbier-type procedure (Fig. 4.29). The reaction tolerated both electron-donating and electron-withdrawing groups on benzylic zinc halides, and moderate to good yields were obtained in the coupled products (54–85% yield).

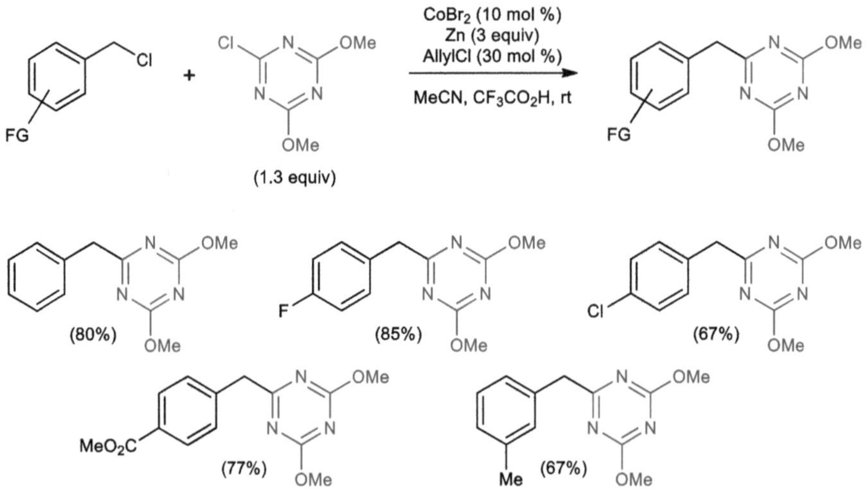

Fig. 4.29: Cobalt-catalyzed benzylation of 2-chloro-4,6-dimethoxy-1,3,5-triazine.

A few years later, Gosmini and Knochel et al. have described a practical strategy to form Csp^3-Csp^2 bond using cobalt-catalyzed cross-coupling reactions using benzylic organozinc reagents and aryl/heteroaryl halides [80] (Fig. 4.30). This reaction was catalyzed by $CoCl_2$ (5 mol %) and isoquinoline (10 mol %) as the ligand, to form valuable polyfunctionalized diaryl- and aryl-heteroaryl methane derivatives. The scope of this reaction showed a good tolerance toward various functional groups on the benzylic zinc chloride (62–82% yield). Furthermore, with substituted heterocycles chlorides such as pyridine or furan, or various aryl halides bearing an ester, nitrile or ketone as

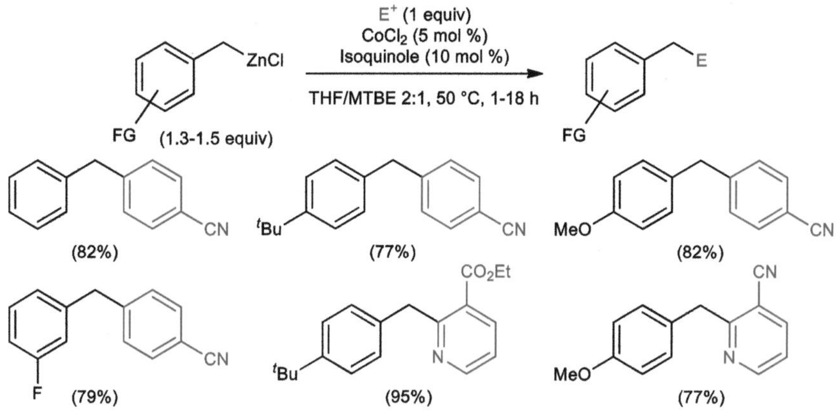

Fig. 4.30: Cobalt-catalyzed cross-coupling of benzylic zinc reagents with aryl and heteroaryl chlorides.

electrophiles, the reaction gave moderate to excellent yields (54–95%). Additionally, it was observed that electron-donating substituents on the aryl halide gave low yields.

More recently, the Knochel group reported the catalyzed Csp^2-Csp^3 cross-coupling reaction of various less reactive functionalized primary and secondary alkylzinc reagents with (hetero)aryl halides [81]. The reaction takes place under mild conditions using $CoCl_2$ (10 mol %) with the bipyridine ligand (20 mol %) (Fig. 4.31). The reaction is tolerated different *N*-heterocyclic halides such as quinoline, isoquinoline, quinazoline, and pyrimidine derivatives, with various substituents. Additionally, linear and cyclic alkylzinc reagents bearing several functionalities such as nitriles, protected amines, acetates, heterocycles or alkynes provided alkylated pyridines in good to high yields (58–87%). Furthermore, the scope of the reaction was extended to electron-deficient aryl halides coupled with alkyzinc reagents leading to moderate to good yields of the corresponding coupling product (66–85%). Finally, the reaction was demonstrated to be efficient by using substituted cyclopropylzinc reagents with heteroaryl halides as coupling partners in a diastereoselective cross-coupling reaction. In this last example, the use of $CoCl_2$ (10 mol %) and 4,4′-di-*tert*-butyl-2,2′-dipyridyl (20 mol %) as the catalytic system in acetonitrile revealed to be the optimal experimental conditions to get the best yield and diastereomeric ratio.

Pyridine moieties are important pharmacophores in bioactive molecules [82, 83]. Especially alkylated pyridines are present in many pharmaceutical compounds due to their chemical properties [84], and they are attractive molecules in organic synthesis [85]. The synthesis of alkylated pyridines was described through C-H functionalizations [86, 87] and by using transition metal-catalyzed cross-coupling reactions between pyridyl(pseudo) halides and alkyl organometallic reagents [2, 3, 88]. The group of McNally recently reported a cobalt-catalyzed cross-coupling reaction between alkylzinc reagents and pyridine phosphonium salts to form the corresponding alkylated pyridines [89] which can be applied to the functionalization of pharmaceutical products (Fig. 4.32). The pyridine

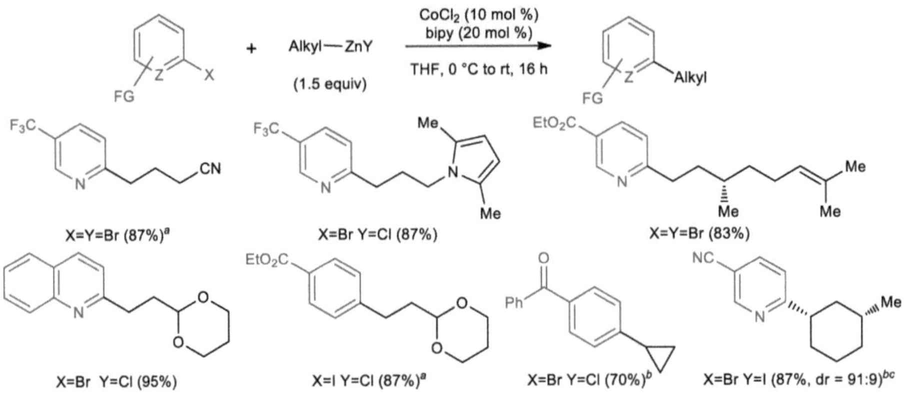

[a]CoCl$_2$ (20 mol %), bipy (40 mol %); [b]Dtbbpy was used instead of bipy; [c]CH$_3$CN was used instead of THF.

Fig. 4.31: Cobalt-catalyzed cross-coupling of functionalized alkylzinc reagents with (hetero)aryl halides.

phosphonium salts reacted with alkylzinc reagents in the presence of Co(acac)$_3$ associated to cyclohexyloxy substituted bipyridine as the ligand and *N*-methylimidazole as the additive. Various linear and branched alkylzinc reagents, bearing a phenyl, chloro, cyano, ester, silyl ether or carbazole group, provided the corresponding coupling product in moderate to good yields (48–78%). Moreover, different azine-containing compounds were employed in this reaction and a good tolerance toward functional groups such as cyano, methoxy, trifluoromethyl groups, esters and boronic esters was observed (37–41%

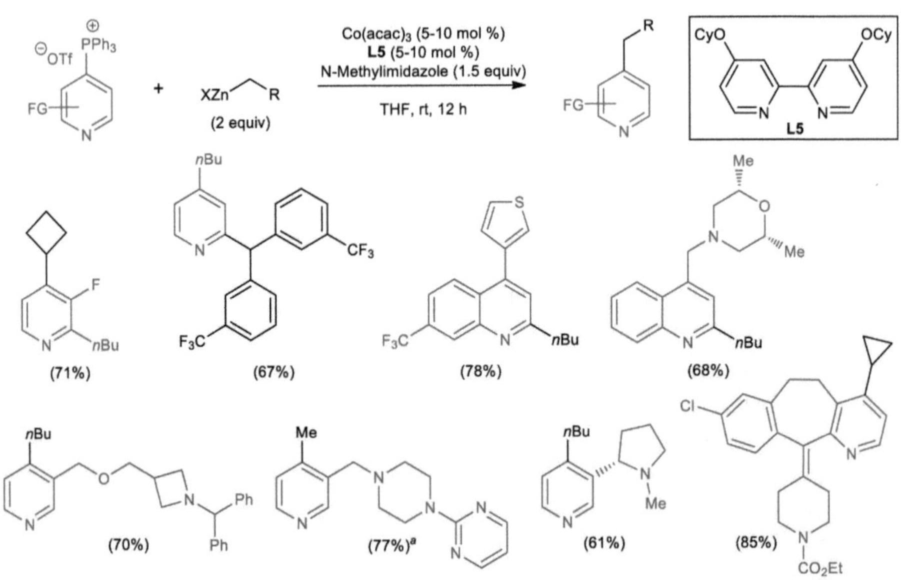

[a]Reaction conducted at 50 °C

Fig. 4.32: Cobalt-catalyzed cross-coupling of heterocyclic phosphonium salts with alkylzinc reagents.

yield). Limitation of pyridine phosphonium salts bearing chlorine, bromine and iodine yielded mixture of alkylated products *via* C –X and C –P bonds as well as bis-alkylation. Furthermore, this reaction was applied to more complex 3-substituted pyridines and the cross-coupling products were isolated with moderate to high yields (37–87%). Finally, the late-stage functionalization of pharmaceutical molecules was examined in the phosphonium-mediated strategy. Chlorphenamine, loratadine, pyriproxyfen, abiraterone acetate and varenicline derivatives were efficiently alkylated (52–85%).

4.6.2 With allyl and vinyl halides

Alkyl organometallic compounds are key intermediates in organic synthesis and can react with electrophiles in the presence of transition metals [90, 91]. Since the discovery of palladium-catalyzed cross-coupling reactions with organozinc halides by Negishi et al., transition metal-catalyzed cross-coupling reactions of organozinc reagents with unsaturated halides have been widely studied [92–94]. The Knochel's group reported in 1996 the cobalt-catalyzed allylation of dialkylzinc compounds [95] (Fig. 4.33). Simple $CoBr_2$ efficiently catalyzed the cross-coupling in THF. The use of various substituted geranyl chlorides led to excellent yields of coupling products with a selectivity

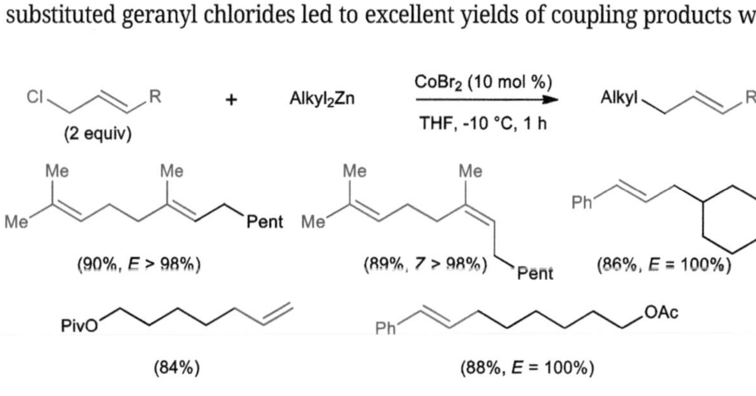

Fig. 4.33: New cobalt-catalyzed reactions of alkylzinc compounds.

of 98% for the S_N2 product. Moreover, a complete retention of the stereochemistry of the double bond was observed.

A few years later, Knochel, Cahiez et al. developed a cobalt-catalyzed cross-coupling of alkylzinc halides with alkenyl halides [96] (Fig. 4.34). Various cobalt salts were evaluated and led to the formation of homocoupling product of the alkenyl halide when $CoBr_2$ was used, while the desired coupling product was efficiently obtained with $Co(acac)_2$. Various functionalized alkylzinc iodides bearing an ester or a nitrile group provided the corresponding alkene product in high yields with retention of the configuration of the double bond.

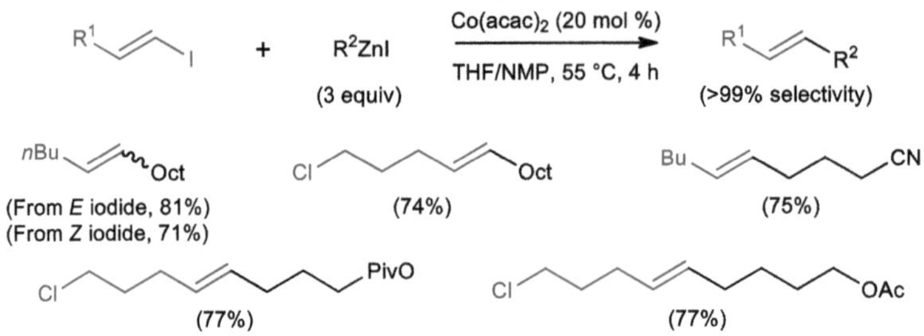

Fig. 4.34: Cobalt-catalyzed alkenylation of alkylzinc compounds.

4.6.3 With alkyl halides

The Fan group has studied the asymmetric reactions of bicyclic alkenes with various benzyl bromides and has demonstrated the efficiency of transition metal and Lewis acid as the co-catalytic system for the asymmetric ring-opening reactions of heterobicyclic alkenes [97, 98]. Within this study, they explored the combination of a chiral cobalt complex with a Lewis acid for the asymmetric reactions of heterobicyclic alkenes with organozinc nucleophiles [69] (Fig. 4.35). The reaction was performed in the presence of $CoCl_2$ (10 mol %) and (2*S*,4*S*)-2,4-bis(diphenylphosphino)pentane (12 mol %) as the catalytic system with $ZnCl_2$ as additive (5 mol %). Various substituents such as halogens and electron-donating groups on the benzylzinc reagents were well tolerated (56–88% yield). The scope of this reaction was extended by using various oxabenzonorbornadienes and led to moderate to high yields and very good enantioselectivity for the corresponding cross-coupling products (61–94%).

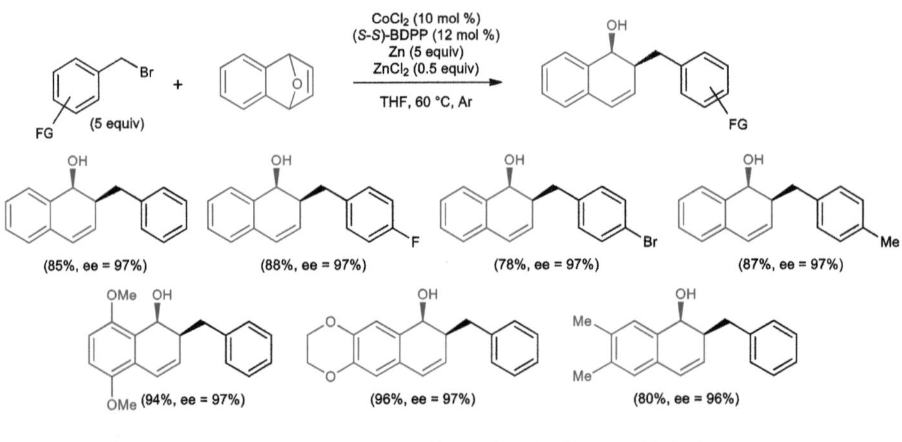

Fig. 4.35: Cobalt-catalyzed asymmetric reactions of heterobicyclic alkenes with *in situ*-generated benzylzinc halides.

A proposed mechanism indicated the formation of a cobalt intermediate which reacts to give the ring-opening product after a β-elimination and a rearrangement (Fig. 4.36).

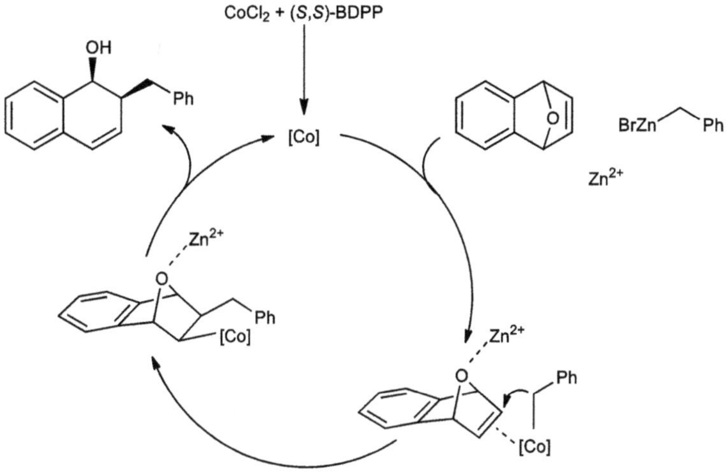

Fig. 4.36: Proposed mechanism for the cobalt-catalyzed asymmetric reactions of heterobicyclic alkenes with *in situ*-generated benzylzinc halides.

Csp^3–Csp^3 cross-coupling reactions are particularly challenging due to the delicate oxidative insertion of the transition-metal catalyst into the alkyl–X bond [99, 100] and the favorable β-hydride elimination of the intermediate alkyl-transition metal species [101, 102]. The use of mild organometallic reagents such as organozinc compounds remains a challenge. Knochel et al. described a mild cobalt-catalyzed Negishi-type Csp^3-Csp^3 cross-coupling between various functionalized dialkylzinc reagents and primary or secondary alkyl iodides [103] (Fig. 4.37). The reaction conditions combined $CoCl_2$ and a chelating nitrogen ligand in acetonitrile. *N,N,N′,N′*-Tetramethylcyclohexane-1,2-diamine (Me_4DACH) was used for primary alkyl iodides and neocuproine for second alkyl iodides. Various functions such as nitrile, aryl or heteroaryl moieties on the primary alkyl iodides are tolerated (61–67%). When cyclic, heterocyclic or acyclic secondary alkyl iodides were employed, functionalities such as alkyl, aryl or heteroaryl esters, TBS-protected iodohydrin and derivative from menthol provided the corresponding coupling products in moderate to high yields (51–91%). Furthermore, functionalized dialkylzinc reagents bearing heterocycles, acetates, carbamates or esters demonstrated an efficient reactivity (65–80%).

The formation of polyfunctionalized quaternary carbons, which are key molecules in organic and medicinal chemistry, still remains a major challenge [104, 105]. Recently, the group of Alcázar reported the formation of quaternary carbons through

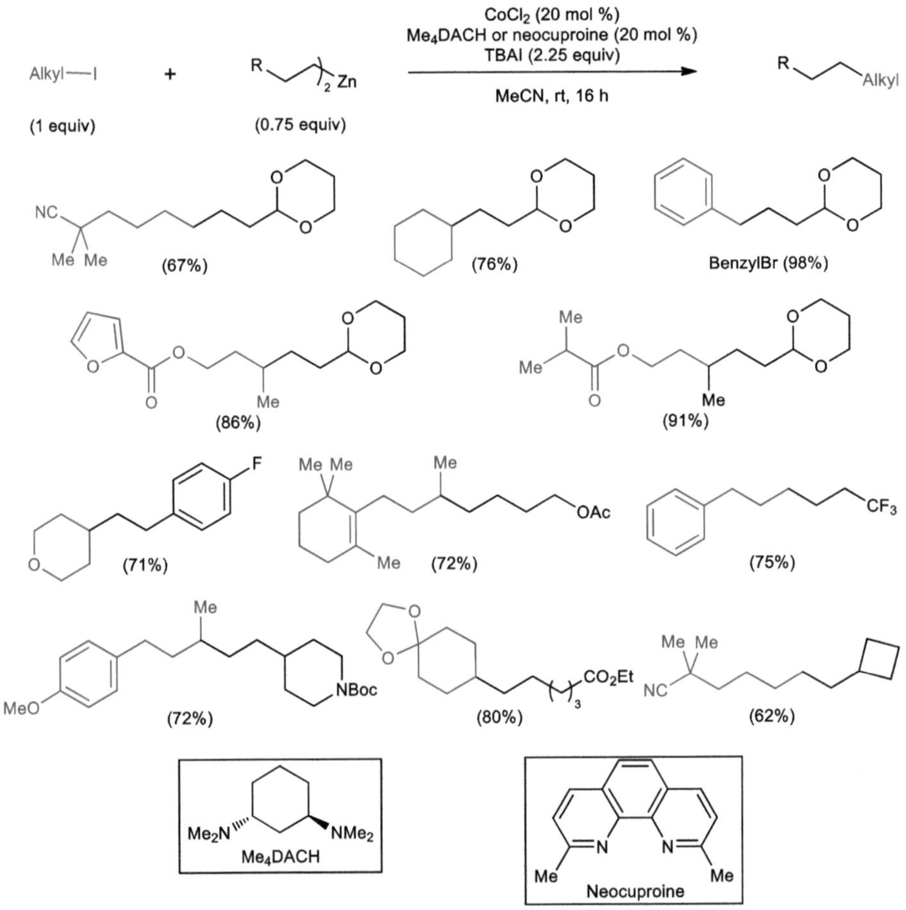

Fig. 4.37: Cobalt-catalyzed Csp^3–Csp^3 cross-coupling of functionalized alkylzinc reagents with alkyl iodides.

cobalt-catalyzed Csp^3-Csp^3 Negishi-type cross-coupling under mild conditions, without additive and using benzylzinc species [106] (Fig. 4.38). The reaction was performed with various cyclic and acyclic α-bromo derivatives bearing various functionalities such as esters, amides, heterocycles carboxylic acids and nitriles, and provided the corresponding coupling products in moderate yields (40–59%). Moreover, halogens, electron-rich and electron-deficient substituents on heterobenzyl- or benzylzinc bromide reagents were tolerated, leading to moderate to good yields in the cross-coupling products (27–65%).

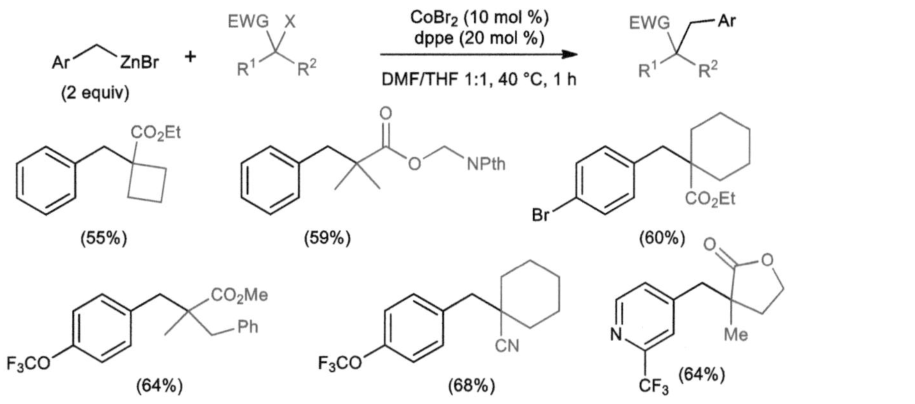

Fig. 4.38: Formation of quaternary carbons through cobalt-catalyzed Csp^3-Csp^3 Negishi-type cross-coupling.

4.6.4 With acyl chlorides

Knochel et al. also reported the acylation of dialkylzinc compounds under cobalt-catalysis in a mixture of NMP/THF (1:1) as the solvent [95] (Fig. 4.39). Various polyfunctionalized ketones were synthesized in excellent yields (78–84%). The reaction tolerated aliphatic and aromatic acid chlorides, as well as oxalyl chloride and trifluoroacetic anhydride.

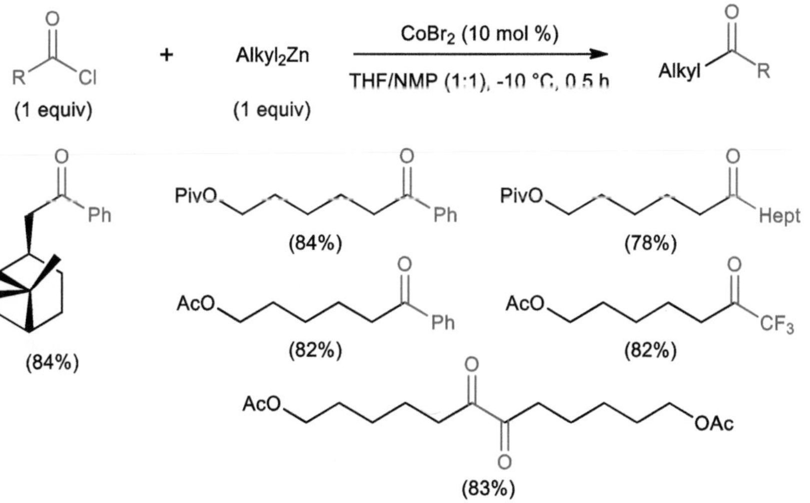

Fig. 4.39: New cobalt-catalyzed reactions of alkylzinc compounds.

4.7 Multicomponent reactions

Multicomponent reactions are important and efficient tools to have a direct access to compounds. Compared to multistep syntheses, these reactions are more atom-economy and step efficient and are part of current environmental issues [107–109].

α-Amino acids, which are key fragments of proteins and important intermediates in metabolism, are widely used in the preparation of peptides and chiral ligand design as well as in the synthesis of drugs [110, 111]. Among the numerous methods employed in the synthesis of α-amino acid derivatives, it is worth mentioning the use of organozinc reagents in one-step multicomponent procedures by Le Gall et al. The authors reported a straightforward cobalt-catalyzed three-component reaction involving organozinc reagents, primary or secondary amines, and ethyl glyoxylate leading to phenylglycine and phenylalanine derivatives [112] (Fig. 4.40). Excellent performances were demonstrated with preformed or *in situ*-generated arylzinc or benzylzinc reagents in a three-component procedure.

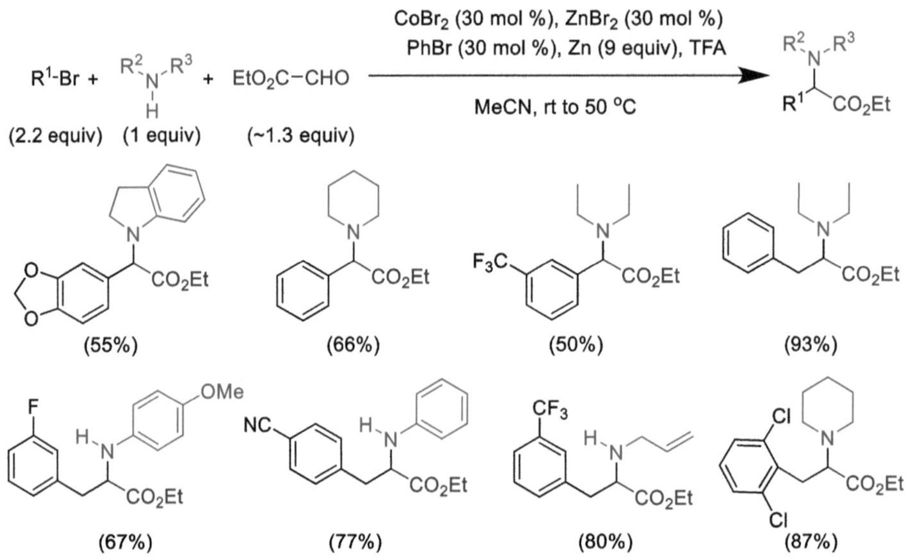

Fig. 4.40: Three-component synthesis of α-amino esters derived from phenylglycine and phenylalanine.

Later on, the same group reported the three-component synthesis of α-branched amines using aldehyde derivatives instead of ethyl glyoxylate [113] (Fig. 4.41). When the reaction was performed with benzyl bromide and allyl halides, no cobalt catalyst was required. However, the use of cobalt bromide was necessary to facilitate the aryl bromide conversion into the corresponding organozinc compound and the reaction was performed at 60 °C.

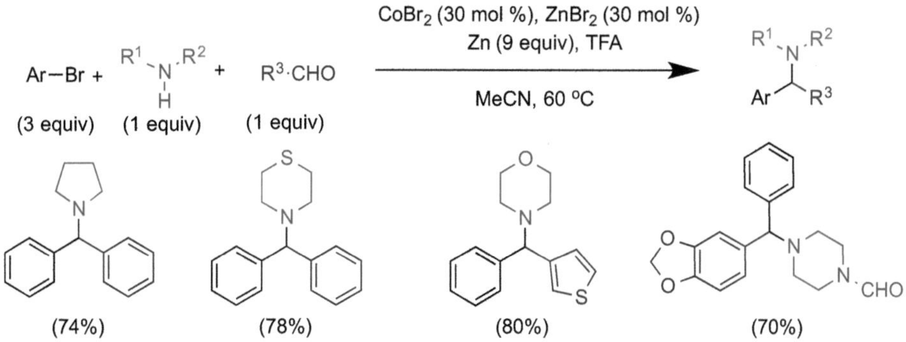

Fig. 4.41: Three-component reaction between an aromatic bromide, an amine and an aldehyde.

The same year, Le Gall et al. extended this three-component reaction to the synthesis of functionalized five-membered ring lactone with *in situ*-generated arylzinc reagents, dimethyl itaconate and aryl aldehydes [114] (Fig. 4.42). This procedure went through the formation of an organozinc reagent, a Michael addition, an aldol coupling and a final cyclization to provide a wide variety of lactones.

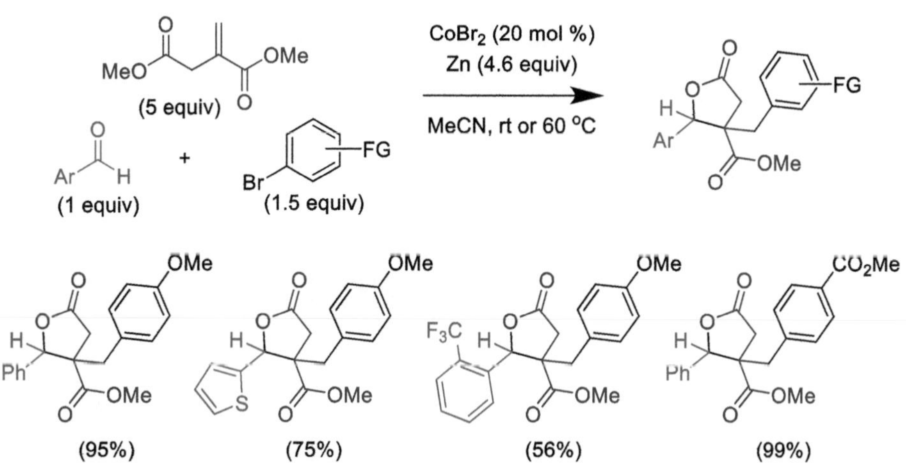

Fig. 4.42: Three-component synthesis of functionalized five-membered ring lactones under Barbier-like conditions.

Later on, the same group extended the reaction to ketones to access five-membered ring lactones. Although yields were generally lower compared to aryl aldehydes, these conditions still provided a convenient way to synthesis novel 2,2,3,3-tetrasubstituted-γ-butyrolactones [115] (Fig. 4.43).

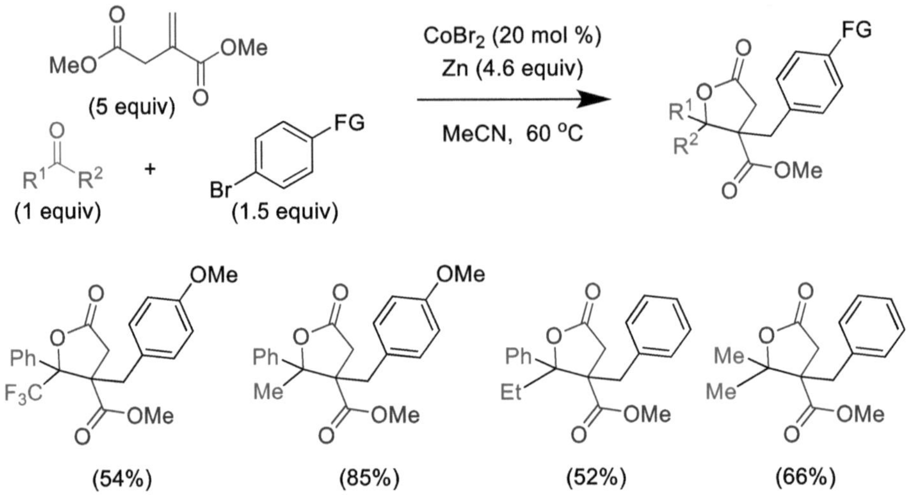

Fig. 4.43: Multicomponent synthesis of γ-butyrolactones.

Benzylamines are of high interest due to their application in functional group transformation and pharmacologically active compounds [116, 117]. Thus, during the development of their three-component reaction between organozinc reagents, amines and aldehyde derivatives [112, 113], Le Gall et al. reported a strategy for preparing a range of functionalized tertiary benzylamines through aromatic halides, amines and paraformaldehyde [118] (Fig. 4.44). Various arylzinc reagents formed *in situ* from different halobenzenes were employed to synthesize functionalized tertiary benzylamines in high yields [118].

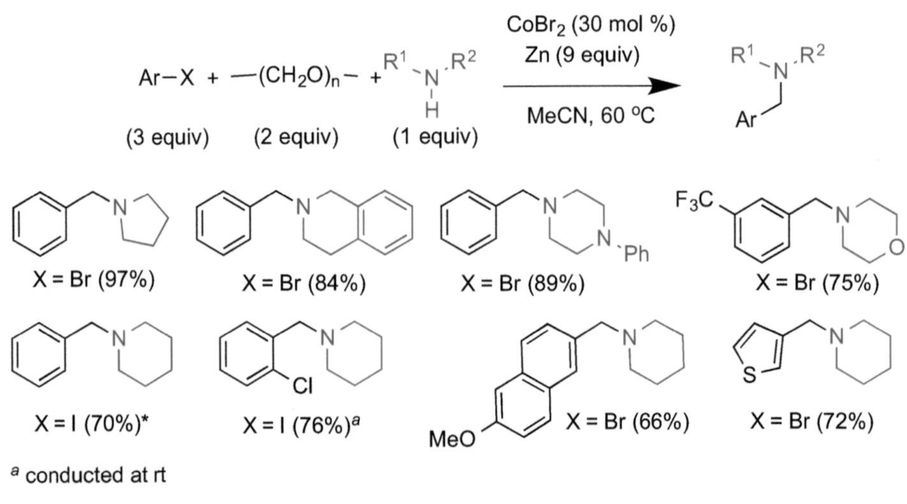

Fig. 4.44: Three-component synthesis of tertiary benzylamines.

In addition, a cobalt-catalyzed multicomponent synthesis of tertiary diarylmethylamines was also reported by Le Gall et al. to produce tertiary amine from aryl bromides, secondary amines and aryl or heteroaryl aldehydes [119] (Fig. 4.45). Contrary to tertiary benzylamines, for this coupling reaction, the pre-formed arylzinc species has to be achieved by using the Gosmini's conditions [12, 13, 120]. Organozinc reagent participated in Mannich-like procedure to form diarylmethylamine by trapping imine cations (or related species) which were formed *in situ* by condensation of an amine with the aldehyde.

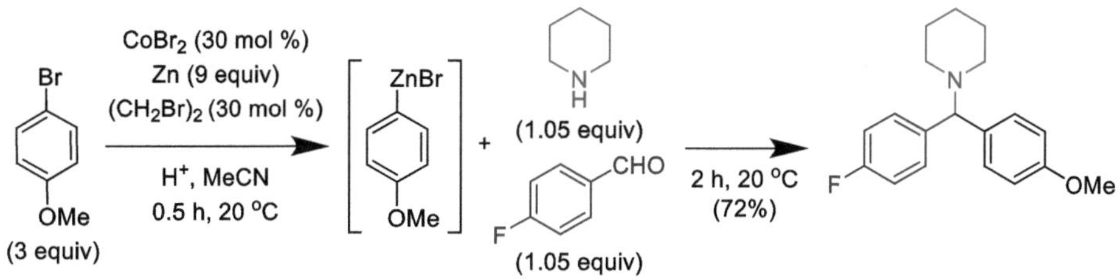

Fig. 4.45: Multicomponent synthesis of tertiary diarylmethylamines: 1-((4-fluorophenyl)(4-methoxyphenyl)methyl)piperidine.

In 2017, Le Gall et al. described a strategy for the straightforward synthesis of β-hydroxycarbonyl compounds by cobalt-catalyzed reductive multicomponent coupling of aldehydes/ketones with aryl halides and Michael acceptors [121] (Fig. 4.46). This strategy could generate various β-hydroxy- and β-aminocarbonyl compounds even from some sensitive imine substrates in good to high yields, through the conjugate addition/aldol or aza-aldol reaction initiated by the *in situ* generation of organozinc species from the organic halide.

The primary mechanism was described for this multicomponent reaction (MCR) (Fig. 4.47). The conjugate addition product **A** produced by ArCo(III) species, generated a zinc enolate **B** and a Co(III) complex though ZnX_2 transmetallation. Enolate **B** was then added to the electrophile to lead to the desired product. On the other hand, the formation of an organozinc intermediate **C** from ArCo(III) species was also described.

More recently, based on the work mentioned above and on the work of Gosmini et al. [122], Le Gall et al. reported the excellent efficiency of cobalt(II)-2,2′-bipyridine and cobalt(II)-1,10-phenanthroline complexes to promote the formal Michael addition/aldol coupling domino reaction between arylzinc species, Michael acceptors and unsaturated electrophiles [123] (Fig. 4.48). Although the role of $CoBr_2$ in furnishing the organozinc compound has been demonstrated, the use of ligand in multicomponent reaction was still necessary. Especially under Barbier's conditions, not only phenanthroline showed slightly higher efficiency than bipyridine as a ligand but high temperature (80 °C) proved to be mandatory. Mechanistic studies showed that the cobalt (bpy) or cobalt(phen) catalyst participated in both the organozinc formation and three-component cross-coupling steps.

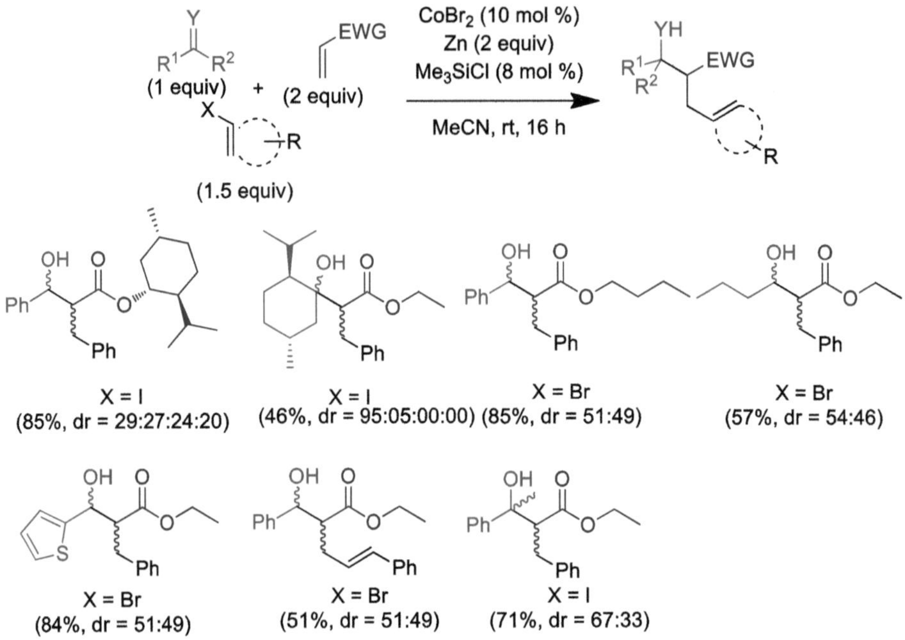

Fig. 4.46: Cobalt-catalyzed reductive multicomponent synthesis of β-hydroxy- and β-aminocarbonyl compounds.

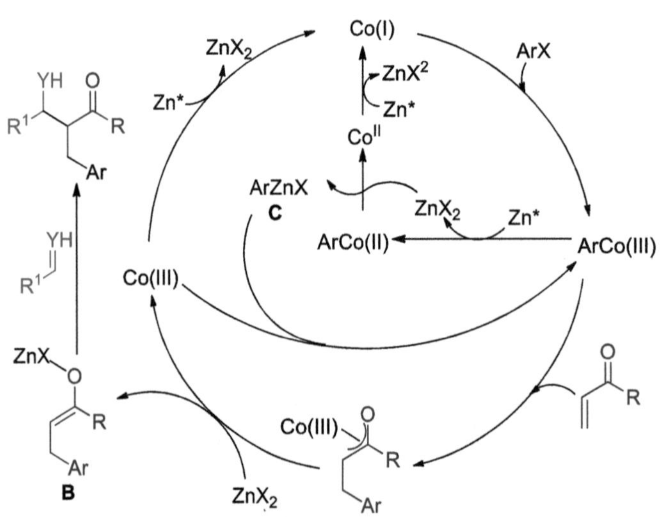

Fig. 4.47: Proposed mechanism for cobalt-catalyzed reductive multicomponent synthesis of β-hydroxy- and β-aminocarbonyl compounds.

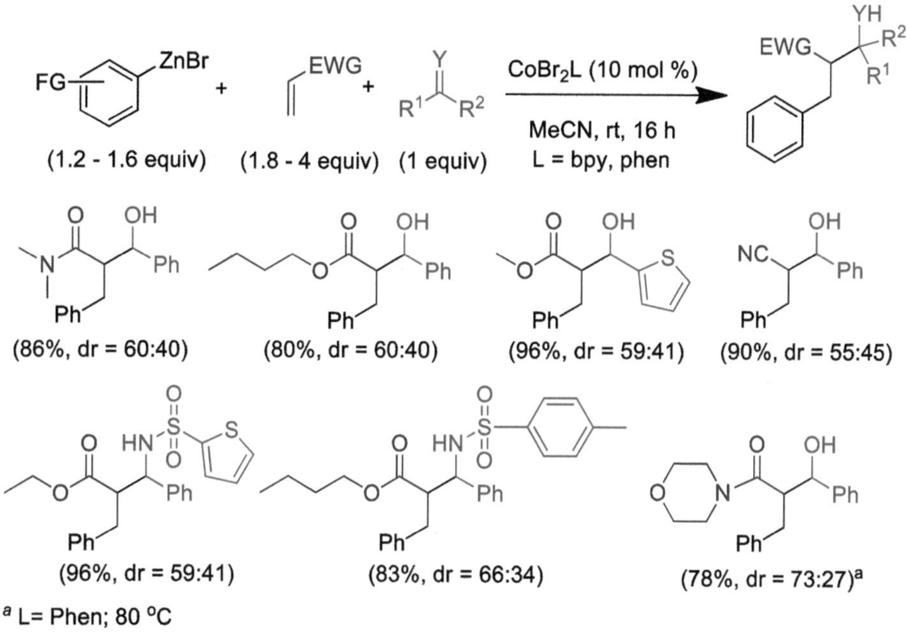

Fig. 4.48: Cobalt-catalyzed three-component coupling of mixed aromatic organozinc species, carbonyl compounds or imines and Michael acceptors.

Multicomponent reactions allow a straightforward access to complex products by mixing together several reagents to generate a multipart molecule in a one-pot process. However, four or more reagents involved in multicomponent reactions were rarely reported due to complex side reactions. In 2013, Le Gall et al. reported the synthesis of α,β-disubstituted β-amino esters by a zinc-mediated, cobalt-catalyzed four-component reaction between organic bromides, alkyl acrylates, amines, and aldehydes in moderate to good yields [124] (Fig. 4.49). This multicomponent method utilized readily available starting compounds to form a range of β-amino esters. As proposed in the mechanism, the role of zinc is important and an organocobalt species initiated the reaction.

A proposed mechanism was described (Fig. 4.50). As shown, the organometallic species **D** formed by a classical route would be involved in two concurrent pathways. The first pathway would be a zinc salt transmetalation to provide an organozinc species **E**. The second pathway would be a conjugate addition on the acrylate to form enolate **F**, and then enolate **F** is reacting with iminium intermediate **G** to produce the corresponding four-component coupling product.

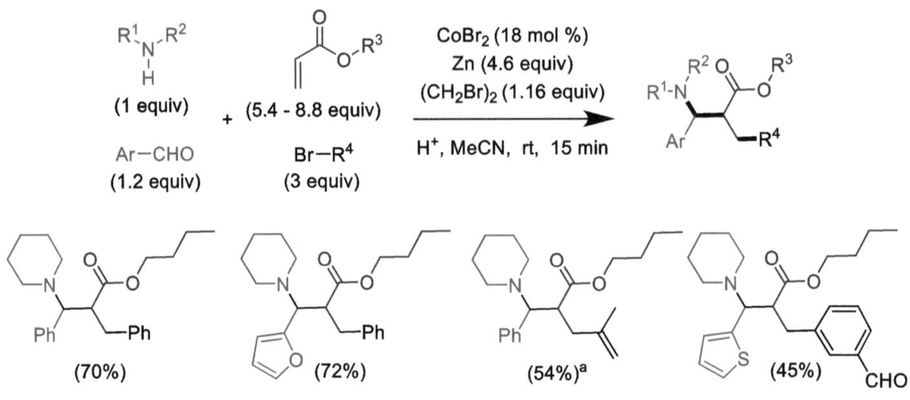

a the amount of Co was lowered by half

Fig. 4.49: Four-component reaction involving *in situ*-generated organometallic reagents.

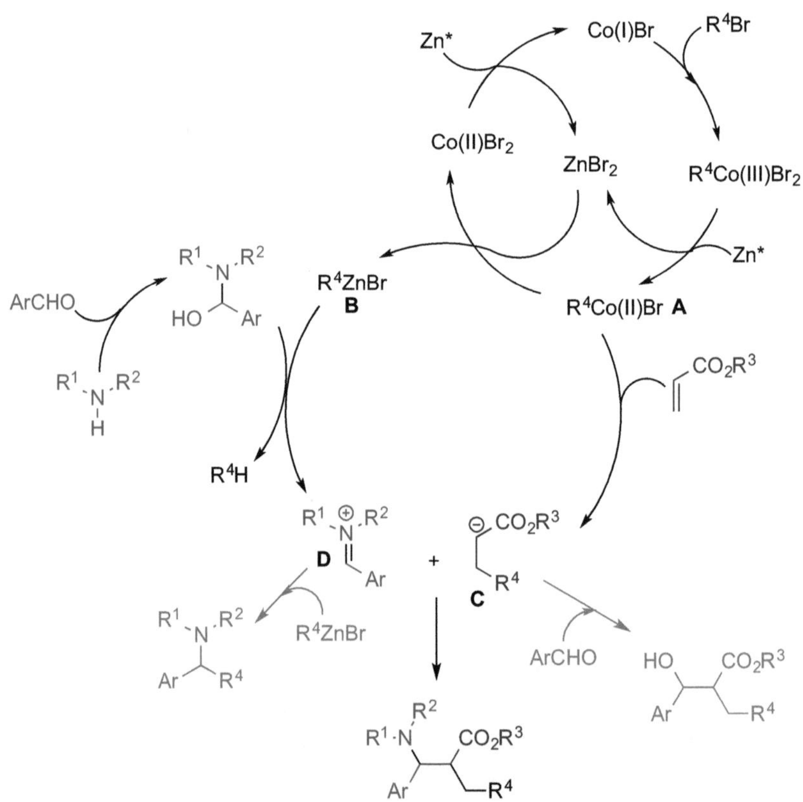

Fig. 4.50: Proposed mechanism for four-component reaction involving *in situ*-generated organometallic reagents.

4.8 Conclusion

The use of cobalt salts is an interesting alternative to other non-economical and toxic transition metals. Some functional groups could be present both on the organozinc reagent and on the coupling partner. Functionalized organozinc reagents are now easy to synthesize and less challenging than the synthesis of functionalized Grignard reagents. Moreover, sometimes the cobalt catalyst used for the generation of the organozinc species can also be used for the cross-coupling reaction without supplementary catalyst. Thereby, since the preliminary Kharash et al. results, and after the Co-catalyzed cross-coupling of Grignard reagents, a wide range of other new cobalt-catalyzed cross-coupling reactions have been studied to increase the scope of the functional groups. It is worth mentioning that at the beginning of the century, by combining alkyl, aryl, alkenyl and alkynyl zinc reagents with different halides, new simple processes were developed. Although at the present time Grignard reagents are the most commonly used organometallic species for performing cobalt-catalyzed cross-coupling reactions, generally satisfying yields of more functionalized compounds are obtained with the system Co/Zn. This promising system is very interesting for future developments. As reported for the Co-catalyzed cross-coupling reaction with aryl Grignard reagent [125], simple ligands are also generally used with cobalt salts compared to bulky, expensive and less commercially available analogues often associated with palladium and nickel catalysts. Most of the reactions described in this chapter proceed in a very simple manner and can be extended to multicomponent reactions. Even if the scope of cobalt-catalyzed cross-coupling reactions involving organozinc species has been significantly expanded, some efforts have to be particularly made for enantioselective transformations.

References

[1] de Meijere A, Diederich F, Metal-catalyzed Cross-coupling Reactions. WILEY-VCH: Weinheim, 2004.
[2] Haas D, Hammann JM, Greiner R, Knochel P, ACS Catal. 2016, 6, 1540–1552.
[3] Jana R, Pathak TP, Sigman MS, Chem. Rev. 2011, 111, 1417–1492.
[4] Gosmini C, Bégouin J-M, Moncomble A, Chem. Comm. 2008, 3221–3233.
[5] Cahiez G, Moyeux A, Chem. Rev. 2010, 110, 1435–1462.
[6] Gosmini C, Moncomble A, Isr. J. Chem. 2010, 50, 568–576.
[7] Knappke CEI, Grupe S, Gärtner D, Corpet M, Gosmini C, Jacobi von Wangelin A, Isr. Chem. Eur. J. 2014, 20, 6828–6842.
[8] Shu T, Cossy J. Chem. Eur. J. 2021, 27, 11021–11029.
[9] Kharasch M, Fuchs CF, J. Am. Chem. Soc. 1943, 65, 504–507.
[10] Gosmini C, Rollin Y, Nédélec JY, Périchon J, J. Org. Chem. 2000, 65, 6024–6026.
[11] Fillon H, Le Gall E, Gosmini C, Périchon J, Tetrahedron Lett. 2002, 43, 5941–5944.
[12] Fillon H, Gosmini C, Périchon J, J. Am. Chem. Soc. 2003, 125, 3867–3870.
[13] Gosmini C, Amatore M, Claudel S, Perichon J, Synlett 2005, 2005, 2171–2174.
[14] Kazmierski I, Gosmini C, Paris J-M, Perichon J, Synlett 2006, 2006, 881–884.

[15] Begouin J-M, Rivard M, Gosmini C, Chem. Comm. 2010, 46, 5972–5974.
[16] Jin M-Y, Yoshikai N, J. Org. Chem. 2011, 76, 1972–1978.
[17] Linke S, Manolikakès SM, Auffrant A, Gosmini C, Synthesis 2018, 50, 2595–2600.
[18] Gosmini C, Périchon J, Org. Biomol. Chem. 2005, 3, 216–217.
[19] Claudel S, Gosmini C, Paris JM, Perichon J. Chem. Comm. 2007, 3667–3669.
[20] Begouin J-M, Gosmini C, J. Org. Chem. 2009, 74, 3221–3224.
[21] Begouin JM, Claudel S, Gosmini C, Synlett 2009, 3192–3194.
[22] Marghad I, Kim DH, Tian X, Mathevet F, Gosmini C, Ribierre J-C, Adachi C, ACS Omega 2018, 3, 2254–2260.
[23] Bourne-Branchu Y, Moncomble A, Corpet M, Danoun G, Gosmini C, Synthesis 2016, 48, 3352–3356.
[24] Haas D, Hammann JM, Lutter FH, Knochel P, Angew. Chem. Int. Ed. 2016, 55, 3809–3812.
[25] Greiner R, Ziegler DS, Cibu D, Jakowetz AC, Auras F, Bein T, Knochel P, Org. Lett. 2017, 19, 6384–6387.
[26] Hammann JM, Lutter FH, Haas D, Knochel P, Angew. Chem. Int. Ed. 2017, 56, 1082–1086.
[27] Lutter FH, Graßl S, Grokenberger L, Hofmayer MS, Chen YH, Knochel P, Chem. Cat. Chem 2019, 11, 5188–5197.
[28] Gomes P, Gosmini C, Périchon J, Org. Lett. 2003, 5, 1043–1045.
[29] Dunet G, Knochel P, Synlett 2007, 1383–1386.
[30] Amatore M, Gosmini C, Chem. Comm. 2008, 5019–5021.
[31] Araki K, Inoue M, Tetrahedron 2013, 69, 3913–3918.
[32] Hammann JM, Haas D, Knochel P, Angew. Chem. Int. Ed. 2015, 54, 4478–4481.
[33] Hofmayer MS, Sunagatullina A, Brösamlen D, Mauker P, Knochel P, Org. Lett. 2020, 22, 1286–1289.
[34] Liu F, Zhong J, Zhou Y, Gao Z, Walsh PJ, Wang X, Ma S, Hou S, Liu S, Wang M, Chem. Eur. J. 2018, 24, 2059–2064.
[35] Zhou Y, Liu C, Wang L, Han L, Hou S, Bian Q, Zhong J, Synlett 2019, 30, 860–862.
[36] Sun X, Wang X, Liu F, Gao Z, Bian Q, Wang M, Zhong J, Chirality 2019, 31, 682–687.
[37] Patai S, The Chemistry of the Carbon-carbon Triple Bond. Wiley: New York, 1978.
[38] Shirakawa H, Angew. Chem. Int. Ed. 2001, 40, 2574–2580.
[39] Liu J, Lam JWY, Tang BZ. Chem. Rev. 2009, 109, 5799–5867.
[40] Corpet M, Bai XZ, Gosmini C, Adv. Synth. Catal. 2014, 356, 2937–2942.
[41] Knochel P, Singer RD, Chem. Rev. 1993, 93, 2117–2188.
[42] Negishi E-i, Bagheri V, Chatterjee S, Luo F-T, Miller JA, Stoll AT, Tetrahedron Lett. 1983, 24, 5181–5184.
[43] Fillon H, Gosmini C, Périchon J, Tetrahedron 2003, 59, 8199–8202.
[44] Kazmierski I, Bastienne M, Gosmini C, Paris J-M, Périchon J, J. Org. Chem. 2004, 69, 936–942.
[45] Rérat A, Michon C, Agbossou-Niedercorn F, Gosmini C, Eur. J. Org. Chem. 2016, 2016, 4554–4560.
[46] Dorval C, Dubois E, Bourne-Branchu Y, Gosmini C, Danoun G, Adv. Synth. Catal. 2019, 361, 1777–1780.
[47] Dorval C, Stetsiuk O, Gaillard S, Dubois E, Gosmini C, Danoun G, Org. Lett. 2022, 24, 2778–2782.
[48] Lutter FH, Grokenberger L, Hofmayer MS, Knochel P, Chem. Sci. 2019, 10, 8241–8245.
[49] Fatiadi AJ, Preparation and Synthetic Applications of Cyano Compounds. Wiley: New-York, 1983.
[50] Larock RC, Comprehensive Organic Transformations. VCH: New York, 1989.
[51] Miller JS, Manson JL, Acc. Chem. Res. 2001, 34, 563–570.
[52] Fleming FF, Wang Q, Chem. Rev. 2003, 103, 2035–2078.
[53] Cai Y, Qian X, Rerat A, Auffrant A, Gosmini C, Adv. Synth. Catal. 2015, 357, 3419–3423.
[54] Liu X-G, Zhou C-J, Lin E, Han X-L, Zhang S-S, Li Q, Wang H, Angew. Chem. Int. Ed. 2018, 57, 13096–13,100.
[55] Hammann JM, Thomas L, Chen Y-H, Haas D, Knochel P, Org. Lett. 2017, 19, 3847–3850.
[56] Thomas L, Lutter FH, Hofmayer MS, Karaghiosoff K, Knochel P, Org. Lett. 2018, 20, 2441–2444.

[57] Paolis MD, Chataigner I, Maddaluno J, Recent Advances in Stereoselective Synthesis of 1, 3-dienes. Springer, 2012.
[58] Li J, Knochel P, Angew. Chem. Int. Ed. 2018, 57, 11436–11440.
[59] McManus HA, Fleming MJ, Lautens M, Angew. Chem. 2007, 119, 437–440.
[60] Webster R, Boyer A, Fleming MJ, Lautens M, Org. Lett. 2010, 12, 5418–5421.
[61] Lautens M, Fagnou K, Yang D, J. Am. Chem. Soc. 2003, 125, 14884–14892.
[62] Lautens M, Dockendorff C, Org. Lett. 2003, 5, 3695–3698.
[63] Long Y, Yang D, Zhang Z, Wu Y, Zeng H, Chen Y, J. Org. Chem. 2010, 75, 7291–7299.
[64] Feng C-C, Nandi M, Sambaiah T, Cheng C-H, J. Org. Chem. 1999, 64, 3538–3543.
[65] Sawano T, Ou K, Nishimura T, Hayashi T, Chem. Comm. 2012, 48, 6106–6108.
[66] Huang Y, Ma C, Lee YX, Huang R-Z, Zhao Y, Angew. Chem. Int. Ed. 2015, 54, 13696–13700.
[67] Arrayás RG, Cabrera S, Carretero JC, Org. Lett. 2005, 7, 219–221.
[68] Ito S, Itoh T, Nakamura M, Angew. Chem. 2011, 123, 474–477.
[69] Li Y, Chen J, He Z, Qin H, Zhou Y, Khan R, Fan B, Org. Chem. Front. 2018, 5, 1108–1112.
[70] Metzger A, Piller FM, Knochel P, Chem. Comm. 2008, 5824–5826.
[71] Metzger A, Schade MA, Knochel P, Org. Lett. 2008, 10, 1107–1110.
[72] Dhainaut A, Regnier G, Atassi G, Pierre A, Leonce S, Kraus-Berthier L, Prost JF, J. Med. Chem. 1992, 35, 2481–2496.
[73] Jarman M, Coley HM, Judson IR, Thornton TJ, Wilman DE, Abel G, Rutty CJ, J. Med. Chem. 1993, 36, 4195–4200.
[74] Mylari BL, Withbroe GJ, Beebe DA, Brackett NS, Conn EL, Coutcher JB, Oates PJ, Zembrowski WJ, Bioorg. Med. Chem. 2003, 11, 4179–4188.
[75] D'Atri G, Gomarasca P, Resnati G, Tronconi G, Scolastico C, Sirtori CR, J. Med. Chem. 1984, 27, 1621–1629.
[76] Zhou C, Min J, Liu Z, Young A, Deshazer H, Gao T, Chang Y-T, Kallenbach NR, Bioorg. Med. Chem. Lett. 2008, 18, 1308–1311.
[77] Maeda S, Kita T, Meguro K, J. Med. Chem. 2009, 52, 597–600.
[78] Porter JR, Archibald SC, Brown JA, Childs K, Critchley D, Head JC, Hutchinson B, Parton TA, Robinson MK, Shock A, Bioorg. Med. Chem. Lett. 2002, 12, 1591–1594.
[79] Yu S-B, Hu X-P, Deng J, Huang J-D, Wang D-Y, Duan Z-C, Zheng Z, Tetrahedron Lett. 2008, 49, 1253–1256.
[80] Benischke AD, Knoll I, Rérat A, Gosmini C, Knochel P, Chem. Comm. 2016, 52, 3171–3174.
[81] Lutter FH, Grokenberger L, Spieß P, Hammann JM, Karaghiosoff K, Knochel P, Angew. Chem. Int. Ed. 2020, 59, 5546–5550.
[82] Baumann M, Baxendale IR, Beilstein J. Org. Chem. 2013, 9, 2265–2319.
[83] Vitaku E, Smith DT, Njardarson JT, J. Med. Chem. 2014, 57, 10257–10274.
[84] Joule JA, Mills K, Smith GF, Heterocyclic Chemistry. CRC Press: London, 1995.
[85] Wurz RP, Chem. Rev. 2007, 107, 5570–5595.
[86] Murakami K, Yamada S, Kaneda T, Itami K, Chem. Rev. 2017, 117, 9302–9332.
[87] Duncton MA, Med. Chem. Comm 2011, 2, 1135–1161.
[88] Yang Y, Niedermann K, Han C, Buchwald SL, Org. Lett. 2014, 16, 4638–4641.
[89] Zhang X, McNally A, ACS Catal. 2019, 9, 4862–4866.
[90] Hegedus LS, Macomber R, Transition Metals in the Synthesis of Complex Organic Molecules. University Science Books: Sausalito, California, 1994.
[91] Kotha S, Meshram M, J. Organomet. Chem. 2018, 874, 13–25.
[92] Negishi E, King AO, Okukado N, J. Org. Chem. 1977, 42, 1821–1823.
[93] King AO, Okukado N, Negishi E-I, J. Chem. Soc. Chem. Commun. 1977, 683–684.
[94] Negishi E, Acc. Chem. Res. 1982, 15, 340–348.
[95] Reddy CK, Knochel P, Angew. Chem. Int. Ed. Engl. 1996, 35, 1700–1701.

[96] Avedissian H, Bérillon L, Cahiez G, Knochel P, Tetrahedron Lett. 1998, 39, 6163–6166.
[97] Zhang W, Chen J, Zeng G, Yang F, Xu J, Sun W, Shinde MV, Fan B, J. Org. Chem. 2017, 82, 2641–2647.
[98] Zeng C, Yang F, Chen J, Wang J, Fan B, Org. Biomol. Chem. 2015, 13, 8425–8428.
[99] Cárdenas DJ, Angew. Chem. Int. Ed. 2003, 42, 384–387.
[100] Frisch AC, Beller M, Angew. Chem. Int. Ed. 2005, 44, 674–688.
[101] Rudolph A, Lautens M, Angew. Chem. Int. Ed. 2009, 48, 2656–2670.
[102] Choi J, Fu GC, Science 2017, 356, eaaf7230.
[103] Lutter FH, Grokenberger L, Benz M, Knochel P, Org. Lett. 2020, 22, 3028–3032.
[104] Christoffers J, Quaternary Stereocenters. Wiley-VCH: 2005.
[105] Quasdorf KW, Overman LE, Nature 2014, 516, 181–191.
[106] Palao E, López E, Torres-Moya I, de la Hoz A, Díaz-Ortiz Á, Alcázar J. Chem. Comm. 2020, 56, 8210–8213.
[107] Sunderhaus JD, Martin SF, Chem. Eur. J. 2009, 15, 1300–1308.
[108] Biggs-Houck JE, Younai A, Shaw JT, Curr. Opin. Chem. Biol. 2010, 14, 371–382.
[109] Dömling A, Chem. Rev. 2006, 106, 17–89.
[110] Kazmaier U, Angew. Chem. Int. Ed. 2005, 44, 2186–2188.
[111] Kaiser J, Kinderman SS, van Esseveldt BC, van Delft FL, Schoemaker HE, Blaauw RH, Rutjes FP, Org. Biomol. Chem. 2005, 3, 3435–3467.
[112] Haurena C, Sengmany S, Huguen P, Le Gall E, Martens T, Troupel M, Tetrahedron Lett. 2008, 49, 7121–7123.
[113] Le Gall E, Haurena C, Sengmany S, Martens T, Troupel M, J. Org. Chem. 2009, 74, 7970–7973.
[114] Le Floch C, Bughin C, Le Gall E, Léonel E, Martens T, Tetrahedron Lett. 2009, 50, 5456–5458.
[115] Le Floch C, Le Gall E, Léonel E, Pure Appl. Chem. 2011, 83, 621–631.
[116] Cecchetto A, Minisci F, Recupero F, Fontana F, Pedulli GF, Tetrahedron Lett. 2002, 43, 3605–3607.
[117] Miyazaki Y, Kato Y, Manabe T, Shimada H, Mizuno M, Egusa T, Ohkouchi M, Shiromizu I, Matsusue T, Yamamoto I, Bioorg. Med. Chem. Lett. 2006, 16, 2986–2990.
[118] Le Gall E, Decompte A, Martens T, Troupel M, Synthesis 2010, 249–254.
[119] Martens T, Le Gall E, Org. Synth. 2012, 89, 283–293.
[120] Kazmierski I, Gosmini C, Paris J-M, Périchon J, Tetrahedron Lett. 2003, 44, 6417–6420.
[121] Paul J, Presset M, Cantagrel F, Le Gall E, Léonel E, Chem. Eur. J. 2017, 23, 402–406.
[122] Amatore M, Gosmini C, Périchon J, J. Org. Chem. 2006, 71, 6130–6134.
[123] Paul J, Presset M, Le Gall E, Leonel E, Retailleau P, Synthesis 2018, 50, 254–266.
[124] Le Gall E, Léonel E, Chem. Eur. J. 2013, 19, 5238–5241.
[125] Rérat A, Gosmini C, Grignard reagents and cobalt. In Phys. Sci. Rev. 2018, 3, 20160021.

Armelle Ouali and Marc Taillefer

5 Organozinc reagents and copper

5.1 Introduction

Among organometallic reagents, organozinc reagents display several assets, including easy preparation from organo halides and zinc as well as tolerance toward a wide variety of functional groups [1]. Therefore, The Negishi-type cross-coupling constitutes one of the most versatile transformation for creating C-C bonds and thus reaching pharmaceuticals, natural products or building blocks for materials. Palladium and nickel are recognized as the metals of choice for catalyzing these couplings [2]. Over the last decades however, copper, an abundant and cheap metal, emerged as a valuable alternative, and catalysts based on this metal have been reported to promote the cross-coupling reactions of various zinc reagents with a wide range of electrophiles.

In this chapter, we will focus on Cu-catalyzed cross-coupling reactions of organozinc reagents with saturated and unsaturated electrophiles. The sections are organized according to the type of zinc derivative involved (Fig. 5.1). Therefore, Sections 5.2, 5.3 and 5.4 are dedicated to aryl, alkenyl and alkynyl zinc reagents respectively, Section 5.5 concerns alkyl zinc derivatives, and Section 5.6 focuses on perfluoroalkyl zinc reagents. In every cases, the starting nucleophilic and electrophilic partners are highlighted in black (R') and in red (R) respectively.

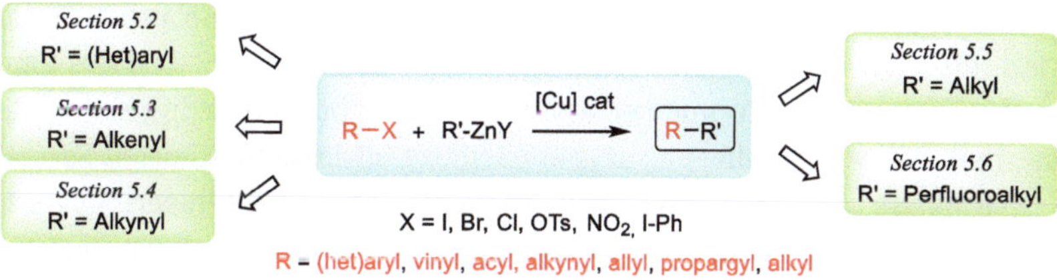

Fig. 5.1: Cu-catalyzed cross-coupling reactions involving organometallic zinc reagents with organic halides and pseudohalides.

It is worth noting that zinc reagents can also react with other unsaturated electrophiles including enones, ynones and nitroolefins, in the presence of copper catalysts [3–8]. These addition reactions do not, strictly speaking, belong to the cross-coupling reactions considered in the scope of this chapter, and they will not be detailed.

Armelle Ouali, ICGM, Univ Montpellier, CNRS, ENSCM, 34296 Montpellier, France, e-mail: armelle.ouali@enscm.fr
Marc Taillefer, ICGM, Univ Montpellier, CNRS, ENSCM, 34296 Montpellier, France, e-mail: marc.taillefer@enscm.fr

https://doi.org/10.1515/9783110728859-005

5.2 Coupling reactions of (hetero)arylzinc reagents

Despite the lower nucleophilicity of the sp^2-hybridized carbon atom as compared to the sp^3-hybridized one, a wide range of cross-coupling reactions involving aryl zinc reagents have been described [9]. Examples have been reported allowing the cross-coupling reaction of aryl zinc reagents with various electrophiles including aryl and heteroaryl halides (see Section 5.2.1), acyl chlorides and carbon dioxide (see Sections 5.2.2 and 5.2.3 respectively). Allyl bromides and trifluoromethyl group using Togni's reagent, have also been used as electrophilic partners (see Sections 5.2.4 and 5.2.5 respectively).

5.2.1 With (hetero)aryl halides

In 2015, Giri et al. reported a general method for the copper-catalyzed Negishi cross-coupling between a wide range of aryl zinc reagents and heteroaryl iodide (Fig. 5.2, eq. 1) [1, 10]. The heteroaryl electrophiles indeed constitute challenging substrates since the heteroatom(s), included in their structure, can act as binding site(s) for the metals catalysts and therefore deactivate or even poison the catalysts. The main contribution of these work concerns the discovery of a ligand-free copper catalyst based on copper iodide. One equivalent of lithium chloride was used to generate more reactive organozinc species and avoided the inhibition of the reaction by ZnI_2, a by-product of the reaction. In the case of aryl zinc reagents, the cross-coupling reactions occurred in DMF within a short time (3 h) as in the case of alkyl zinc partners, although harsher temperature conditions were required (100 °C instead of room temperature). The biaryl compounds were isolated in fair to very high yields (41–87%). Concerning the functional group tolerance, the method was compatible with electron-neutral or electron-rich aryl zinc reagents bearing methyl, amino (dimethylamine, morpholine, piperidine) or oxygen (methoxy, silyloxy) substituents. Several heteroaryl electrophiles can be used, eventually containing chloro substituents. Interestingly, the cross-coupling reactions gave rise to bidentate pyridylamine ligands which nicely illustrate the synthetic potential of the method.

Later on, the same catalytic system was shown to promote the coupling of aryl zinc reagents with aryl iodides (Fig. 5.2, eq. 2) [11]. In this related work, a diaryl zinc derivative was used as the nucleophile and the cross-coupled biaryl compound was isolated in good to very good yield (51–84%) within 12 h, in DMF, at 100 °C. The method was compatible with a wide range of electrophilic and nucleophilic partners bearing halides (Br, Cl) as well as electron-neutral (Me), -withdrawing (CF_3, CO_2Me, CN) or -donating (OMe) substituents. From a mechanistic point of view (Fig. 5.3), the formation of diaryl zincates in solution resulting from the binding of lithium chloride to the diaryl zinc reagent is proposed as the first step. The diaryl zincate would next undergo transmetallation with Cu(I) iodide to give rise to an arylcopper(I) (ArCu) intermediate. This intermediate would react in the last step with the aryl iodide to produce the expected biaryl compound and the catalytic active species (CuI) is regenerated [11, 12].

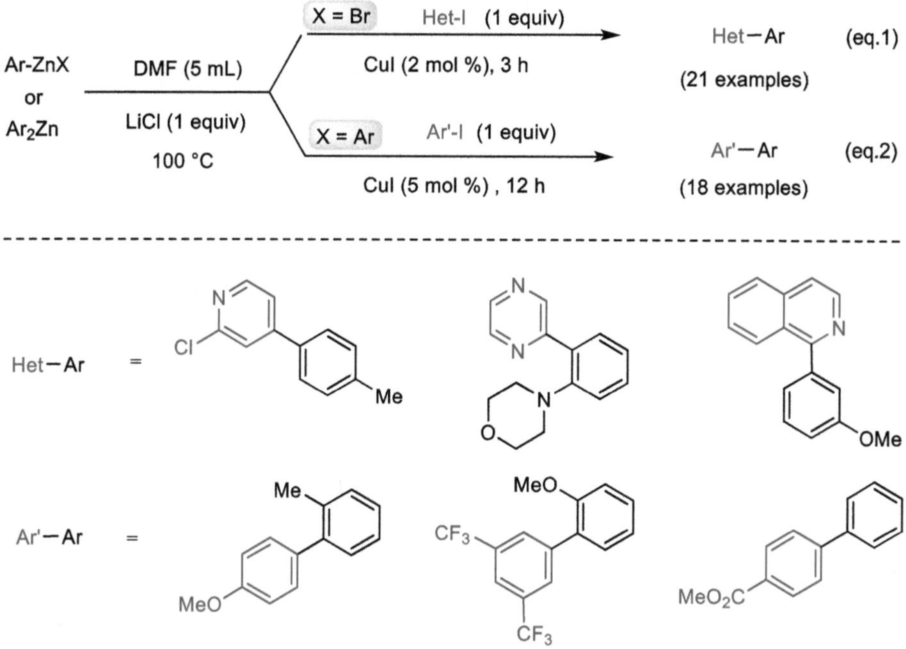

Fig. 5.2: Cu-catalyzed cross-coupling reactions between aryl zinc reagents and heteroaryl- [10] or aryl-iodides [11].

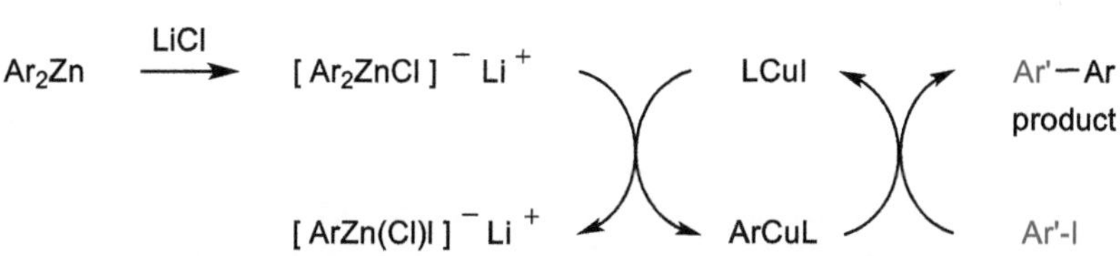

Fig. 5.3: Mechanism proposed by Giri et al. for the Cu-catalyzed cross-coupling reactions between diaryl zinc reagents and aryl iodides [11, 12].

Recently, Burungale et al. have shown that $Cu(CH_3CN)_4BF_4$ (15 mol %) can promote the cross-coupling reactions of electron-poor aryl zinc reagents with aryl bromides [13]. The reaction occurred in refluxing THF and the biaryl compounds were obtained in good yields (62–82%) within 6 h.

The homocoupling of aryl zinc derivatives under oxidative conditions has been reported as well. It constitutes a powerful synthetic method to create carbon-carbon bonds and to reach, for example, biphenylenes, tetraphenylenes or even more complex structures. The latter are valuable building blocks to prepare organic materials. Along these lines, diaryl zinc reagents are formed from the corresponding bromo or iodo derivatives in two steps – (1) *n*-BuLi and (2) $ZnCl_2$ (Fig. 5.4) – and have been shown by Iyoda et al. to give rise to bi- and tetraphenylene in 81% and 5% yield respectively in the presence of 3 equiv of copper(II) chloride [14]. The method was applied to methyl- and fluoro-functionalized diaryl zinc derivatives to yield the expected functionalized biphenylenes in moderate to good yields. Later on, the method was extended to heteroaryl zinc reagents to give rise, for example, to tricyclic thiophenes and cyclic tetrathiophenes (Fig. 5.4) [15]. Octamethoxytetraphenylene derivatives can be prepared under analogous conditions as well (Fig. 5.4) [16]. It is noteworthy that the methods of construction of such polycyclic compounds are limited, which highlights the synthetic potential of the reported approach.

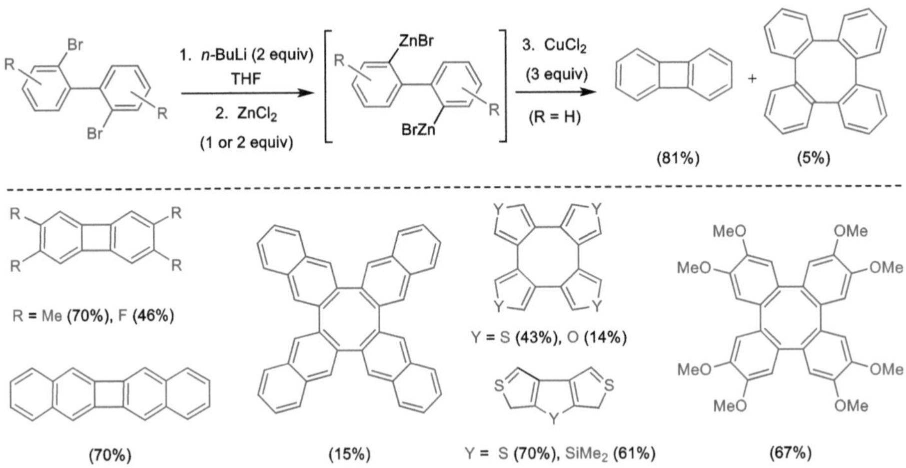

Fig. 5.4: Cu-catalyzed homocoupling reaction of aryl zinc intermediates for the synthesis of biphenylenes, tetraphenylenes, tricyclic thiophenes and cyclic tetrathiophenes [14–16].

Methods involving copper catalysts have been next proposed by Spring et al. for promoting oxidative homocoupling of aryl zinc derivatives (Fig. 5.5) [17]. The organozinc reagents were formed *in situ* from the corresponding iodo-, bromo- or chloro-arenes and Rieke zinc. In the presence of $CuBr.SMe_2$ (10 mol %) and a dinitroarene as the oxidant, the expected biaryl compounds were formed under very mild conditions (room temperature) and isolated in good to excellent yields (76–95%, some examples are reported in Fig. 5.5). The conditions were compatible with various functional groups (ester, nitrile, ketone). The copper-catalyzed method could be applied in the total synthesis of sanguiin H-5, a natural product [18, 19]. In this transformation, the key step involved the intramolecular oxidative coupling of an aryl zinc intermediate leading to a diastereoselective biaryl bond with the concomitant formation of a medium ring (structure of sanguiin H-5 is given in Fig. 5.5). Later on, other related cyclic analogues were synthesized by the same method and their cytotoxicity was compared [20].

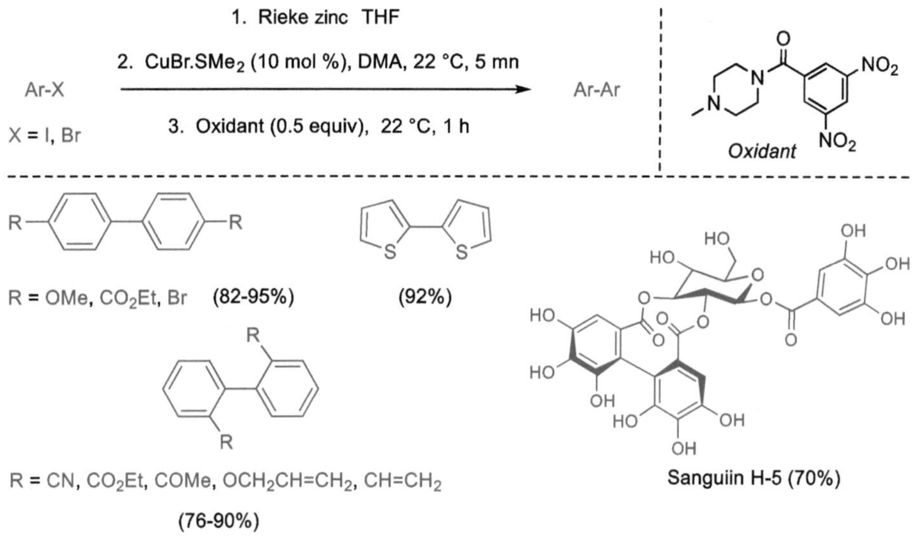

Fig. 5.5: Cu-catalyzed oxidative homocoupling of organozinc halides [17–20].

From a mechanistic point of view, the authors proposed the formation of an organocuprate intermediate from the organozinc reagent in the presence of the copper catalyst CuBr (Fig. 5.6). This latter catalyst would next be oxidized in the corresponding copper(II) complex. The subsequent reductive elimination of the expected biaryl compound (0.5 equiv) would lead to a copper(I) intermediate further converted into the organocuprate by reaction with the organozinc reagent. In this way, the copper could be used in a catalytic manner [17].

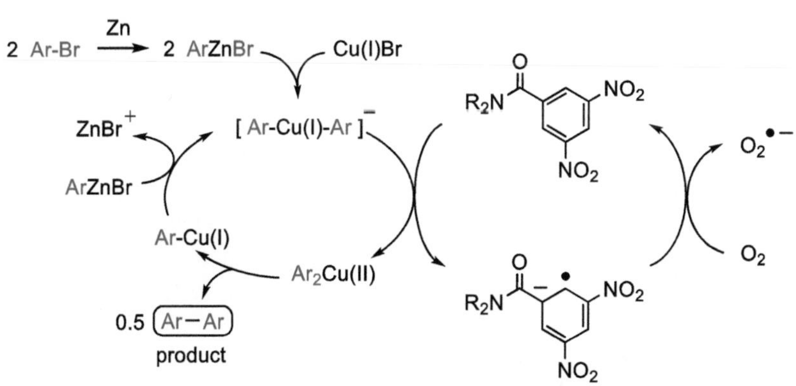

Fig. 5.6: Mechanism proposed by Spring et al. for the Cu-catalyzed oxidative homocoupling of organozinc halides [17].

5.2.2 With acyl chlorides

The couplings of aryl zinc reagents with non-aromatic electrophiles were widely explored by Knochel et al [21]. Aryl- and heteroaryl zinc iodides, prepared from the insertion of Zn in the presence of LiCl, were cross-coupled with aromatic and aliphatic acyl chlorides. The catalytic system involved 20 mol % CuCN.2LiCl, and the resulting ketones were obtained in very good yields (81–90%, some examples are reported in Fig. 5.7) under mild conditions (25–50 °C).

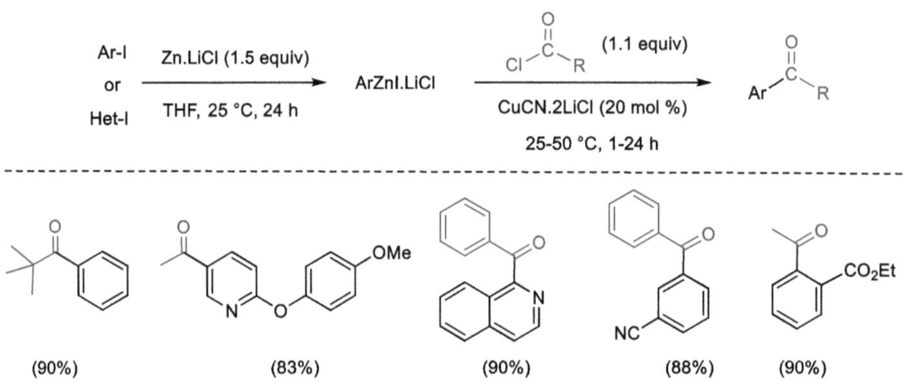

Fig. 5.7: Cross-coupling reactions of aryl zinc iodides with acyl chlorides [21].

5.2.3 With carbon dioxide

Interestingly, Daugulis et al. demonstrated that a copper catalyst, composed of CuI and *N,N,N',N'*-tetramethylethylediamine (TMEDA) or *N,N'*-dimethylethylediamine (DMEDA), was able to promote the carboxylation of aryl iodides with carbon dioxide in the presence of diethylzinc in excess (Fig. 5.8) [22]. The reactions were performed in polar solvents, DMSO being preferred for electron-rich aryl iodides, whereas for electron-poor substrates *N,N*-dimethylacetamide (DMA) was preferred. The corresponding carboxylic acids were isolated in fair to very high yields (40–88%) under mild temperature conditions (20–70 °C). From a mechanistic point of view (Fig. 5.8), the authors proposed a first reduction step of the Cu(I) by diethylzinc to produce the catalytic active species Cu(0). This latter species would next react with the aryl iodide to generate an Ar-Cu(I) intermediate able to undergo CO_2 insertion to yield copper carboxylate. This copper carboxylate would react with diethylzinc in the last step to deliver the corresponding zinc carboxylate and regenerate the Cu(0) active species (acid quench of the reaction allows the generation of the product). In this transformation, the ArCu(I) intermediate is thought to be in equilibrium with the corresponding ArZnX species, a reversible aryl transfer between copper and zinc being likely to occur [23]. Due to the generation of aryl zinc

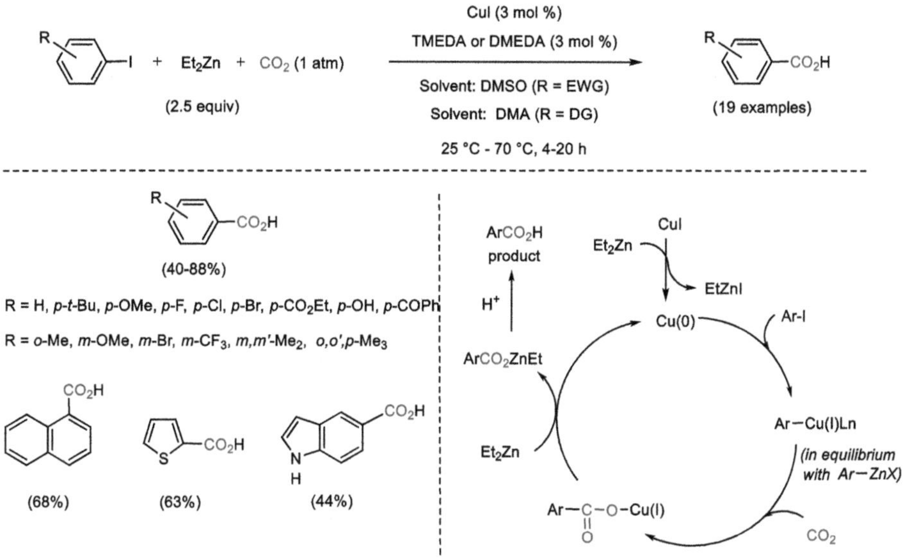

Fig. 5.8: Mechanism proposed for the Cu-catalyzed/ZnEt$_2$-mediated carboxylation of aryl iodides with CO$_2$ [22].

species in the medium, and even if this transformation is not, strictly speaking, a “cross-coupling reaction,” it has been included in this chapter.

5.2.4 With allyl bromides

The copper catalytic system reported by Knochel et al. allowing the cross-coupling reaction of (hetero)aryl zinc reagents with acyl chlorides (see Section 5.2.2) (Fig. 5.7) was also shown to promote the coupling with allyl bromides as electrophiles, thus leading to various olefins in excellent yields (eight examples, 85–96%) [21]. It is worthy to note that the catalyst loading could be decreased to 2 mol % (as compared to 20 mol % when acyl chlorides are used). Later on, the same authors reported a related method involving aromatic and heteroaromatic zinc pivalates as the coupling partners (Fig. 5.9) [24–26]. These air-stable and easy-to-handle powders are prepared by selective metallation of Ar-H derivatives in the presence of an hindered zinc amide base. Sensitive functions (e.g. nitro, aldehyde) were tolerated. While zinc pivalates smoothly underwent cross-coupling reaction with acyl chlorides using stoichiometric amounts of CuCN.2LiCl, catalytic quantities of this copper complex were sufficient to efficiently promote the cross-coupling reaction with allyl bromides. The expected products were obtained in very good yields under very mild conditions (Fig. 5.9). One isolated example describing the cross-coupling reaction with a vinyl iodide was also reported.

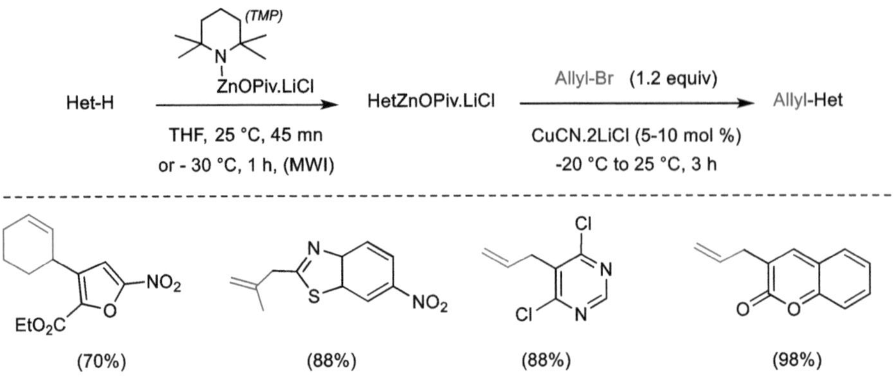

Fig. 5.9: Cu-catalyzed cross-coupling reactions between heteroaryl zinc pivalates with allyl bromides [24].

5.2.5 With a trifluoromethyl group using Togni's reagent

An efficient copper-catalyzed trifluoromethylation of aryl zinc reagents with Togni's reagent was described by Yao et al. (Fig. 5.10) [27]. The reaction proceeded under mild temperature conditions (room temperature, 20 min, in DMF) and trifluoromethylarenes substituted by various functional groups such as ester, cyano, protected ketone and amide were prepared in moderate to good yield (32–68%). Initial mechanistic studies suggested that the reaction of CuI with Togni's reagent lead in a first step to a cationic intermediate. Transmetallation of the aryl group from organic zinc reagent to this cationic intermediate would then generate a trisubstituted hypervalent iodine species. The latter species would next undergo a reductive elimination through a concerted bond formation to afford the trifluoromethylated arene. For the last step, a radical route was also considered. Some specific examples involving alkyl- and vinyl zinc reagents were also reported to illustrate the scope of the method.

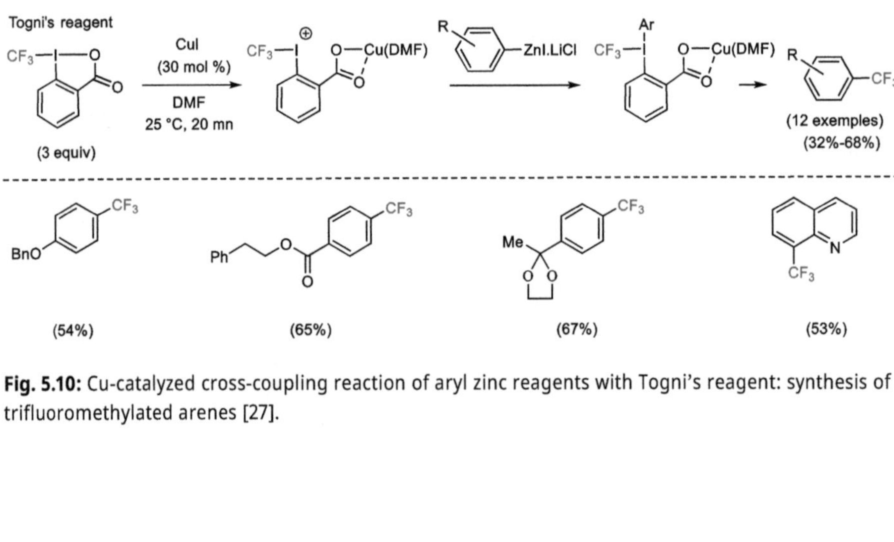

Fig. 5.10: Cu-catalyzed cross-coupling reaction of aryl zinc reagents with Togni's reagent: synthesis of trifluoromethylated arenes [27].

5.3 Coupling reactions of alkenylzinc reagents

In 2015, Knochel et al. reported a general and efficient method to access functionalized cyclic and acyclic alkenyl zinc reagents bearing sensitive functional groups (aldehydes, bromide, thiophenol, ketones or esters) by magnesium insertion performed in the presence of $ZnCl_2$ associated to LiCl as additive. Subsequent copper-catalyzed cross-coupling reaction with 2-(bromomethyl)acrylate, bromoacetylene or bromocyclohexene electrophiles furnished the corresponding polyfunctional compounds in excellent yield (76–96%), proceeding at low temperature in short time (Fig. 5.11) [28]. In most of the cases, not represented in Fig. 5.11, a stoichiometric amount of copper was employed. Some acyclic alkenyl zincs, owing to the chelating effect of the vicinal carbonyl group, were *in situ* generated from the alkenyl bromides without loss of stereochemistry, thus allowing the synthesis of the corresponding trisubstituted olefins with high *Z*-selectivity. It is noteworthy that the post-functionalization was not limited to copper-catalyzed or -mediated reactions and, for example, the palladium-catalyzed Negishi-type cross-coupling involving electron-poor aryl or heteroaryl halides was also successfully reported.

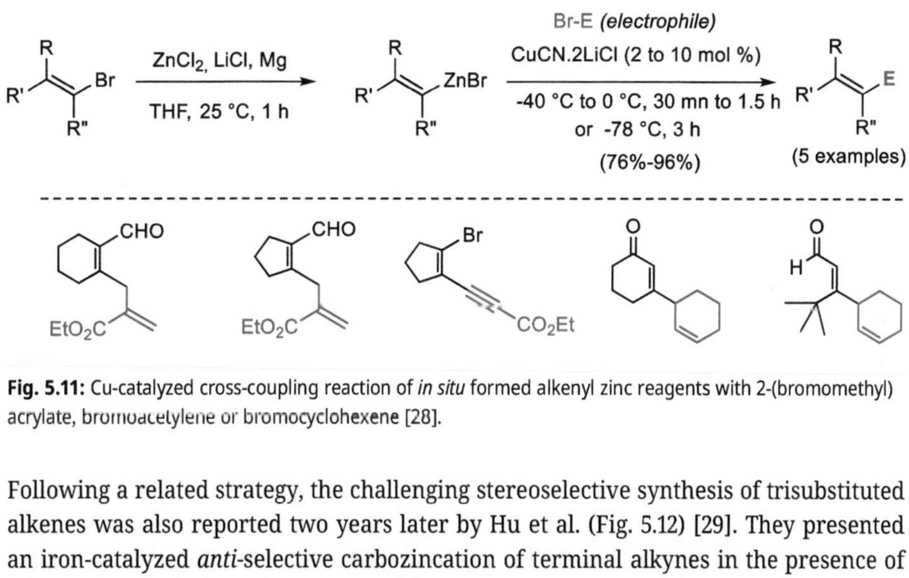

Fig. 5.11: Cu-catalyzed cross-coupling reaction of *in situ* formed alkenyl zinc reagents with 2-(bromomethyl) acrylate, bromoacetylene or bromocyclohexene [28].

Following a related strategy, the challenging stereoselective synthesis of trisubstituted alkenes was also reported two years later by Hu et al. (Fig. 5.12) [29]. They presented an iron-catalyzed *anti*-selective carbozincation of terminal alkynes in the presence of alkyl iodides, leading to the *in situ* formation of alkenyl zinc bromides [30]. By combining this procedure with copper-catalyzed Neigishi-type cross-coupling with *sp*-hybridized carbon electrophiles (bromoalkynes), various functionalized conjugated *E*-enynes which are rarely described molecules, were prepared with high regio- and stereoselectivity. This one-pot method, involving inexpensive and low toxic iron and copper catalysts, has a good functional group tolerance and a broad substrate scope (primary, secondary and tertiary alkyl iodides). This system was also extended to nickel and cobalt Negishi

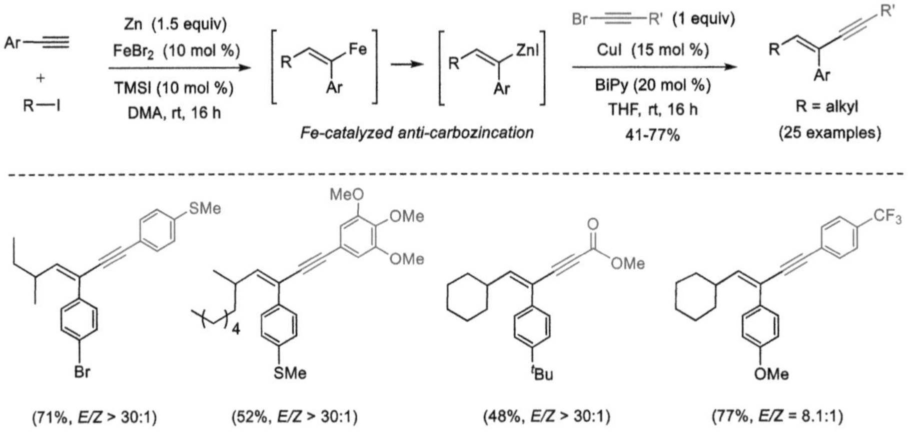

Fig. 5.12: Cu-catalyzed cross-coupling reaction of *in situ* formed alkenyl zinc reagents with 2-(bromomethyl) acrylate, bromoacetylene or bromocyclohexene [29].

cross-coupling performed in the presence of sp^2-, and sp^3-hybridized carbon electrophiles, respectively.

5.4 Coupling reactions of alkynylzinc reagents

A general method reported by Giri et al., also reported previously, was extended to the ligand free copper-catalyzed cross-coupling reaction of alkynyl zinc reagents with heteroaryl iodides (Fig. 5.13). Some alkynyl-quinolines, substituted on the triple bond by aromatic or alkyl groups were thus obtained at 100 °C in DMF, in fair to good yield, in the presence of LiCl (1 equiv) as an additive [10].

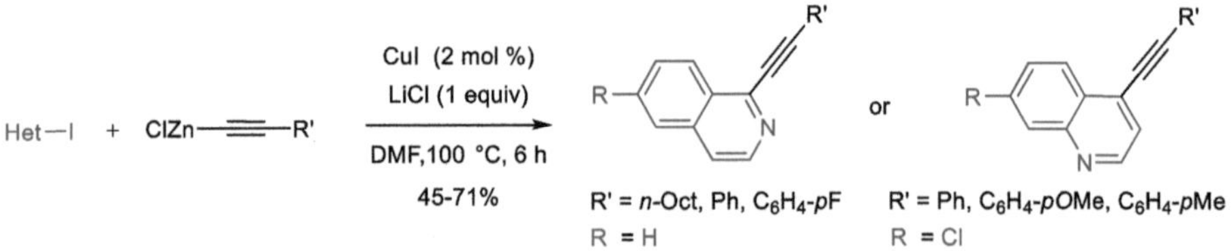

Fig. 5.13: Cu-catalyzed cross-coupling reaction of in alkynyl zinc reagents with heteroaryl iodides [10].

5.5 Coupling reactions of alkylzinc reagents

Alkyl zinc reagents are widely used substrates in Negishi cross-couplings [2]. This section is dedicated to this class of derivative and their cross-coupling reactions with various electrophilic partners including aryl/heteroaryl halides (Section 5.5.1), vinyl halides and pseudohalides (Section 5.5.2), acyl chlorides (Section 5.5.3), allyl halides or propargyl pseudohalides (Sections 5.5.4 and 5.5.5), and α-chloroketones (Section 5.5.6). Lastly, the oxidative homocoupling of alkyl zinc reagents is also included in section 5.5.7.

5.5.1 With aryl and heteroaryl halides

In a pioneering work, Nakamura, Kuwajima et al. reported the coupling of a chiral zinc homoenolate of isobutyrate with iodobenzene catalyzed by $CuBr.SMe_2$ (5 mol %) [31]. The reaction proceeded smoothly (room temperature) to yield the expected coupling product in high isolated yield (79%) and enantiomeric excess (ee = 95%). Two decades later, Giri et al. proposed a general method involving heteroaryl halides using copper iodide as the catalyst (Fig. 5.14) [10]. The cross-coupling reactions of primary, secondary or tertiary alkyl zinc reagents, possibly bearing β-hydrogen atoms, with heteroaryl halides occurred in DMF at room temperature within a short time (3 h). The expected cross-coupling products were obtained in satisfying yields in most of the cases (45–92%). It is worth mentioning that alkyl zinc reagents possessing β-hydrogen atoms are challenging substrates since they can undergo side reactions (β-hydride elimination) leading to undesired compounds. In addition, heteroaryl derivatives are challenging partners since they may act as ligands and consequently poison the corresponding

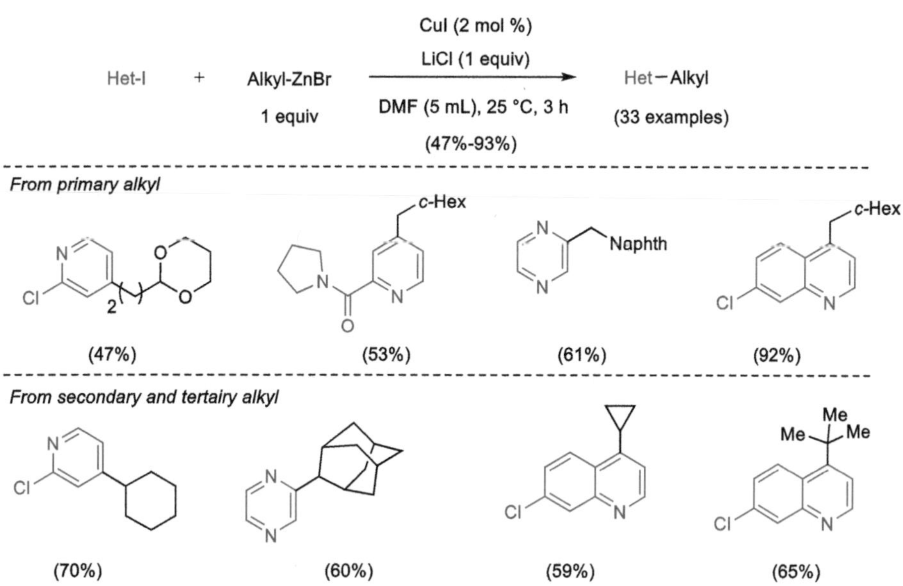

Fig. 5.14: Cu-catalyzed cross-couplings between alkyl zinc reagents and heteroaryl iodides [10].

catalysts as shown with palladium systems [10, 32]. The copper catalyst enabled circumvention of both issues of potential β-hydride elimination and coordinating heteroaryl electrophiles. The use of one equivalent of lithium chloride allowed generating more reactive organozinc species to avoid the inhibition of the reaction by ZnI_2 which is a by-product of the reaction.

5.5.2 With vinyl halides and vinyl pseudohalides

Nakamura, Kuwajima et al. reported the copper-catalyzed cross-coupling reaction of the chiral zinc homoenolate of isobutyrate with 1-bromocyclooctene [33]. Performed at room temperature with 5 mol % of CuBr.SMe, the reaction led to the corresponding cross-coupling product in very good isolated yield (85%) and enantiomeric excess (ee = 90%).

As reported by Wang et al., vinyl iodonium salts revealed to be efficient electrophilic partners in copper-catalyzed cross-coupling reactions involving organozinc reagents (Fig. 5.15) [34]. The latter were generated *in situ* from substrates combining functional groups (phosphonate, sulfoxide, sulfone, ester, amide, cyano), and a C(sp^3)-H bond that underwent direct zincation using bis(2,2,6,6-tetramethylpiperinyl)zinc as the base (Fig. 5.15). The method gave rise to the corresponding functional olefins in one step with fair to excellent yields (38–99%). These compounds are of interest for fine chemistry as well as for functional materials design. The cross-coupling reactions proceeded at room temperature or 50 °C. Interestingly, in these transformations, the copper(II) triflate revealed to be the most efficient catalyst (10 mol %) among copper(I) precursors usually employed in cross-coupling reactions involving organozinc derivatives.

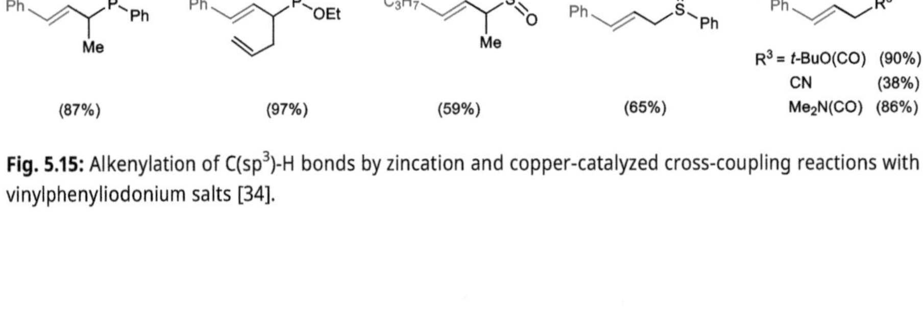

Fig. 5.15: Alkenylation of C(sp^3)-H bonds by zincation and copper-catalyzed cross-coupling reactions with vinylphenyliodonium salts [34].

5.5.3 Coupling with acyl chlorides

The cross-coupling reactions of acyl chlorides with organometallic derivatives are an attractive method to reach ketones. Along these lines, organozinc reagents have been used as nucleophilic partners. Therefore, Knochel et al. reported the coupling of adamantyl zinc reagents, generated from the corresponding bromides in the presence of magnesium, lithium and zinc chlorides, with aromatic acid chlorides (Fig. 5.16) [35, 36]. Using a copper(I) catalyst (20 mol % of CuCN.2LiCl), the expected ketones were obtained in high yields (70–89%) with electron-rich and electron-poor aroyl chloride derivatives. For pyridine derived acyl chloride, the performances were found to be moderate (44% yield). In all cases, the cross-coupling reaction proceeded under mild conditions (– 40 °C to 25 °C) within a couple of hours. It is worth mentioning that the adamantyl moiety is found in many compounds of interest for various applications including pharmaceutical industry and nanotechnologies.

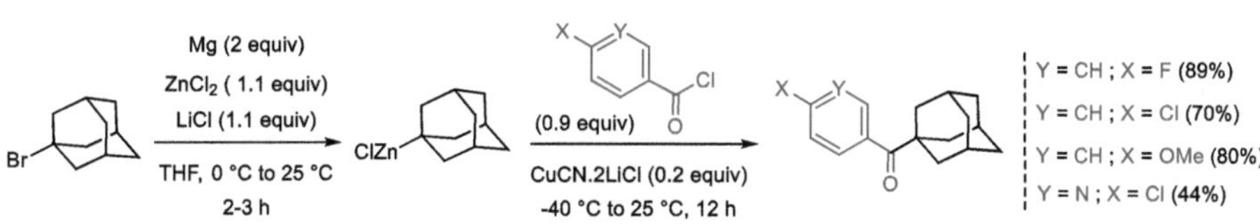

Fig. 5.16: Cu-catalyzed cross-coupling reactions between adamantyl-zinc reagents and aroyl chlorides [35].

In addition to adamantyl zinc derivatives, zinc enolates have been shown to act as partners in cross-coupling reactions. Thus, Nakamura et al. reported the addition of zinc homoenolates (possibly chiral) to benzoyl chloride using copper iodide (2 examples) (Fig. 5.17) [37] and next $CuBr.SMe_2$ (one example) [31] as the catalysts [38]. The corresponding ketones were formed in good to excellent yields.

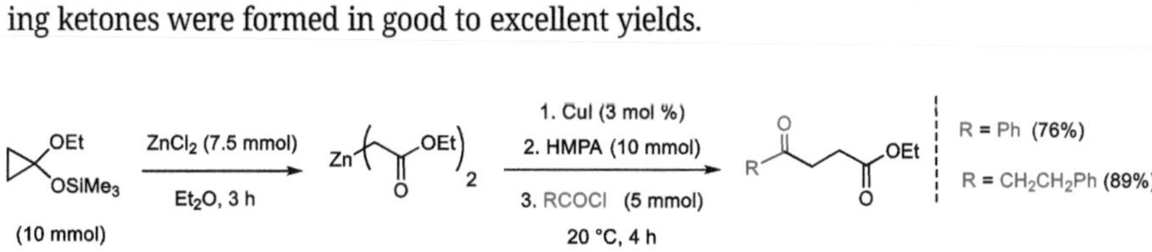

Fig. 5.17: Cu-catalyzed acylation of zinc homoenolate [37].

5.5.4 With allyl halides

In their pioneering work, Nakamura, Kuwajima et al. demonstrated the ability of $CuBr.SMe_2$ (5 mol %) to catalyze the cross-coupling reaction of chiral zinc homoenolate of isobutyrate with 3-bromocyclohexene (Fig. 5.18) [31]. The expected ester was obtained in 59% yield and 96% ee after 4 h at room temperature (Fig. 5.18, **left**). The same year, a general method for promoting allylation of organozinc esters has been reported by Yoshida et al. [39]. This method involved copper cyanide CuCN as the catalyst (15 mol %), and the cross-coupling reactions occurred under mild conditions (1 h at 60 °C or overnight at room temperature) (Fig. 5.18). The corresponding coupling products were generally obtained in very good yields (50–99%). For unsymmetrical substrates, the allylation preferentially occurred on the most substituted position (γ-isomer) (Fig. 5.18, **right**), with α/γ selectivities ranging from 28:72 to 0:100. Similar regioselectivity for S_N2' products (α/γ< 1:99) was observed in the case of allylic halides and mesylates using a copper(II) catalyst (5–10 mol % of $Cu(OTf)_2$, $CuBr_2$ or $Cu(acac)_2$). This method was applied to prepare (*E*)-alkene dipeptide isosteres in high yields (75–96%) [40]. Another related example concerns the S_N2' regioselective allylation of cyclobutylmethylzinc bromide with an allyl bromide derivative bearing an ester function [21]. The expected coupling product was obtained in 82% yield, using only 1 mol % of CuCN.2LiCl as the catalyst. Interestingly, the regioselectivity in copper catalysis differs from palladium catalysis for which a sterically control was observed (S_N2 process; preference for the α-isomer).

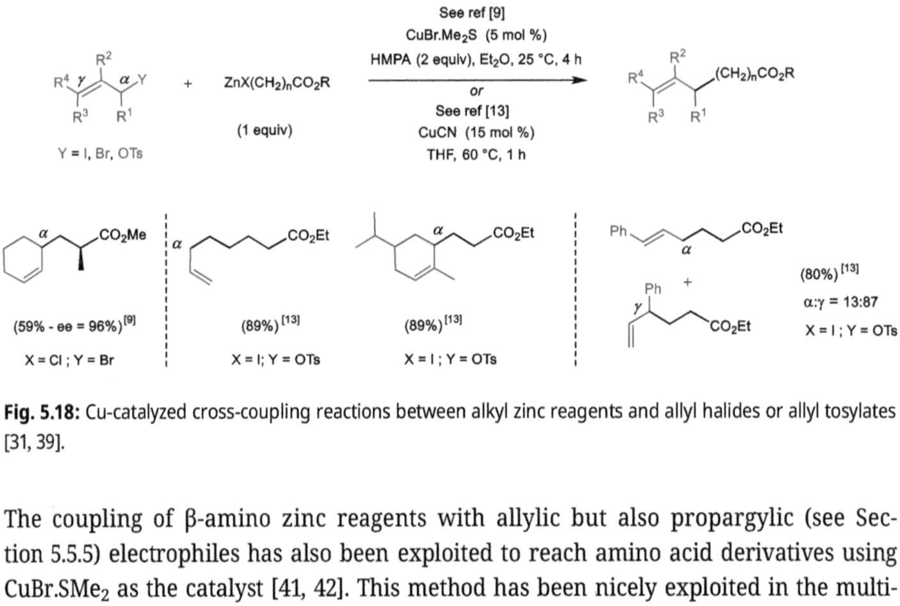

Fig. 5.18: Cu-catalyzed cross-coupling reactions between alkyl zinc reagents and allyl halides or allyl tosylates [31, 39].

The coupling of β-amino zinc reagents with allylic but also propargylic (see Section 5.5.5) electrophiles has also been exploited to reach amino acid derivatives using $CuBr.SMe_2$ as the catalyst [41, 42]. This method has been nicely exploited in the multistep synthesis of tabtoxinine-β-lactam, a phytopathogenic compound [43].

Knochel et al. reported another example that clearly illustrates the relevance of copper-catalyzed cross-coupling reactions involving allyl zinc reagents to reach molecules of great interest (Fig. 5.19) [44]. In this work, a chiral ferrocenyl amine-copper(I) complex was shown to catalyze the enantioselective cross-coupling reaction of unsymmetrical allyl chlorides with dialkylzinc derivatives. The substitution occurred with high S_N2' regioselectivities, the α/γ selectivities ranging from 10:90 to 1:99. The desired chiral compounds were obtained with moderate to very good yields (41–72%) and enantioselectivities (ee up to 87%). The scope included allyl chlorides possibly functionalized with phenyl, naphthyl, cyclohexyl, thienyl, trifluoromethyl or silyl ether substituents. Interestingly, the use of bulky organozinc reagents enabled reaching the highest enantioselectivities.

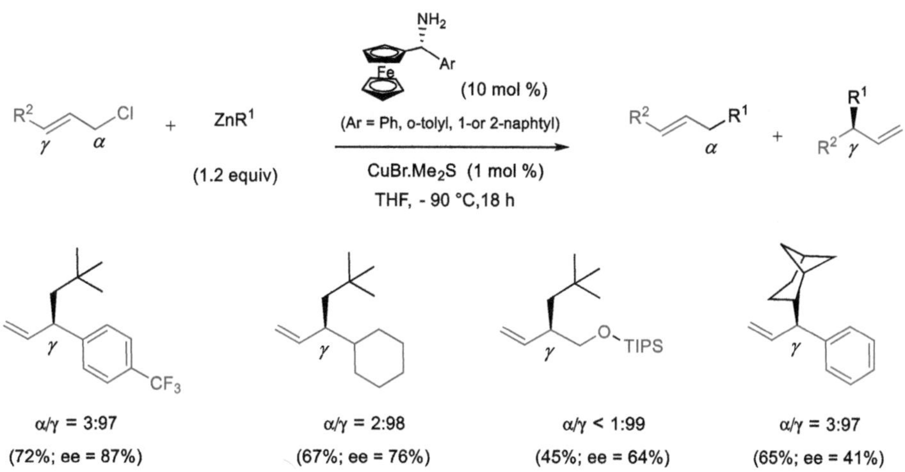

Fig. 5.19: Cu-catalyzed enantioselective cross-coupling reactions between alkyl zinc reagents and allyl chloride [44].

5.5.5 With propargyl pseudohalides

Hiemstra et al. reported a general and elegant copper-catalyzed S_N2' displacement of propargylic tosylates using organozinc derivatives to prepare highly substituted allenes (Fig. 5.20) [45]. This work nicely extended the pioneering work realized by Knochel et al., describing analogous cross-coupling reactions mediated by stoichiometric amounts of copper salts [46]. In Hiemstra's method (Fig. 5.20), the functionalized and substituted expected allenic oxazolidinones and lactams were obtained in yields ranging from 49% to 59% using $CuBr.SMe_2$ as the catalyst (5 mol %). As before, the reaction occurred under mild conditions (from – 20 °C to rt). The allenes were shown to subsequently undergo cyclization in the presence of iodobenzene under palladium catalysis to yield bicyclic enamides. In some other isolated examples, propargyl halides and sulfonates have also been employed as electrophilic partners of alkyl zinc reagents to synthesize allenes in the presence of a copper catalysis ($CuBr.SMe_2$, 5 mol %) [42].

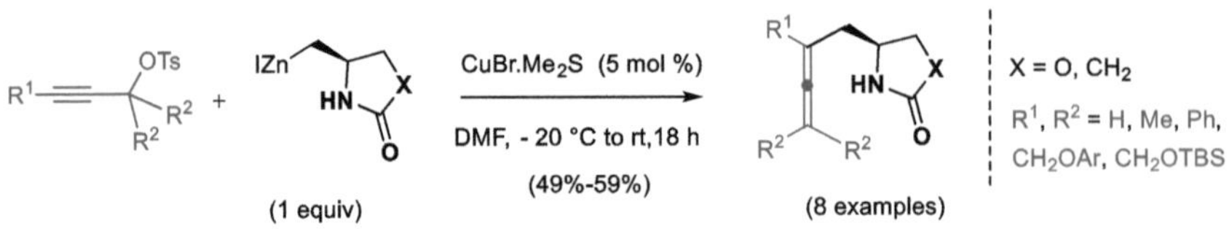

Fig. 5.20: Cu-catalyzed cross-coupling reactions between pyroglutamic acid derived zinc reagents and propargylic tosylates [45].

5.5.6 With α-halogeno carbonyl derivatives

In general, enolates alkylation is a powerful method to create C-C bonds. This approach is fully relevant for introducing unhindered alkyl groups, but overalkylation can occur and the control of regioselectivity may be challenging. To circumvent these issues, an alternative strategy has been proposed by Ready et al. [47]. In this strategy, the preparation of substituted ketones is achieved by a cross-coupling reaction of α-chloroketones with organozinc halides (Fig. 5.21). The potential side reactions of dehalogenation or dimerisation of α-chloroketones can be limited thanks to a proper choice of the copper catalyst and organometallic reagent. Along these lines, it was found that the transmetallation of a Grignard reagent with $ZnCl_2$ was the most efficient way to generate the alkyl zinc halide. The reaction conditions are mild (room temperature) and the scope wide. The introduction of primary and secondary (cyclic or acyclic) alkyl groups on cyclic or acyclic electrophilic α-haloketones occurs in fair to high yields (46–96%). The mechanism of this cross-coupling reaction has been studied, and the experiments realized are consistent with the direct transfer of the alkyl group from the metal (Cu, Zn or Mg) to the electrophilic alpha carbon of the α-chloroketones (Fig. 5.21). This pathway is in line with the inversion of the stereochemistry observed in the cross-coupling reaction of an optically active α-chloroketone.

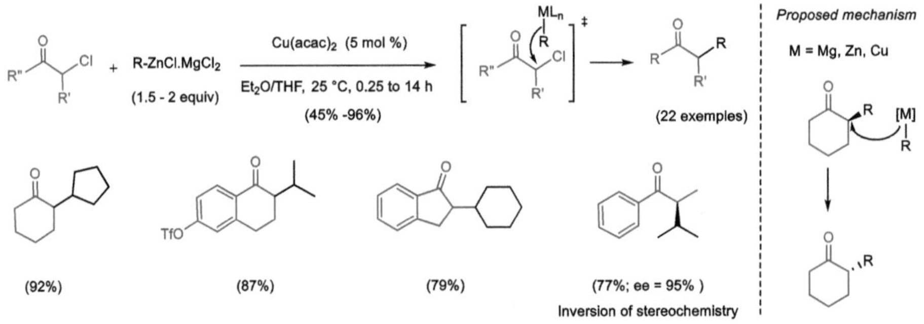

Fig. 5.21: Cu-catalyzed cross-coupling reactions between α-chloroketones and alkyl zinc reagents [47].

5.5.7 Oxydative homocoupling of dialkylzinc reagents

Catalytic methods based on copper have also been proposed by Spring et al. for promoting the oxidative homocoupling of alkyl zinc derivatives (Fig. 5.22) [17]. The organozinc reagents are formed *in situ* from the corresponding bromo- or chloro-alkyl derivatives and Rieke zinc. In the presence of $CuBr.SMe_2$ (10 mol %) and a dinitroarene as the oxidant, the desired coupling products were obtained in high yields (85–95%) at room temperature (Fig. 5.22). It is worthy of note that the conditions are compatible with ketones which differentiates organozinc reagents from their magnesium counterparts.

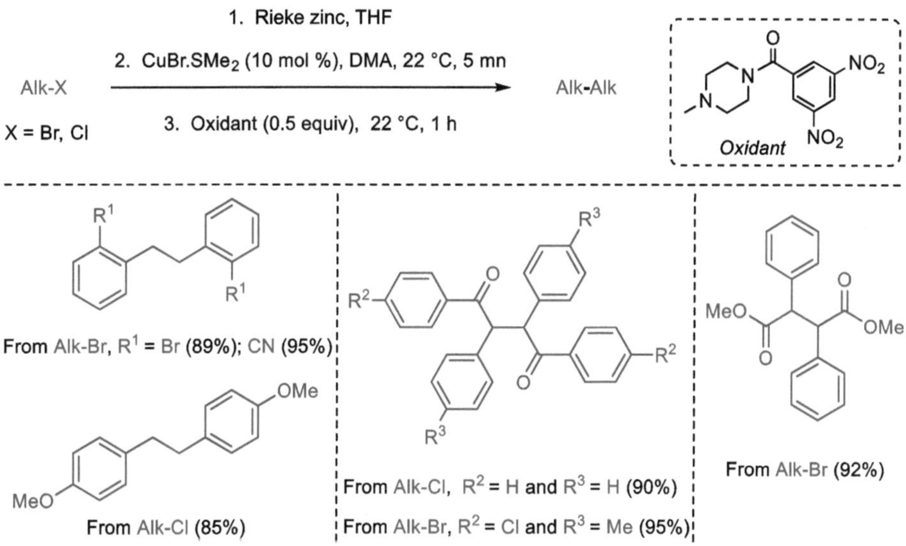

Fig. 5.22: Cu-catalyzed oxidative homocoupling of alkyl zinc halides [17].

5.6 Coupling of perfluoroalkylzinc reagents

In recent years, organofluorine chemistry has become a strategic research topic for the pharmaceutical and agrochemical industries. This is obviously linked to the unique and remarkable properties of the fluorine atom, whose presence on an organic fragment allows, among other factors, to increase its lipophilicity and metabolic stability [48, 49]. Along these lines, synthetic pathways to molecules bearing fluorine atoms, trifluoromethyl (CF_3) groups or perfluoroalkyl (C_nF_{2n+1}) chains have been designed. Although the synthesis of difluoromethylated molecules (CF_2R, R ≠ F) has been less explored, it represents also a very important field of research because the $-CF_2$ moiety is a bioisostere of a carbonyl group or an oxygen atom. The following paragraphs show that copper-catalyzed cross-coupling reactions involving organozinc halides play an important role in the synthesis of these key families of fluorinated compounds.

5.6.1 With (hetero)aryl halides

5.6.1.1 Introduction of CF_3 and perfluoroalkyl chains

In their pioneering work, Kremlev, Tyrra et al. described the preparation of trifluoromethyl and pentafluoroethyl copper from $Zn(CF_3)Br.2DMF$ [50]. The organocopper reagents were next shown to react with iodoarenes to form the expected cross-coupling products with yields ranging from 20% to 70%. Still using stoichiometric amount of copper, Vicic et al. reported the synthesis of perfluoroalkyl linked arenes from 1,2-diiodobenzene or 2,3-diiodopyridine and dizinc reagents $[(MeCN)_2Zn((CF_2)_3Zn(MeCN)_2]$ [51, 52].

To go one step further, Daugulis et al. reported a catalytic version of related cross-coupling reactions (Fig. 5.23) [53]. The catalytic system, associating copper chloride (10 mol %) and 1,10-phenanthroline as the ligand (20 mol %), promoted the cross-coupling reactions of a wide range of electron-rich and electron-poor (hetero)aryl iodides with zinc perfluoroalkyls (Fig. 5.23, eq. 1). The latter were generated from various 1*H*-perfluoroalkanes and bis(2,2,6,6-tetramethylpiperinyl)zinc ($ZnTMP_2$) acting as the base. The couplings occurred at 90 °C and the best performances were obtained using dimethylpropyleneurea (DMPU) as the solvent. Perfluorinated aromatics were isolated in good to excellent yields (51–96%). The authors achieved mechanistic investigations and the first step consisted in the deprotonation of the 1*H*-perfluoroalkanes by $ZnTMP_2$ (Fig. 5.23, eq. 2). The resulting organozinc reagent would next undergo transmetallation with the copper catalyst to give rise to a mixture of anionic copper species. The latter were subsequently proposed to react with the electrophilic partner (aryl iodide) to produce the expected coupling product.

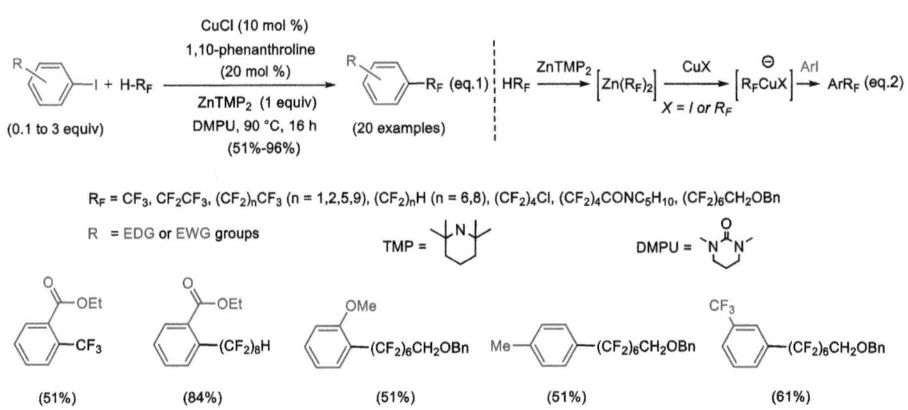

Fig. 5.23: Copper-catalyzed arylation of *1H*-perfluoroalkanes in the presence of TMP_2Zn reagent [53].

Later on, Mikami et al. reported that an analogous Cu(I)-phenanthroline catalyst promoted the trifluoromethylation of various aryl iodides using a trifluoromethylzinc reagent : $Zn(CF_3)I$ [54]. Interestingly, the latter was synthesized *in situ* from Zn dust and trifluoromethyl iodide in the same solvent as previously e.g. DMPU. After 24 h at 50 °C, the cross-coupling products were isolated in yields ranging from 5% to 96%. The mechanistic pathway proposed for this transformation was consistent with the one previously described by Daugulis et al. [53]. In a next report, Mikami et al. enlarged the scope by designing and synthesizing air-stable and easy-to-handle bis(perfluoroalkyl) zinc reagents $Zn(Rf)_2(DMPU)_2$, prepared from perfluoroalkyl iodide and diethylzinc [55]. Copper iodide was shown to promote the cross-coupling reactions of these organozinc derivatives with aryl iodides in DMPU (50–120 °C), no additional ligand being required. The desired perfluorinated compounds were generally isolated in very good yields (67–99%). In addition, this study demonstrated that the rate of the transmetallation of the perfluoroalkyl chain Rf from $Zn(Rf)_2(DMPU)_2$ to copper depends on its length and the following order of reactivity was found: $CF_2 > C_2F_5 > nC_3F_7 > nC_6F_{13}$. Uchiyama et al. reported a related cross-coupling reaction in which the bis(perfluoroalkyl)zinc reagents were generated *in situ* from the corresponding perfluoroalkyl iodides and diethylzinc [56]. With CuI (10 mol %)/1,10-phenanthroline (20 mol %) as the catalyst and DMPU as the solvent (90–120 °C), the expected coupling products obtained from aryl iodides and bromides were isolated in fair to excellent yields (29–95%) (more than 40 examples).

5.6.1.2 Introduction of the CF_2H moiety

As mentioned earlier, the introduction of a CF_2 moiety is of great interest to access bioactive molecules. Along these lines, Mikami et al. applied their know-how in preparing bis(perfluoroalkyl)zinc reagents to synthesize $Zn(CF_2H)_2(DMPU)_2$ [57]. The latter was shown to act as an efficient partner in cross-coupling reactions involving aryl halides (Fig. 5.24). The reaction conditions were very similar to those employed in the case of (trifluoromethyl) and (perfluoroalkyl)zinc derivatives and, therefore, CuI (10 mol %) was used as the catalyst and DMPU as the solvent. While electron-poor aryl and heteroaryl iodides were converted in good yields (46–92%), electron-rich substrates were poorly reactive (Fig. 5.24). This work can be seen as a breakthrough since it corresponds to the first catalytic method for copper-mediated difluoromethylation of aryl halides [58].

5.6.1.3 Introduction of the $-CF_2P(O)(OR)_2$ moiety

Among the difluoromethylene groups, one of the most promising group is the difluorophosphonate ($-CF_2P(O)(OR)_2$), which is considered as a non-hydrolysable mimetic of enzymatic phosphates. The corresponding aryldifluoromethylphosphonates ($Ar-CF_2PO(OEt)_2$) have found important applications in medicinal chemistry [59] but, until recently, only stoichiometric methods have been developed to reach these valuable structures. In

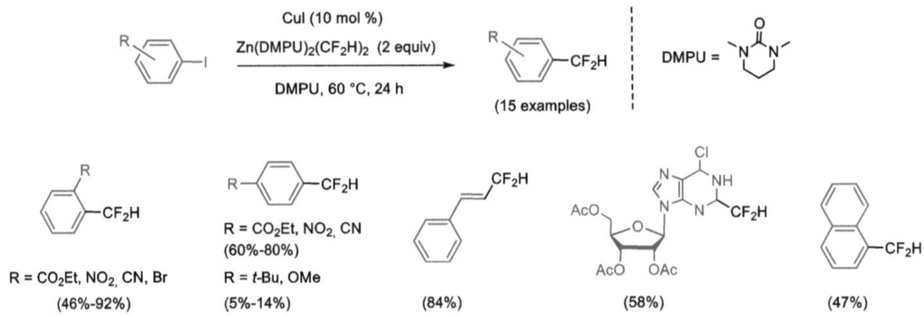

Fig. 5.24: Copper-catalyzed difluoromethylation of aryl iodides with zinc reagents [57].

1996, Shibuya et al. reported the cross-coupling reaction of bromozinc difluoromethylphosphonate ($BrZnCF_2PO(OEt)_2$) with aryl iodides, carried out in presence of more than stoichiometric amount of copper [60, 61]. As part of a mechanistic study, the authors described only one isolated example with a copper catalytic loading (10 mol %), involving β-bromostyrene as a coupling partner which produced the corresponding styril difluoromethylphosphonate (41% yield). Some years after, Burton et al. presented the copper-mediated synthesis of α,α-difluoropropargylphosphonate from bromoalkyl or bromophenylacetylene using a related method [62]. The first catalytic route in this field was reported in 2012 by Zhang et al. with the first example of a copper-catalyzed Negishi-type cross-coupling involving iodobenzoates with bromozinc-difluorophosphonates (Fig. 5.25) [63, 64]. The presence of an *ortho*-benzoate ester group on the aryl iodides was shown to be necessary for reaching good performances. The method is compatible with many functional groups and a wide range of aryl difluorophosphonates could be prepared in good to excellent yields (53–95%) in dioxane at 60 °C. The proposed mechanism, supported by DFT calculations undertaken by Jover [65], is based on a Cu(I/III) catalytic cycle (Fig. 5.25). The first step involves the initial generation of an organocopper-zinc intermediate, able to activate the electrophile by acting as a linker between the catalyst and the carboxylate directing group (DG). The oxidative addition of the *ortho*-iodobenzoate to copper, which is facilitated, is followed by a reductive elimination which closes the cycle by generating the aryl difluorophosphonate while regenerating the catalyst.

The authors also applied their conditions to the cross-coupling reaction of bromozinc-difluorophosphonates with iodo- or bromoaryltriazenes [66]. The triazene group acted as a nitrogenous directing and chelating group, instead of the carboxylate previously involved. By using this method, the first example of a copper-catalyzed difluoroalkylation with aryl bromides was reported in the literature. The reaction conditions are quite similar to those reported previously: 10 mol % CuCN precatalyst, 3,4,7,8-tetramethyl -1,10-phenanthroline ligand (20 mol %), Zn (3 equiv), dioxane at 60 °C, 48 h reaction time. A series of aryl difluoromethylphosphonates, difficult to synthesize by any other ways,

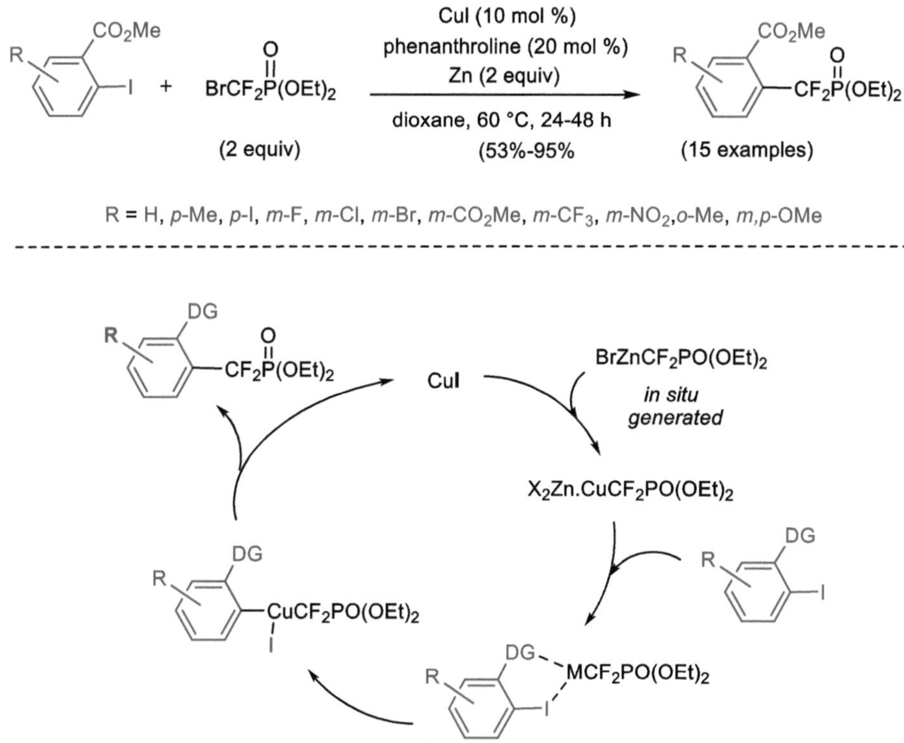

Fig. 5.25: Copper-catalyzed cross-coupling reaction involving *ortho*-iodobenzoates and the bromozinc-difluorophosphonate [63–65].

were obtained in excellent yields *via* functionalization of the arene-bearing position, demonstrating the synthetic potential of this method, particularly in medicinal chemistry.

5.6.2 With vinyl iodides and β-nitrostyrenes

In their work reporting the copper-catalyzed cross-coupling reaction of bis(perfluoroalkyl)zinc reagents $Zn(Rf)_2(DMPU)_2$ with aryl iodides (Section 5.5.1) [55], Mikami et al. also reported the cross-coupling reaction with vinyl iodides and bromides, performed in similar experimental conditions. The corresponding perfluorinated compounds were obtained in very good yields (67–97%).

The same year Dilman et al. reported an original method for the synthesis of *gem*-difluorinated *E*-alkenes, based on the copper(I)-catalyzed cross-coupling reaction of the corresponding *gem*-difluorinated organozincs with β-nitrostyrenes (Fig. 5.26) [67]. The reaction proceeds by substitution of the nitro group providing alkenes in fair to good yields, under mild reaction conditions (DMF, rt, 18 h).

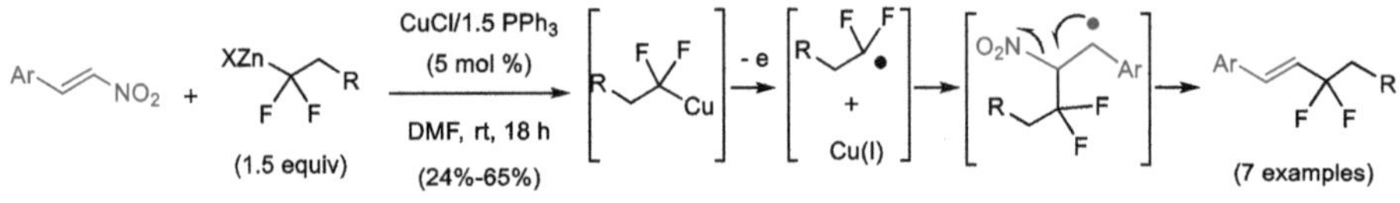

R = Ph, H, 4-MeO$_2$C-C$_6$H$_4$

Ar = 4-MeO$_2$C-C$_6$H$_4$, 4-F-C$_6$H$_4$, 4-MeO-C$_6$H$_4$, 1-Naphthyl, 2-Furyl, Styryl

Fig. 5.26: Copper-catalyzed coupling reaction of *gem*-difluorinated organozinc derivatives with β-nitrostyrenes [67].

Although they do not rule out a non-radical pathway, the authors believe that the mechanism is initiated by a transmetallation between zinc and copper (Fig. 5.26). The resulting organocopper intermediate would undergo a one-electron oxidation to give an organofluorine radical, which after reaction with β-nitrostyrenes, would provide a new radical stabilized by conjugation with the phenyl group. β-Nitrostyrenes could play the role of oxidant. Finally the formation of the *gem*-difluorinated *E*-alkenes would occur with the removal of NO_2.

5.6.3 With acyl chlorides

A copper-catalyzed cross-coupling method involving acyl chlorides and difluorinated organozinc reagents (prepared as described in Section 3.2) was also proposed by Dilman et al. (Fig. 5.27) [68]. Acyl chlorides are previously transformed with potassium dithiocarbamate in *S*-acyl dithiocarbamates which serve as electrophilic intermediates. Various α,α-difluorinated ketones were synthesized under mild conditions (–25 °C to rt, 18 h) in the presence of CuCl (10–20 mol %) associated with a triphenylphosphine ligand.

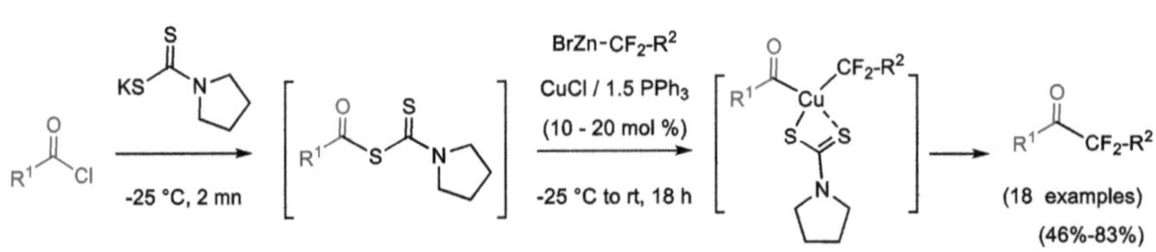

R^1 = Aryl, Vinyl, 2-Thiophenyl, Alkyl, 2-Furyl, Vinyl, 2-Furyl, Styryl

R^2 = Benzyl, CH$_2$-Ar (*p*-CO$_2$Me, *o*-Br, *p*-Br), CH$_2$-Naphthyl, Me, -(CH$_2$)$_3$-CO$_2$Et

Fig. 5.27: Copper-catalyzed cross-coupling reaction of *gem*-difluorinated organozinc reagents with acyl chlorides, *via* acyl dithiocarbamates [68].

A Cu(I)/Cu(III) catalytic cycle was proposed, involving a zinc/copper transmetallation step, the oxidative addition of the acyl dithiocarbamate, and the reductive elimination of copper(III), leading to α,α-difluorinated ketones while regenerating the copper(I) species.

In the copper(III) intermediate resulting from the oxidative addition step, the stabilizing interaction between the metal and the thioamide moiety of the dithiocarbamates probably plays an important role in the efficiency of the system.

5.6.4 With bromoalkynes

Dilman et al. described a copper-catalyzed ligand free method in which α,α-difluoro-substituted organozinc derivatives couple with 1-bromoalkynes provided the corresponding gem-difluorinated alkynes under mild conditions (5 mol % of CuI, DMF, 0 °C to rt, 18 h) (Fig. 5.28) [69]. The *gem*-difluoroorganozinc solutions were prepared by reacting organozinc derivatives with difluorocarbene, this latter was generated *in situ* from bromodifluoromethyl)trimethylsilane or potassium bromodifluoroacetate.

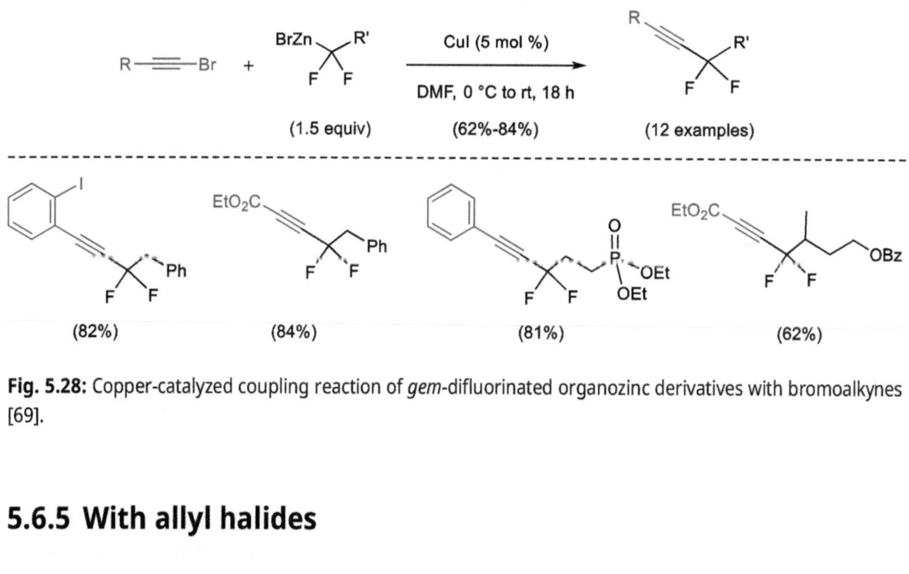

Fig. 5.28: Copper-catalyzed coupling reaction of *gem*-difluorinated organozinc derivatives with bromoalkynes [69].

5.6.5 With allyl halides

Dilman et al. also described a method for the cross-coupling reaction of organozinc reagents with difluorocarbene and allylic electrophiles, leading to the synthesis of *gem*-difluorinated allylic compounds) (Fig. 5.29) [70]. This process, which takes place *via* the initial *in situ* insertion of difluorocarbene into the carbon–zinc bond, involved the presence of 10 mol % of copper iodide and 1,10-phenanthroline ligand. Ester, nitrile and bromide functionalities are well tolerated.

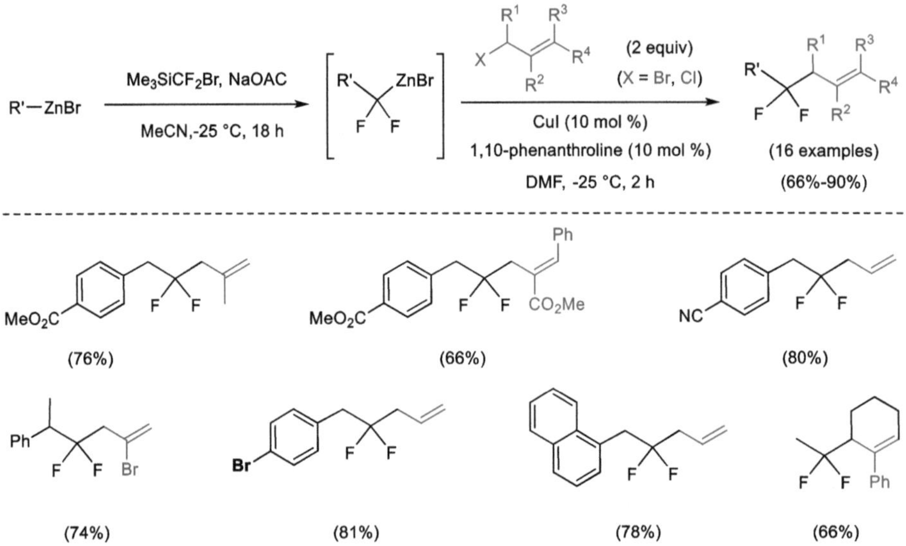

Fig. 5.29: Copper-catalyzed coupling reaction of *gem*-difluorinated organozinc derivatives with allyl halides [70].

5.6.6 With propargyl halides

In a related system involving propargyl halides as the electrophiles, Dilman et al. also succeeded in synthesizing 4,4-difluoro-substituted terminal allenes [71]. Although only three examples have been reported, this method is promising because these allenic derivatives, which are difficult to synthesize, were obtained in excellent yield using copper cyanide (CuCN 10 mol %) without any additional ligand.

5.6.7 With alkyl halides

Recently Liu et al. presented an original strategy taking advantage of the halogen absorption capacity of aryl radicals to allow the copper-catalyzed cross-coupling of difluoromethyl zinc reagents with unactivated alkyl iodides (Fig. 5.30) [72]. This unprecedented Negishi-type cross-coupling was performed at room temperature in DMSO using the 2,4,6-trimethylbenzene diazonium as an additive and a terpyridine (20 mol %) as the copper ligand. The method showed high tolerance to functional groups allowing the synthesis of a wide range of alkyl difluoromethane products, including late-modified pharmaceutical agents and natural products.

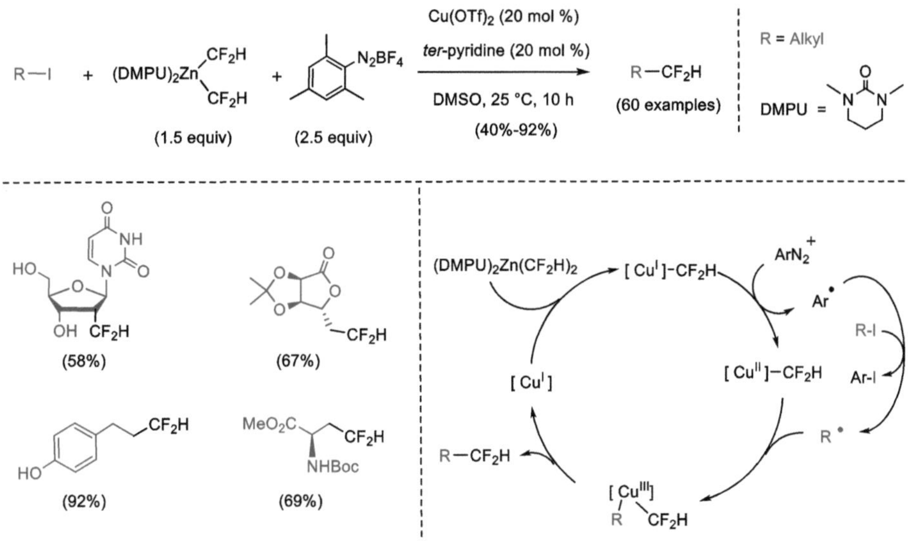

Fig. 5.30: Aryl radical-enabled, copper-catalyzed difluoromethylation of alkyl iodides [72].

A mechanism with a first transmetallation step of a [Cu(I)] catalyst with $(DMPU)_2$-Zn-$(CF_2H)_2$ was proposed, leading to a [Cu(I)]-CF_2H species supposed to be a strong reductant (Fig. 5.30) [73]. A copper(II) intermediate and an aryl radical are then formed by a Single Electron Transfer (SET) involving the aryl diazonium salt. In the next step, the alkyl iodide exchanges the halide with the aryl radical to form an alkyl radical which oxidatively recombines with the Cu(II). From the resulting copper(III) intermediate, the alkyl difluoromethane is expelled and the catalyst is regenerated by reductive elimination.

5.7 Conclusion

Organozinc reagents are powerful partners in cross-coupling reactions due to their easy preparation, their functional group tolerance and their ability to be activated under very mild conditions. Besides, although palladium and nickel are often considered as the metals of choice to promote a lot of carbon-carbon cross-coupling reactions, it should be recalled that organocopper reagents were firstly exploited in these reactions, using stoichiometric amounts of metal [74, 75]. With respect to organozinc reagents, efficient copper catalytic versions later emerged and have been widely developed as illustrated throughout this chapter.

It is worth mentioning that copper is by far much cheaper and more available than palladium and nickel. In addition, it plays a crucial role for human and living organisms and is less toxic with a significantly higher oral permissible daily exposure value: 2500 μg for copper as compared to 250 µg for nickel and 100 µg for palladium [76]. Copper-catalyzed cross-coupling reactions involving organozinc reagents therefore constitute highly valuable reactions to reach functional and complex molecules, in a limited number of steps and possibly in a stereoselective and/or enantioselective manner.

References

[1] Haas D, Hammann JM, Greiner R, Knochel P, ACS Catal. 2016, 6, 1540–1552.
[2] Jana R, Pathak TP, Sigman MS, Chem. Rev. 2011, 111, 1417–1492.
[3] Feringa BL, Naasz R, Imbos R, Arnold LA, Modern organocopper chemistry. In: Krause N Ed., Wiley-VCH, 2002, vol. 7, 224–258.
[4] Mori S, Nakamura E, Modern Organocopper Chemistry. Krause N Ed., Wiley-VCH, 2002, vol. 10, 315–346.
[5] Alexakis A, Benhaim C, Eur. J, Org. Chem. 2002, 3221–3236.
[6] Hajra A, Yoshikai N, Nakamura E, Org. Lett. 2006, 8, 4153–4155.
[7] Jarugumilli GK, Zhu C, Cook SP, Eur. J, Org. Chem. 2012, 1712–1715.
[8] Thaler T, Knochel P, Angew. Chem. Int. Ed. 2009, 48, 645–648.
[9] Frisch AC, Beller M, Angew. Chem. Int. Ed. 2005, 44, 674–688.
[10] Thapa S, Kafle A, Gurung SK, Montoya A, Riedel P, Giri R, Angew. Chem. Int. Ed. 2015, 54, 8236–8240.
[11] Thapa S, Vangala AS, Giri R, Synthesis. 2016, 48, 504–511.
[12] Gurung SK, Thapa S, Vangala AS, Giri R, Org. Lett. 2013, 15, 5378–5381.
[13] Gawade RB, Pise AS, Burungale AS, Emerg J, Technol. Innov. Res. 2020, 7, 390–393.
[14] Iyoda M, Kabir SMH, Vorasingha A, Kuwatani Y, Yoshida M, Tetrahedron Lett. 1998, 39, 5393–5396.
[15] Kabir SMH, Miura M, Sasaki S, Harada G, Kuwatani Y, Yoshida M, Iyoda M, Heterocycles. 2000, 52, 761–774.
[16] Kabir SMH, Hasegawa M, Kuwatani Y, Yoshida M, Matsuyama H, Iyoda M, J. Chem. Soc. Perkin Trans. 2001, 1, 159–165.
[17] Su X, Fox DJ, Blackwell DT, Tanaka K, Spring DR, Chem. Commun. 2006, 3883–3885.
[18] Su X, Surry DS, Spandl RJ, Spring DR, Org. Lett. 2008, 10, 2593–2596.
[19] Su X, Thomas GL, Galloway W, Surry DS, Spandl RJ, Spring DR, Synthesis. 2009, 22, 3880–3896.
[20] Zheng S, Laraia L, O' Connor C, Sorrell D, Tan YS, Xu Z, Venkitaraman AR, Wu W, Spring DR, Org. Biomol. Chem. 2012, 10, 2590–2593.
[21] Krasovskiy A, Malakhov V, Gavryushin A, Knochel P, Angew Chem. Int. Ed. 2006, 45, 6040–6044.
[22] Tran-Vu H, Daugulis O, ACS Catal. 2013, 3, 2417–2420.
[23] Williams CM, Johnson JB, Rovis T, Am J, Chem. Soc. 2008, 130, 14936–14937.
[24] Stathakis CI, Manolikakes SM, Knochel P, Org. Lett. 2013, 15, 1302–1305.
[25] Chen J-Q, Dong Z-B, Synthesis. 2020, 52(24), 3714–3734.
[26] Manolikakes SM, Ellwart M, Stathakis CI, Knochel P, Chem. Eur. J. 2014, 20, 12289–12297.
[27] Wang C-S, Wang H, Yao C, RSC Adv. 2015, 5, 24783–24787.
[28] Sämann C, Schade MA, Yamada S, Knochel P, Angew. Chem. Int. Ed. 2013, 52, 9495–9499.
[29] Cheung CW, Hu X, Chem. Eur. J. 2015, 21, 18439–18444.
[30] Cheung CW, Zhurkin FE, Hu X, J. Am. Chem. Soc. 2015, 137, 4932–4935.
[31] Nakamura E, Sekiya K, Kuwajima I, Tetrahedron Lett. 1987, 28, 337–340.
[32] Slagt VF, de Vries A, de Vries JG, Kellogg RM, Org. Process Res. Dev. 2010, 14, 30–47.
[33] Nakamura E, Sekiya K, Kuwajima I, Tetrahedron Lett. 1987, 28, 337–340.
[34] Liu C, Wang Q, Angew. Chem. Int. Ed. 2018, 57, 4727–4731.

[35] Sämann C, Dhayalan V, Schreiner PR, Knochel P, Org. Lett. 2014, 16, 2418–2421.
[36] Ghouse S, Sreenivasulu C, Kishore DR, Satyanarayana G, Tetrahedron. 2022, 105, article 132580.
[37] Nakamura E, Kuwajima I, J. Am. Chem. Soc. 1984, 106, 3368–3370.
[38] Rilatt I, Caggiano L, Jackson RFW, Synlett. 2005, 18, 2701–2719.
[39] Ochiai H, Tamaru Y, Tsubaki K, Yoshida Z-I, J. Org. Chem. 1987, 52, 4418–4420.
[40] Fujii N, Nakai K, Habasbita H, Yosbizawa H, Ibuka T, Garrido F, Mann A, Chounan Y, Yamamoto Y, Tetrahedron Lett. 1993, 34, 4227–4230.
[41] Deboves HJC, Grabowska U, Rizzo A, Jackson RFW, J. Chem. Soc. Perkin Trans. 2000, 1, 4284–4292.
[42] Hunter C, Jackson RFW, Rami HK, J. Chem. Soc. Perkin Trans. 2001, 1, 1349–1352.
[43] Rilatt I, Caggiano L, Jackson RFW, Synlett. 2005, 18, 2701–2719.
[44] Dübner D, Knochel P, Angew. Chem. Int. Ed. 1999, 38, 379–381.
[45] Karstens WFJ, Stol M, Rutjes F, Hiemstra H, Synlett. 1998, 10, 1126–1128.
[46] Duddu R, Eckhardt M, Furlong M, Knoess H, Berger S, Knochel P, Tetrahedron. 1994, 50, 2415–2432.
[47] Malosh CF, Ready JM, J. Am. Chem. Soc. 2004, 126, 10240–10241.
[48] Purser S, Moore PR, Swallow S, Gouverneur V, Chem. Soc. Rev. 2008, 37, 320–330.
[49] Wang J, Sánchez-Roselló M, Aceña JL, Del Pozo C, Sorochinsky AE, Fustero S, Soloshonok VA, Liu H, Chem. Rev. 2014, 114, 2432–2506.
[50] Kremlev MM, Tyrra W, Mushta AI, Naumann D, Yagupolskii YL, J. Fluorine Chem. 2010, 131, 212–216.
[51] Kaplan PT, Xu L, Chen B, McGarry KR, Yu S, Wang H, Vicic DA, Organometallics. 2013, 32, 7552–7558.
[52] Kaplan PT, Chen B, Vicic DA, J. Fluorine Chem. 2014, 168, 158–162.
[53] Popov I, Lindeman S, Daugulis O, Am J, Chem. Soc. 2011, 133, 9286–9289.
[54] Nakamura Y, Fujiu M, Murase T, Itoh Y, Serizawa H, Aikawa K, Mikami K, Beilstein J, Org. Chem. 2013, 9, 2404–2409.
[55] Aikawa K, Nakamura Y, Yokota Y, Toya W, Mikami K, Chem. Eur. J. 2015, 21, 96–100.
[56] Kato H, Hirano K, Kurauchi D, Toriumi N, Uchiyama M, Chem. Eur. J. 2015, 21, 3895–3900.
[57] Serizawa H, Ishii K, Aikawa K, Mikami K, Org. Lett. 2016, 18, 3686–3689.
[58] Prakash GKS, Ganesh SK, Jones JP, Kulkarni A, Masood K, Swabeck JK, Olah GA, Angew. Chem. Int. Ed. 2012, 51, 12090–12094.
[59] Zhang Z-Y, Acc. Chem. Res. 2003, 36, 385–392.
[60] Yokomatsu T, Suemune K, Murano T, Shibuya S, J. Org. Chem. 1996, 61, 7207–7211.
[61] Yokomatsu T, Murano T, Suemune K, Shibuya S, Tetrahedron. 1997, 53, 815–822.
[62] Zhang X, Burton DJ, Tetrahedron Lett. 2000, 41, 7791–7794.
[63] Feng Z, Chen F, Zhang X, Org. Lett. 2012, 14, 1938–1941.
[64] Feng Z, Xiao Y-L, Zhang X, Acc. Chem. Res. 2018, 51, 2264–2278.
[65] Jover J, Organometallics. 2018, 37, 327–336.
[66] Feng Z, Xiao Y-L, Zhang X, Org. Chem. Front. 2014, 1, 113–116.
[67] Kondratyev NS, Levin VV, Zemtsov AA, Struchkova MI, Dilman A, J. Fluorine Chem. 2015, 176, 89–92.
[68] Ashirbaev SS, Levin VV, Struchkova MI, Dilman A, J. Org. Chem. 2018, 83(1), 478–483.
[69] Zemtsov AA, Volodin AD, Levin VV, Struchkova MI, Dilman AD, Beilstein J, Org. Chem. 2015, 11, 2145–2149.
[70] Zemtsov AA, Kondratyev NS, Levin VV, Struchkova MI, Dilman AD, J. Org. Chem. 2014, 79, 818–822.
[71] Zemtsov AA, Kondratyev NS, Levin VV, Struchkova MI, Dilman AD, Russ. Chem. Bull., Int. Ed. 2016, 11, 2760–22.
[72] Cai A, Yan W, Wang C, Liu W, Angew. Chem. Int. Ed. 2021, 60, 27070–27077.
[73] Zeng XJ, Yan WH, Zacate SB, Chao TH, Sun XD, Cao Z, Bradford KGE, Paeth M, Tyndall SB, Yang KD, Kuo TC, Cheng MJ, Liu W, J. Am. Chem. Soc. 2019, 141, 11398–11403.
[74] Beletskaya IP, Cheprakov AV, Coord. Chem. Rev. 2004, 248, 2337–2364.
[75] Lipschutz BH, Acc. Chem. Res. 1997, 30, 277–282.
[76] Abernethy DR, DeStefano AJ, Cecil TD, Zaidi K, Williams RL, The USP Metal Impurities Advisory Panel. Pharm. Res. 2010, 27, 750–755.

List of contributors

Mae Féo
CNRS
Chimie ParisTech
i-CleHS, CNRS
11 rue Pierre et Marie Curie
75005 Paris
France

Guillaume Lefèvre
CNRS
Chimie ParisTech
i-CleHS, CNRS
11 rue Pierre et Marie Curie
75005 Paris
France

Olivier PIVA
Université Claude Bernard LYON 1
ICBMS, CNRS
69622 villeurbanne cedex
France

Julien Legros
Normandie Univ, UNIROUEN
INSA Rouen
COBRA, CNRS
1 rue Lucien Tesnière
76821 Mont-Saint-Aignan
France

Bruno Figadère
Univ Paris-Saclay
BioCIS, CNRS
Bat. Henri Moissan
17 av. des Sciences
91400 Orsay
France

Pauline Schiltz
Ecole Polytechnique
Laboratoire de Chimie Moléculaire
CNRS
Route de Saclay
91128 Palaiseau Cedex
France

Mengyu Gao
Ecole Polytechnique
Laboratoire de Chimie Moléculaire
CNRS
Route de Saclay
91128 Palaiseau Cedex
France

Corinne Gosmini
Ecole Polytechnique
Laboratoire de Chimie Moléculaire
CNRS
Route de Saclay
91128 Palaiseau Cedex
France

Armelle Ouali
Univ Montpellier, ENSCM
ICGM, CNRS
34296 Montpellier
France

Marc Taillefer
Univ Montpellier, ENSCM
ICGM, CNRS
ENSCM
34296 Montpellier
France

https://doi.org/10.1515/9783110728859-006

Index

1,3,5-triazine 145
1,5-hydrogen transfer 133
1-bromoalkyne 201
2,2,3,3-tetrasubstituted-γ-butyrolactones 169
2,4,6-trimethylbenzene diazonium 202
2-aryldiazines 145
2-chloropyrazines 145
2-chloropyrimidines 145
(2*S*,4*S*)-2,4-bis(diphenylphosphino)pentane 164
2-thiophen-1,3,5-triazine 145
4,4′-di-*tert*-butyl-2,2′-dipyridyl 161
4,4′-di-*tert*-butyl-2,2′-dipyridyl 154
4,6-bis-phenyl phenothiazine 145

acetonitrile 142
acetylene 41
acid anhydrides 154
acrylate 173
acyl chloride 153, 180
acyl chlorides 118
adamantylzinc 191
air 146
aliphatic carboxylic acids 156
Alkenyl 148
alkenyl acetates 158
alkenyl bromide 187
alkenyl zinc 187
alkenylation 87
alkyl halides 27
alkyl phosphates 149
alkyl radical 157, 203
alkylzinc 154
alkylzinc reagent 100
alkyne 47, 152
alkynyl zinc 179
alkynyl zinc pivalates 156
alkynyl-substituted glycosides 158
alkynylzinc 126
alkynylzinc pivalate 88
alkynylzinc pivalates 157
allene 193
allenyl acetates 136
Allyl bromide 180
allyl chloride 193
allyl halide 189
allylation 159
allylic acetates 149

Amphos 12
antilipidemic drug fenofibrate 154
Aromatic ketones 153
aryl bromide 181
aryl halides 142
aryl iodide 180
aryl iodides 144
aryl nitriles 155
aryl radical 203
aryl zinc 180
arylboronates 145
aryldifluoroacetates 149
arylketones 153
arylstannanes 145
arylzinc 65, 120–124, 127, 142
arylzinc pivalates 148
asymmetric 151
azaborine 54

Barbier 50, 149
Barbier-type procedure 160
Benzonitriles 155
benzoquinoline 127
benzyl chlorides 149
Benzylamines 170
benzylic zinc halides 160
biaryl 146, 181
bicyclic alkenes 164
bimetallic 48
bioactive molecules 159
bipyridine 161
bis(oxazoline) 77
bisoxazoline 151
bite angle 55
boron 53
bromides 144
bromoalkene 26
bromoalkynes 153
Buchwald 12

carbometallation 117, 122, 136
carbon dioxide 180
carboxamides 131, 133–136
carboxylation 184
catalytic cycle 142
chlorides 144
chloroarenes 57

https://doi.org/10.1515/9783110728859-007

Chlorphenamine 163
$Co(acac)_2$ 149, 163
$Co(acac)_3$ 162
cobalt 141
cobalt(II)-1,10-phenanthroline complexes 171
$CoBr_2(Phen)$ 153
Conjugated dienes 158
continuous flow 48
copper 153, 179
copper chloride 196
copper cyanide 192
copper iodide 191
copper-catalyzed 180
C-Phos 31
cross-coupling 1, 141
Csp^2-Csp^2 157
Csp^2-Csp^3 157
cyanation 155
cyclic iodides or bromides 158
cyclohexyloxy substituted bipyridine 162
cyclopropylzinc 161
C–X and C–P bonds 163

decarboxylation 83
decarboxylative 156
dialkenyl zinc 156
dialkylzinc 192
dialkylbiarylphosphines 53
Dialkynyl zinc 156
diaryl ketones 154
diarylmethanes 149
diarylmethylamines 171
di-arylzinc 120, 123
diarylzinc 149
diarylzincs 118
diastereoselective 151
diazine 145
dienes 25
diethylzinc 117, 130–131
difluorocarbene 201
difluoromethane 203
difluoromethyl 195
difluorophosphonate 197
diheteroarylzinc 150
dimethyl itaconate 169
dimethylzinc 130, 132, 134–135
diphenylzinc 122–124, 127–128, 130, 133–134
discodermolide 37

electrochemical 142
electron-donating 149
electrophile 153
enantiomeric excess 189
enantioselectivities 151
enyne 187
enynes 126, 135
ethyl bromodifluoroacetate 149
ethyl chloroformate 154
ethyl glyoxylate 168

$Fe(acac)_3$ 117–119, 128, 131, 134
$FeCl_3$ 117, 120–122, 125–126, 130
five-membered ring lactone 169
Fritsch-Buttenberg-Wieschell 42
Fu 16
functionalized 147

gem-difluorinated *E*-alkene 200
geranyl chlorides 163
glycosides 120, 125
gram-scale 151
Grignard 117–118, 122, 127, 132, 137
Guram 12

haloalkynes 153
halogenopyridines 119
heteroaryl 146
heteroaryl halide 180
heteroaryl iodide 180
heteroaryl zinc 184
heteroarylzinc 121
heteroarylzinc pivalates 147, 154
heterobicyclic alkenes 159
homocoupling 9, 195
hydrozincation 30

indenones 135
iron(II) chloride 120, 125
isoquinoline 160–161

ketones 167
Knochel 15

Leaching 48
Lewis acid 159
ligand 147
Lipshutz 12

loratadine 163
lpyridine 127
lynestrenol 157

Mannich 171
mechanism 142
methylthio-substituted *N*-heterocycles 144–145
micellar conditions 49
Michael addition, 169
multi-component 168
Multicomponent reactions 168

N,N,N′,N′-Tetramethylcyclohexane-1,2-diamine 165
N-(1-(pyridin-2-yl)ethyliden)quinolin-8-amine 144
naphthyridines 147
natural products 153
N-benzoyl glutarimides 154
N-cyano-*N*-phenyl-*p*-methylbenzenesulfonamide 155
Negishi 1, 150
Negishi cross-coupling 180
Negishi type cross-coupling 179
Negishi-type reaction 63
neocuproine 165
N-Heterocyclic Carbenes 13
N-hydroxyphtalidimide 85
N-hydroxyphthalimide 157
N-methylimidazole 162
N-methylpyrrolidone 149

Organ 13
organozinc 179
organozinc chloride 73
organozincates 6
ortho-magnesiation 21
oxabenzonorbornadienes 159, 164
oxalyl chloride 167
oxidative addition 6, 201
oxidative homocoupling 146

palladium 1, 153
p-benzoquinone 146
$PCyp_3$ 28
Pd 153
PEPPSI 14
perfluoroalkyl 195
perfluoroalkyl zinc 179
pharmaceutical molecules 153
phenylacetylene 152
phosphates 22
phosphines 11
piperidine *N*-hydroxyphthalimide esters 156
polyfunctionalized quaternary carbons 165
PPh_3 151
primary or secondary amines 168
propargyl halide 193, 202
propargyl pseudohalide 189
propargylic tosylate 193
PyBox 101
pyrazole 127
pyridine 137
pyridine phosphonium salts 161
pyridylzinc 144
pyrimidine 127, 161
pyrimidines 119
pyriproxyfen 163

quinazoline 161
quinoline 161
quinolylbenzamide 128–130

radical 84
radiolabelling 55
(*R*)-cinacalcet 151
redox active aliphatic esters 156
reductive elimination 6, 97
Reformatsky 31
ring-opening 122, 131, 159, 165
(*R*)-xanthorrhizol 151

Schlenk 5
secondary alkyl iodides 121
S_N2' 122–123
(*S*)-preclamol 151
S-pyridyl thioesters 154
Stille 26
supported catalyst 48

TADF 145
TBS-protected iodohydrin 165
tetrachloro-substituted *N*-hydroxyphthalimide ester 156
thienyl chloride 142
three-component 168
Togni's reagent 180
tosylates 121–122, 131
trans-1,2-bis(dimethylamino)-cyclohexane 149
transmetallation 5, 65, 157, 180

triazine halides 145
trifluoroacetic anhydride 167
trifluoromethyl 180, 195
trifluoromethylzinc 197

vinyl halide 189
vinyl iodide 37, 185

water 49
-withdrawing 149

XanPhos 144
XantPhos 55
xerulin 43

ylide 52

zinc enolate 191
zinc pivalate 73, 185

α,α-difluorinated ketone 201
α,β-disubstituted β-amino esters 173
α-Amino acids 168
α-arylalkanoic esters 151
α-branched amines 168
α-bromoesters 151
α-bromolactones 151
α-chiral ketones 154
α-chloroketone 189
β-amino esters 173
β-elimination 165
β-hydride elimination 9
β-hydroxycarbonyl compounds 171
β-nitrostyrene 199
(η^2–allylOAc)-Co(I) complex 149

Bei Fragen zur Produktsicherheit wenden Sie sich bitte an:
If you have any questions regarding product safety,
please contact:

Walter de Gruyter GmbH
Genthiner Straße 13
10785 Berlin
productsafety@degruyterbrill.com